Lecture Notes in Mathematics

T0238106

Editors:
J.-M. Morel, Cachan
F. Takens, Groningen
B. Teissier, Paris

George Osipenko

Dynamical Systems, Graphs, and Algorithms

 Springer

Author

Prof. George Osipenko
Sevastopol National Technical University
99053 Sevastopol
Ukraine
e-mail: george.osipenko@mail.ru

Library of Congress Control Number: 2006930097

Mathematics Subject Classification (2000): 37Bxx, 37Cxx, 37Dxx, 37Mxx, 37Nxx, 54H20, 58A15, 58A30, 65P20

ISSN print edition: 0075-8434
ISSN electronic edition: 1617-9692
ISBN-10 3-540-35593-6 Springer Berlin Heidelberg New York
ISBN-13 978-3-540-35593-9 Springer Berlin Heidelberg New York

DOI 10.1007/3-540-35593-6

Springer is a part of Springer Science+Business Media
springer.com
© Springer-Verlag Berlin Heidelberg 2007

Typesetting by the author and SPi using a Springer LaTeX package
Cover design: WMXDesign GmbH, Heidelberg

Printed on acid-free paper   SPIN: 11772033   VA41/3100/SPi   5 4 3 2 1 0

The book is dedicated to my three sons — Valeriy, Sergey, Egor
and my wife — Valentina.

Preface

The book presents constructive methods of symbolic dynamics and their applications to the study of continuous and discrete dynamical systems. The main idea is the construction of a directed graph which represents the structure of the state space for the investigated dynamical system. The book contains a sufficient number of examples of concrete dynamical systems from illustrative ones to systems of current interest. Results of their numerical simulations with detailed comments are presented. For an understanding of the book matter, it is sufficient to be acquainted with a general course of ordinary differential equations. The new theoretical results are presented with proofs; the most attention is given to their applications. The book is designed for senior students and researches engaged in applications of the dynamical systems theory.

The base of the presented book is the course of lectures given during the Youth Workshop "Computer Modeling of Dynamical Systems" (June 2004, St. Petersburg) initiated and supported by the UNESCO-ROSTE. Parts of these lectures were presented in ETH, Zurich, 1992; Pohang University of Technology, South Korea, 1993; Belmont University, USA, 1996; St. Petersburg University, Russia, 1999; Suleyman Demirel University, Turkey, 2000; Augsburg University, Germany, 2001; Kalmer University, Sweden, 2004.

Symbolic image, coding, pseudo-orbit, shadowing property, Newton method, attractor, filtration, structural graph, entropy, projective space, Lyapunov exponent, Morse spectrum, hyperbolicity, structural stability, controllability, invariant manifold, chaos.

St. Petersburg – Sebastopol   *George Osipenko*

2005 – 2006

George Osipenko at 1952, Sebastopol, Crimea.

Contents

1

Introduction

1.1 Dynamics

In order to investigate some physical phenomenon usually one constructs its mathematical model. The model is a system of equations which describe a process under study in mathematical terms. Equations involved in a system may be of different nature. The dependence between quantities involved in equations may be linear, i.e. this dependence is represented by a linear function, or nonlinear. Parameters may be included in equations, and in this case we have the equations with parameters. Equations may contain both functions sought for and their derivatives – differential equations. Such models are commonly known, e.g. a model of the pendulum motion, a model of the fluid motion, a model of the heat diffusion, a model of the bacteria reproduction, and other. By the process we mean the observed parameters variables which depend on the time t. Parameter values at a time t determine the state of a process. The set of process states constitutes the phase space of a system. Thus, a system of equations describing a given process is determined on the phase space.

For an example, the law of radioactive decay can be stated as: the rate of the decay at a given moment is proportional to an amount of a substance remaining at this moment. In this case the state of a process is determined by the amount of a substance. The process of bacteria reproduction under wide enough amount of a nutritive material can be stated as: the rate of population reproduction is proportional to the population size. In this case the state of a process is determined by the bacteria quantity. In the cases just discussed above, the phase space is one-dimensional and constitutes the set of positive real numbers.

Let us consider a mechanical system that describes the motion of a mass point. The state of the mass point is specified by two quantities: coordinates and velocity. In order to determine uniquely the state of the mass point one needs different number of characteristics depending on where the movement occurs. If the mass point moves along the straight line, one needs two quantities: line coordinate and velocity. Thus, the phase space is the plane $\mathbb{R}^2$ or its part. If the mass point moves in the plane, the point position is determined by its two coordinates and by two components of the velocity vector. Hence, the phase space is four-dimensional Euclidean space $\mathbb{R}^4$. Similarly, to describe the motion of a mass point in the three-dimensional space one needs six quantities that determine the point state at a given time, and the phase space is $\mathbb{R}^6$.

A system of equations governs changes in the object state that occurs with time via some law. If this law is expressed by a system of differential equations then one says that a continuous-time system is given. If equations that govern a system determine changes of the object state through a fixed time interval then the system is called a discrete-time system. A length of the time interval is determined by a problem at hand. Thus, we can became aware of the behavior of an object at hand by treating the movement of points in a phase space at given instants of time with the law of this movement governed by the system of equations.

One of the mostly known classes of systems is that describing so-called determinate processes. This means that there exists a rule in terms of a system of equations that uniquely determines the future and the past of the process on the basis of knowledge of its state at present. The systems describing radioactive decay and bacteria reproduction as well as mechanical systems of a mass point motion outlined above are determinate, i.e. the process progress is uniquely determined by initial conditions and equations. Needless to say that there exist also indeterminate systems, e.g. the process of heat propagation in a medium is semi-determinate as the future is determined by the present whereas the past is not. It is well known that the motion of particles in quantum mechanics is an indeterminate process.

It should be noted that whether or not a process is determinate can be established only experimentally, hence with a certain degree of accuracy. In the subsequent discussion we will return to this subject, but now we suppose that a mathematical model reflects closely a given physical process, i.e. the model is sufficiently accurate. In what follows we will treat both discrete and continuous dynamical systems.

A discrete system is given by a mapping (a difference equation) of the form

$$x_{n+1} = f(x_n),$$

where each subsequent system state x_{n+1} is uniquely determined by its previous state x_n and the mapping f, n can be viewed as the discrete time. Thus, the evolution of the system is governed by the sequence $\{x_n, n \in \mathbb{Z}\}$ in the phase space. A continuous dynamical system is generally given by an equation of the form

$$\frac{dx}{dt} = F(x)$$

or by a system of such equations. Let $\Phi(t, x_0)$ be a solution of the equation, where x_0 is an initial state at $t = 0$, t is viewed as the time. In this case, the system evolution is governed by the curve $\{x = \Phi(t, x_0), \ t \in \mathbb{R}\}$ in the phase space. Fundamental theorems of the differential equations theory ensure the existence of the solution Φ under some reasonable conditions posed on the mapping F, however, its explicit finding (integration of a system) is a sufficiently challenging task. Moreover, solutions of the most part of differential equations cannot be expressed in elementary functions. In practice, when solving an actual problem, Φ is often constructed numerically.

At this point of view, discrete dynamical systems are more favored for the study as the mapping f is similar to the solution Φ and the integration of a system does not complicate understanding of the system evolution. Computer modeling allows to construct easily a trajectory of the system on each finite-time interval that gives a possibility to solve many problems. If we simulate an orbit of a dynamical system for a given initial condition we reach to an attractor of this system and in general, we are not be able to locate any other objects existing in the state space. Although several coexisting attractors might be detected by variation of initial conditions, it is not possible to find unstable objects like, for instance, unstable limit cycles. In this context we need methods that studies the global structure of dynamical system rather than tracing single orbits in the state space.

The method presented approaches this task. It provides a unified framework for the acquisition of information about the system flow without any restrictions concerning the stability of specific invariant sets.

1.2 Order and Disorder

Since the behavior of the process described by a determinate system is uniquely determined by a given initial state, it is reasonable to assume that the behavior of such a system is sufficiently regular, i.e. it obeys a certain law. This mode of thought prevailed in the 19th century. However, with the advance of science our concepts on outward things have been changed. In the 20th century, theory of relativity, quantum mechanics, and theory of chaos have been created.

The theory of relativity dispelled Newton's ideas about the absolute nature of time and space. The quantum mechanics showed that many physical phenomena cannot be considered determinate. The theory of chaos proved that many determinate systems can exhibit irregularity, i.e. they obey solutions that depend on the time in an unpredictable way. One example of chaotic dependence is the decimal representation of an irrational number, where each subsequent digit may be arbitrary independently of preceding digits, i.e. being aware of the first n digits one cannot predict the next one.

The term "chaos" was likely introduced by J. Yorke in 60th. However, H. Poincaré is recognized a pioneer in the study of chaotic behavior of trajectories [117]. In 1888, H. Poincaré [116] revealed strongly unstable trajectories in the three-body problem. For this work, in 1889 he was awarded a prize of Swedish King Oscar II. More precisely, H. Poincaré proved the existence of so-called doubly asymptotic orbits in the three-body problem. Now these orbits are called homoclinic. The main property of such an orbit is that it starts and ends near the same periodic orbit. It should be noted that in this case chaotic trajectories appear in a fully determinate mechanical system that obeys Newton's laws.

In 1935, G. Birkhoff [13] applied symbolic dynamics for coding trajectories near a homoclinic orbit. The same technique was used by S. Smale [136] in construction of the so-called "horseshoe" – a simple model of the chaotic dynamics. Smale's "horseshoe" influenced very much on the theory of chaos as this example is typical and the symbolic dynamics methods turned out to be just an instrument that allows to describe the nature of chaos.

The systematic study of chaos begins in 1960, when researches perceived that even very simple nonlinear models can provide as much disorder as the most violent waterfall. Minor distinctions between initial conditions produce considerable difference in results that is called a "sensitive dependence on initial conditions". One of the pioneer investigators of chaos, E. Lorenz, called this phenomenon a "butterfly effect": trembling of the butterfly wings may cause a tornado in New York within a month. However, the majority of researches continue to hold the viewpoint of Laplace, a philosopher and mathematician of the 18th century, who reasoned that there exists formulas that describe the motion of all physical bodies and hence there is nothing indeterminate neither in the future nor in the past. They believe that by adding complexity to a mathematical model and by increasing accuracy of calculations on can achieve an absolute determinate description of a system, the chaos in a model is viewed as a weakness of the model and the work of investigator is negatively appreciated. If in the course of investigation or in the performance of experiment it emerges that instability or chaos are inherent characteristics of an object of study then this is explained by extraneous "noise", unaccounted perturbations, or bad quality of the experiment performance. It is reasonable that biologists, physiologists, economists and others desire to decompose systems investigated into "elements" and then to construct their determinate models. However, it should be remembered the following:

1) the absolute accuracy of calculations cannot be achieved;
2) the more complicated mathematical models, the greater is the dependence on initial conditions.

In addition, many of system parameters are known with a certain degree of accuracy, e.g. the acceleration of gravity. Moreover, every model describes a real system only approximately and an initial state is also not known precisely. An attempt to achieve a closer description of a system implies a complication of a mathematical model which generally becomes nonlinear. This inevitably leads to systems admitting indeterminate or chaotic solutions (trajectories). Hence, we cannot circumvent chaotic behavior of systems and must foresee the chaos and control it. A practical implementation of such an approach is a solution of the problem of transmitting information. It is known that the transmission of information (in computers, telephone nets, etc.) is attended with interference or noise: intervals of pure transmission alternate with intervals with noise. The unexpected appearance of noise was believed to be associated with a "human element". Costly attempts to improve characteristics of nets or to increase signal power did not lead to solution of the problem of noise. Intervals of pure transmission and intervals of noise are arranged highly chaotic both in duration and in order. However, it turned out that in the chaos of noise and pure intervals there is a certain regularity: the mean ratio of the summarized time of pure transmission and the summarized time of noise is kept constant and, in addition, this ratio is independent of the scale, i.e. it is the same both for an hour and for a second. This means that the problem of noise is not a local problem and is associated not only with a "human element". The way out from this seemingly hopeless situation is very simple: it is reasonable to use a rather weak and inexpensive communication network but duplicate it for correcting errors. This strategy of communicating information is applied now in computer networks.

Economics also provides examples of the chaotic behavior. Studying the variation diagram of prices of cotton within eight years, Hautxacker, a professor of economics at the Harvard university, revealed that there were too many big jumps and that the frequency curve did not correlate with the normal distribution curve. He consulted B. Mandelbrot who worked in the IBM research center. A computer analysis of the variation of prices showed that the points which do not fall on the normal distribution curve form a strange symmetry. Each individual jump of the price is random, but the sequence of such jumps is independent of the scale: day's and month's jumps correspond well to each other under appropriate scaling of the time. Such a regularity persists during the last sixty years with two world wars and many crises. Thus, a striking regularity appears within chaotic dynamics.

Chaotic behavior can be viewed not only in statistic processes but in determinate ones. Let us consider a pendulum built up from two or more rigid components. The first component is secured at a fixed point, to the end of the first component is secured the second component, and so forth. This mechanical system is entirely determinate and described by a collection of differential equations. If one actuates the pendulum in such away that it highly rotates then a chaotic motion can be observed: The pendulum will change the direction of rotation in a chaotic manner. In addition, it is impossible to repeat exactly the motion in subsequent experiments. Thus, we can observe chaos in fully determinate mechanical systems. An explanation is very simple: the system offers the property of sensitive dependence on initial conditions [136], [21].

1.3 Orbit Coding

The modern theory and practice of dynamical systems require the necessity of studying structures that fall outside the scope of traditional subjects of mathematical analysis — analytic formulas, integrals, series, etc. An important tool that allows to investigate such complicate phenomena as chaos and strange attractors is the method of symbolic dynamics. The name reflects the main idea of the method — the description of system dynamics by admissible sequences (admissible words) of symbols from a finite symbol collection (alphabet). We explain this idea by the following hypothetic sample.

Assume that a "device" (realizable or hypothetic) note a system state (a position of the phase point) by some values. These values are obtained with certain accuracy. For example, an electronic clock displays the value t_i, when the exact time t lies in the interval $[t_i, t_i + h)$, where $h > 0$ depends on clock's design. It is convenient to suppose that the phase space M of the system studied is covered by a finite number of cells $\{M_i\}$ and the "device" marks the cell number (index) i when the point x is in the cell M_i. The cells M_i and M_j can intersect when the device indicator is exactly on the boundary between M_i and M_j. In the last case any of i and j are accepted as correct. For simplicity we suppose that the device marks indices of cells through equal time intervals and the trajectory (the sequence of phase points under the action of a system) is coded by the sequence of indices of the system $\{z(k), \ k \in \mathbb{Z}\}$. As indices, we can use symbols of different nature: numbers, letters, coordinates etc. If symbols are letters of some alphabet then the number of letters coincides with the number of cells and trajectories are coded by sequences of letters named admissible words. For transmission of communications by telegraph, as an example, an alphabet with two symbols ("dot" and "dash") is usually used.

Thus, the set of potential system states (phase space) is divided into a finite number of cells. Each cell is coded by a symbol and the "device" in every unit of time "displays" a symbol which corresponds to that cell where the system occurs. Notice that given a sequence of symbols, we can uniquely restore the sequence of cells a trajectory passes through. Clearly, the smaller are cells, the closer is the description of dynamics. The transition from an infinite phase space to a finite collection of symbols can be viewed as a discretization of the phase space.

Thus, the behavior of a system is "coded" with a specially constructed language; in so doing there is a certain correspondence between sequences of symbols and the system dynamics. For example, to a periodic orbit there corresponds a sequence formed by repeated blocks of symbols. The property of orbit recurrence is expressed in repetition of a symbol in an admissible word. Thus, the system dynamics is determined not by values of symbols but by their order in the sequence. Notice that the system dynamics specifies the permissibility of transition from one cell to another and, hence, from one symbol to other symbol; the transition from one symbol to several ones is not excluded. In this case the set of all admissible words is infinite. As an illustration, if the alphabet is formed by the symbols $\{0, 1\}$ and transitions from each symbol to an each one are allowed then we obtain the set of infinite binary sequences with continuum cardinality. If the transition from 1 to 0 is forbidden, we obtain sequences that differ where the transition from 0 to 1 occurs; such sequences form a denumerable set. The first system has the infinite number of periodic orbits, whereas the second one has only two periodic orbits: $\{\ldots 0 \ldots\}$ and $\{\ldots 1 \ldots\}$.

G. Hadamard was the first who used coding of trajectories. In 1898 he applied coding of trajectories by sequences of symbols to obtain the global behavior of geodesics on surfaces of negative curvature [50]. M. Morse [89] is recognized as a founder of symbolic dynamics methods. The term "symbolic dynamics" was introduced by M. Morse and Hedlund [90] who laid the foundations of its methods. They described the main subject as follows.

"The methods used in the study of recurrence and transitivity frequently combine classical differential analysis with a more abstract symbolic analysis. This involves a characterization of the ordinary dynamical trajectory by an unending sequence of symbols termed symbolic trajectory such that the properties of recurrence and transitivity of the dynamical trajectory are reflected in analogous properties of its symbolic trajectory."

These ideas led in the 1960's an 1970's to the development of powerful mathematical tools to investigate a class of extremely non-trivial dynamical systems. R. Bowen [14, 15] made an essential contribution to their development. Smale's "horseshoe" mentioned above influenced very much the advancement of the theory. In 1972 V.M. Alekseev [3] applied symbolic dynamics to investigate some problems of celestial mechanics. He put into use the term "symbolic image" to name the space of admissible sequences in coding trajectories of a system. For theoretical background and applications of symbolic dynamics we refer the reader to the lectures by V.M. Alekseev [4].

In an attempt to find an approach to computer modeling of dynamical systems, C. Hsu [57] elaborated the "cell-to-cell mapping" method. This method performs well in studying the global structure of dynamical systems with chaotic behavior of trajectories. The idea of the method is to approximate a given mapping by a mapping of "cells"; the image of the cell M_i is considered to coincide with the cell M_j provided the center of M_i is mapped by f to some point of M_j. The method suggested by C. Hsu is computer-oriented and admits a straightforward computer implementation. One of the weaknesses of the method is its insufficient theoretical justification. That is why results and conclusions of simulation require detailed analysis and verification. It is also known a generalized version of the method when the image $f(M_i)$ of M_i may consists of several cells $\{M_j\}$ with probability proportional to the volume (the measure) of the intersection $f(M_i) \cap M_j$. Such approach leads to finite Markov's chains which theory is well developed. In this case the computer implementation is rather complicate and presents certain difficulties. A detailed description of these methods can be found in [57].

In 1983 G.S. Osipenko [95] introduced the notion of symbolic image of a dynamical system with respect to a finite covering. A symbolic image is an oriented graph with vertices i corresponding to the cells M_i and edges $i \to j$; the edge $i \to j$ exists if and only if there is a point $x \in M_i$ whose image $f(x)$ lies in M_j. By transforming the system flow into graph we are able to formulate investigation methods as graph algorithms. The following relations between an initial system and its symbolic image hold:

trajectories of a system agree with admissible paths on the graph;
symbolic image reflects the global structure of a dynamical system;
symbolic image can be considered as a finite approximation of a system;
the maximal diameter of cells control an accuracy of approximation.

We notice that there exist several other approaches which use concepts similar to the construction of the symbolic image graph. In Mischaikow [84], a symbolic image-like graph, called a *multivalued mapping*, is constructed in order to compute isolated blocks in the context of the Conley Index Theory [28]. The *set-oriented* methods of Dellnitz, Hohmann and Junge [7, 31, 33, 36] use a scheme similar to our graph and apply a subdivision technique which is also used slightly modified in our implementation. Hruska [56] makes a *box chain construction* to get a directed graph with the aim to compute an *expanding* metric for dynamical systems. An analogous tool for discretization of dynamical systems was applied by F.S. Hunt [58] and Diamond et al [38]. Furthermore, there are many other constructive and computer-oriented methods, of this kind [29, 30, 46, 48, 78, 134, 135].

M. Dellnitz et al [32, 33, 36] elaborated a subdivision technique for the numerical study of dynamical systems. The main point of this method is as follows: a studied domain is covered by boxes or cells, according to certain rules, a part of cells is excluded from consideration while the remainder part is subdivided, then this procedure is repeated. This approach was used in construction of algorithms localizing various invariant sets, in particular, a numerical method for construction of stable and unstable invariant manifolds was obtained [32]. Algorithms for calculating approximations of the invariant measure and the Lyapunov exponent were also created [35, 36]. Based on the algorithms just mentioned, the package GAIO (available at http://math-www.uni-paderborn.de/agdellnotz/gaio/) was elaborated.

A general scheme of the symbolic analysis proposed is as follows. By a finite covering of the phase space of a dynamical system we construct a directed graph (symbolic image) with vertices corresponding to cells of the covering and edges corresponding to admissible transitions. A symbolic image can be viewed as a finite discrete approximation of a dynamical system; the fine is the covering, the closer is the approximation. A process of adaptive subdivision of cells allows to construct a sequence of symbolic images and in so doing to refine qualitative characteristics of a system. The method described above can be used to solve the following problems:

1. Localization of periodic orbits with a given period,
2. Construction of periodic orbit,
3. Localization of the chain recurrent set,
4. Construction of positive (negative) invariant sets,
5. Construction of attractors and domains of attraction,
6. Construction of filtrations and fine sequence of filtrations,
7. Construction of the structural graph,
8. Estimation of the topological entropy,
9. Estimation of Lyapunov exponents,
10. Estimation of the Morse spectrum,
11. Verification of hyperbolicity,
12. Verification of structural stability,
13. Verification of controllability,
14. Construction of isolating neighborhoods of invariant sets.
15. Calculation of the Conley index.

We remark that the symbolic image construction opens the door to applications of several new methods for the investigation of dynamical systems. Quite a lot of information can be gathered by this, and there might be even some more techniques, yet undiscovered, which could be built around symbolic image in the future.

1.4 Dynamical Systems

Let M be a subset in the q-dimensional Euclidean space $\mathbb{R}^q$. In what follows we assume that M is a closed bounded set (a compact) or a smooth manifold in $\mathbb{R}^q$. Let $\mathbb{Z}$ and $\mathbb{R}$ stand for the sets of integers and real numbers, respectively. By a dynamical system we mean a continuous mapping $\Phi(x,t)$, where $x \in M$, $t \in \mathbb{Z}$ ($t \in \mathbb{R}$), such that $\Phi : M \times \mathbb{Z} \to M$ ($\Phi : M \times \mathbb{R} \to M$) and

$$\Phi(x,0) = x,$$
$$\Phi(\Phi(x,t),s) = \Phi(x,t+s),$$

for all $t, s \in \mathbb{Z}$ ($t, s \in \mathbb{R}$). The variable t is thought of as the time and M is named the phase space. If $t \in \mathbb{Z}$ then we have a discrete time system called, for brevity, discrete system (cascade). Discrete dynamical systems result generally from iterative processes or difference equations $x_{n+1} = f(x_n)$. In the case when $t \in \mathbb{R}$ we deal with a continuous time system called, for brevity, continuous system (flow). Continuous dynamical systems result generally from autonomous systems of ordinary differential equations $\dot{x} = f(x)$, i.e. from systems with right hand sides independent of time.

Example 1. Linear equation.
    Consider the linear differential equation $\dot{x} = ax$ on the straight line R. The solution with initial conditions (x_0, t_0) is of the form $F(x_0, t - t_0)$

$= x_0 \exp a(t - t_0)$. In this case the continuous dynamical system is given by the mapping $F(x, t)$, i.e.

$$\Phi(x, t) = x \exp at.$$

If $a < 0$ then $x \exp at \to 0$ as $t \to +\infty$. If $a > 0$ and $x \neq 0$ then $x \exp at \to \pm\infty$ as $t \to +\infty$. By fixing the time t of the shift along trajectories, e.g. $t = 1$, we reach to the discrete dynamical system

$$x_{n+1} = bx_n$$

where $b = \exp a$ is a positive constant. The discrete system $x_{n+1} = bx_n$ can be considered independently of the differential equation and, as this holds, the constant b may be negative. In the last case the mapping $\Phi(x) = bx$ is said to reverse orientation.

Example 2. The Lotka-Volterra equations.

The Lotka-Volterra equations are a system of differential equations of the form

$$\begin{aligned} \dot{x}_1 &= (a - bx_2)x_1 \\ \dot{x}_2 &= (-c + dx_1)x_2, \end{aligned} \tag{1.1}$$

where a, b, c, and d are positive parameters. The Lotka-Volterra equations are one of the mostly known examples that present dynamics of two interacting biological populations. In (1.1) x_1 and x_2 stand for quantities of preys and predators, respectively, a is the reproduction rate of predators in the absence of preys, the term $-bx_2$ means losses via preys. Thus, for predators the population growth per one predator $\dot{x}_1/x_1$ equals $a - bx_2$. In the absence of predators the population of preys decreases, so that $\dot{x}_2/x_2 = -c, c > 0$ provided $x_1 = 0$. The term dx_1 compensates this decrease in the case of "lucky hunting".

1.4.1 Discrete Dynamical Systems

Assume that a continuous mapping $f : M \to M$ has the continuous inverse f^{-1}, i.e. f is a homeomorphism. Then f generates a discrete dynamical system of the form $\Phi(x, n) = f^n(x)$, $n \in \mathbb{Z}$. The mapping $f^m(x)$ is an m-times composition of the function f for $m > 0$ and an m-times composition of the function f^{-1} for $m < 0$; if $m = 0$ then f is the identity mapping.

Thus, we study the dynamics of the cascade

$$x_{k+1} = f(x_k), \ x_k \in M \subset \mathbb{R}^q, \ k \in \mathbb{Z}.$$

Sometimes we will require a homeomorphism f to be a diffeomorphism. This means that there exists continuous partial derivatives of f and f^{-1}.

The trajectory (or the orbit) of the point x_0 is an infinite two-sided sequence

$$T(x_0) = \{x_k = f^k(x_0), \ k \in \mathbb{Z}\}.$$

A point x_0 is called fixed point if $f(x_0) = x_0$. The trajectory of a fixed point consists of a single point $T(x_0) = \{x_0\}$. A point x_0 is called p-periodic point if $f^p(x_0) = x_0$; a least positive integer p with this property is called the least period. For example, a fixed point is a p-periodic point for each positive integer p but its least period is 1. The trajectory of a periodic point x_0 with the least period p consists of p distinct points $T(x_0) = \{x_0, x_1, ..., x_{p-1}\}$.

Example 3. Consider the mapping of the plane $\mathbb{R}^2$ into itself:

$$f : (x, y) \to (ay + bx^2, -ax).$$

Since $f(0,0) = (0,0)$ the origin $(0,0)$ is a fixed point with trajectory $T(0,0) = \{(0,0)\}$. If $b \neq 0$ there exists one more fixed point (x_0, y_0), where $x_0 = (1 + a^2)/b$, $y_0 = -a(1 + a^2)/b$, with trajectory $T(x_0, y_0) = \{(x_0, y_0)\}$. If $b = 0$ then the mapping f is a composition of two linear mappings: $f = L_1 \circ L_2$, where L_1 is a multiplication by a and $L_2 = (y, -x)$ is a rotation through the angle $\alpha = -90°$. When $a = 1$, f is reduced to a rotation; each point $(x, y) \neq (0,0)$ generates the periodic trajectory with least period $p = 4$, i.e. $f^4(x, y) = (x, y)$. As an example, the trajectory of the point $(1, 1)$ is of the form $T(1, 1) = \{(1, 1), (1, -1), (-1, -1), (-1, 1)\}$. It turns out that under certain values of a and b the dynamical system posses infinitely many periodic trajectories with unbounded least periods (see [57]).

1.4.2 Continuous Dynamical Systems

To describe a continuous dynamical system given by ordinary differential equations we use the shift operator along its trajectories defined as follows. Consider the system of differential equations

$$\dot{x} = F(t, x),$$

where $x \in M$, $F(t, x)$ is a C^1 vector field periodic in t with period ω. Let $\Phi(t, t_0, x_0)$ be the solution of the system with initial conditions $\Phi(t_0, t_0, x_0) = x_0$. The investigation of the global dynamics of the system can be performed by studying the Poincaré mapping $f(x) = \Phi(\omega, 0, x)$ of the system which is nothing that the shift operator along trajectories through the period ω.

Example 4. Duffing equation with forcing.
    Consider the damped Duffing equation with forcing

$$\ddot{x} + k\dot{x} + \alpha x + \beta x^3 = B \cos(ht),$$

where t is an independent variable, k, α, β, B, and $h \neq 0$ are parameters, x is a function sought for. Setting $y = \dot{x}$ we get an equivalent system of the form

$$\dot{x} = y,$$
$$\dot{y} = -ky - \alpha x - \beta x^3 + B \cos(ht).$$

If $B \neq 0$ then the system is periodic in t with least period $\omega = \frac{2\pi}{h}$. Let $(X(t,x,y), Y(t,x,y))$ be its solution with initial conditions (x,y) at $t = 0$. If we put, say, $h = 2$ then the Poincaré mapping takes the form

$$f : (x,y) \rightarrow (X(\pi, x, y), Y(\pi, x, y)).$$

If the system is autonomous (i.e. the vector field F is independent of t), an arbitrary $\omega \neq 0$ can be reasoned as a period. For example, without loss of generality we may take 1. The shift operator takes the form $f(x) = \Phi(\omega, x)$, where $\Phi(t, x)$ is the solution of autonomous system such that $\Phi(0, x) = x$. When differential equations are solved numerically, for instance, by the Runge-Kutta or the Adams methods, we get the shift operator approximately.

Example 5. Duffing equation without forcing.
   Consider the damped Duffing equation without forcing

$$\ddot{x} + k\dot{x} + \alpha x + \beta x^3 = 0.$$

The corresponding system

$$\begin{aligned} \dot{x} &= y, \\ \dot{y} &= -ky - \alpha x - \beta x^3, \end{aligned}$$

is autonomous and the shift operator may be written as

$$f : (x,y) \rightarrow (X(1,x,y), Y(1,x,y)).$$

To study the systems listed above methods of computer modeling are widely applied. For example, the use of the MAPLE yields good results. Obtained with the Runge-Kutta method, the phase portrait of the system

$$\begin{aligned} \dot{x} &= y, \\ \dot{y} &= x - 0.27x^3 - 0.48y, \end{aligned}$$

is depicted in Fig. 1.1.
   The system has three equilibriums O, A, and B. There are two trajectories that approach O as $t \rightarrow +\infty$. These trajectories are called stable separatrices and denoted by $W^s(O)$. Thus, for each $x \in W^s(O)$ the omega limit set (ω-limit set) of x coincides with O. There are also two trajectories called unstable separatrices and denoted by $W^u(O)$ that approach O as $t \rightarrow -\infty$. Similarly, for each $x \in W^u(O)$ the alpha limit set (α-limit set) of x is O. Other trajectories, except for $W^s(O)$ approach equilibriums A and B as $t \rightarrow +\infty$.

Relationship between discrete and continuous dynamical systems. Historically, in the dynamical systems theory continuous dynamical systems governed by ordinary differential equations have been the main object of investigation. However, recent trends are to give much attention to

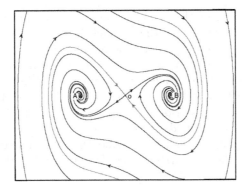

Fig. 1.1. The phase portrait of Duffing's equation

discrete systems governed by diffeomorphisms. Let us show that there is a connection between continuous and discrete systems. We will convince that each continuous system generates a discrete system and vice versa, moreover there is a natural correspondence between trajectories of the systems. The most simple way to obtain a discrete system from a continuous one is to consider the shift mapping (shift operator) at a fixed time along trajectories. The method for constructing the shift mapping was discussed above. By the theorems of existence of ODE solutions and differentiability of solutions with respect to initial data, the shift mapping is a diffeomorphism provided the original system is smooth. In connection with this an inverse problem of including a diffeomorphism in a flow arises: for a given diffeomorphism one needs to find a vector field whose shift operator coincides with the diffeomorphism. However, as M.I. Brin [16] showed, most of diffeomorphisms cannot be included in flows as shift operators. For example, if a diffeomorphism is orientation revising, i.e. its Jacobian is negative, it cannot be included in a flow since the shift operator is always continuously transformed into the identity mapping with positive Jacobian. Thus, diffeomorphisms constitute essentially wide class than flows generated by differential equations on the same manifold. However, using the notion of a section mapping introduced by Poincaré one can construct the correspondence where the opposite situation appears. As an example, consider the section of a torus. A torus can be viewed as the product of two circles $T = S \times S$ with the coordinates $(x, y), 0 \le x, y \le 1$. Let a vector field F on T be such that its trajectories intersect transversally the circle $S \times 0$, which called a section of the flow on a torus. Suppose that the trajectory which starts from the point $(x, 0), x \in S$ returns back to S in a unit time at the point $(f(x), 0)$. In this manner the diffeomorphism $f : S \mapsto S$ called a first return mapping arises. Poincaré was the first who applied this construction to study the system dynamics near a periodic trajectory. In this case, the section is a surface transverse to a periodic trajectory and the return time depends on an initial point. Consider now the inverse passage from a diffeomorphism to a vector field. Let $f : M \mapsto M$ be a diffeomorphism of a manifold M. First of all we define the new manifold M^* by identifying the points $(x, 1)$ and $(f(x), 0)$ in the product $M \times [0, 1]$. Clearly, for the unit vector field $F = (0, 1)$ on $M \times [0, 1]$, the manifold $M \times 0 \cong M$ is a section. The field F generates the vector field F^* on M^* such that its trajectories intersect transversally M and take the point x to $f(x)$ in a unit time. Thus, the diffeomorphism f on M generates the vector field F^* on M^* for which the shift mapping on the zero section M coincides with f, $\dim M^* = \dim M + 1$. Both of the methods discussed for correlation of flows and diffeomorphisms indicate that the qualitative theory of smooth flows (differential equations) and the theory of discrete systems develop in parallel though can differ in details.

2

Symbolic Image

2.1 Construction of a Symbolic Image

Let us consider a discrete dynamical system generated by a homeomorphism $f : M \to M$ on a compact manifold M. We have to note that in practice M is a compact domain in $\mathbb{R}^d$ and f maps from M in $\mathbb{R}^d$. Let $C = \{M(1), ..., M(n)\}$ be a finite covering of the domain M by closed sets. The set $M(i)$ is named a cell (or a box) of the index i. For each cell $M(i)$ we consider its image $f(M(i))$ and set the covering $C(i)$ to consist of cells $M(j) \in C$ whose intersections with $f(M(i))$ are not empty :

$$C(i) = \{M(j) : M(j) \cap f(M(i)) \neq \emptyset\}.$$

Let

$$c(i) = \{j : M(j) \cap f(M(i)) \neq \emptyset\}$$

be a collection of indices corresponding to cells from $C(i)$. The cells of $C(i)$ are called image cells of $M(i)$ under f. Let G be a directed graph with vertices $\{i\}$ corresponding to cells $\{M(i)\}$. Two vertices i and j of G are connected by the directed edge $i \to j$ if and only if $j \in c(i)$, i.e., the cell $M(j)$ is included in the covering of the image $f(M(i))$.

Definition 6. *The graph G is called the symbolic image of f with respect to the covering C.*

Example 7. Constructing a symbolic image.
There are many methods for constructing a symbolic image. Let us consider one of them by an example of the forced Duffing system

$$x' = y,$$
$$y' = -0.1y - (x + x^3) + \cos 2t$$

on the domain $[-2, 2] \times [-2, 2]$. The system is π-periodic in t. The covering consists of the boxes $M(i)$ of the size 0.25×0.25. So we have $16 \times 16 = 256$ cells.

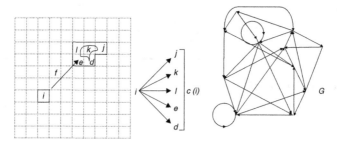

Fig. 2.1. Construction of a symbolic image

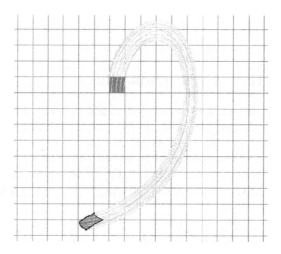

Fig. 2.2. Image of a cell

The numbering of the sells starts from the left-upper corner and finishes to the right-down corner. Let $m(i)$ be a finite set of points in $M(i)$. The placement of points may be systematic or random. Let f be the shift operator along the trajectories on the period $T = \pi$ (the Poincaré mapping). The image $f(m(i))$ is an approximation of $f(M(i))$. We can check the inclusions $f(x) \in M(j)$, $x \in m(i)$ and fix the edge $i \to j$ if the inclusions hold. In Fig. 2.1 the image of the cell $M(87)$ through the period $T = \pi$ is shown. The image $f(M(87))$ intersects the cells $M(213)$, $M(214)$, $M(229)$, and $M(230)$. So we have the edges $87 \to 213$, $87 \to 214$, $87 \to 229$, and $87 \to 230$ in the graph G The same way we can construct each edge and, hence, the symbolic image.

A symbolic image is a geometric tool to describe the quantization process. An other method to get the quantization is to use a matrix of transitions, see Subsection 2.4. We can consider the symbolic image as a finite approximation of the mapping f. It would appear natural that this approximation describes dynamics more precise if the mesh of the covering is smaller. By investigating the symbolic image we can analyze the evolution of a system. It is easily to note that there is a correspondence between orbits of a system and the paths on G. The investigation of the symbolic image permits to get valuable information about the global structure of a system and to obtain such characteristics as the entropy or the Lyapunov exponents. The symbolic image depends on the covering C. By varying C we can change the symbolic image of the mapping f. The existence of the edge $i \to j$ guarantees the existence of a point x in the cell $M(i)$ such that its image $f(x) \in M(j)$. In other words, the edge $i \to j$ is a trace of the matching $x \to f(x)$, where $x \in M(i)$, $f(x) \in M(j)$. If the edge $i \to j$ does not exist then there is no point $x \in M(i)$ such that $f(x) \in M(j)$.

2.2 Symbolic Image Parameters

Definition 8. *An infinite in both directions sequence $\{z_k\}$ of vertices of the graph G which is called an admissible path (or simply a path) if for each k the graph G contains the directed edge $z_k \to z_{k+1}$. A path $\{z_k\}$ is said to be p-periodic if $z_k = z_{k+p}$ for each $k \in Z$.*

A finite path is defined in the same way. For the finite path $\omega = \{z_0, ..., z_m\}$ $|\omega| = m$ is called the length of the path. Denote by Ver the set of vertices of G. The graph G can be considered as a multi-correspondence $G : Ver \to Ver$ between the vertices defined as $G(i) = c(i)$. There is a natural multi-valued projector h, which maps the manifold M on the vertices Ver: $h(x) = \{i : x \in M(i)\}$. By definition it follows that the diagram

$$
\begin{array}{ccc}
M & \xrightarrow{f} & M \\
\downarrow h & & \downarrow h \\
Ver & \xrightarrow{G} & Ver
\end{array}
\tag{2.1}
$$

is commutative in the following sense

$$h(f(x)) \subset G(h(x)). \tag{2.2}$$

In fact, $h(x) = \{i : x \in M(i)\}$ and $h(f(x)) = \{j : f(x) \in M(j)\}$. As $M(j) \cap f(M(i)) \neq \emptyset$ then there are the edges $\{i \to j, \ i \in h(x), \ j \in h(f(x))\}$. The last leads to the inclusion (2.2). Of course, we can not guarantee the equality $h(f(x)) = G(h(x))$. However, the inclusion (2.2) is sufficient to state that orbits of f are transformed by h on paths of the symbolic image. Theorem 14 given below states the properties of this transformation.

Now we introduce some parameters of a symbolic image. Let

$$diam M(i) = \max(\rho(x,y) : x, y \in M(i))$$

be the diameter of the cell $M(i)$, and $d = diam(C)$ be the largest of diameters of cells from C. The parameter d is called a diameter of the covering C. Denote a union of the cells $M(j)$ belonging to the covering $C(i)$ by R_i :

$$R_i = \bigcup_{j \in c(i)} M(j).$$

By construction, R_i contains the image $f(M(i))$ and is in a closed d-neighborhood of the image $f(M(i))$:

$$f(M(i)) \subset R_i \subset \{x : \rho(x, y) \le d, \ y \in f(M(i))\}.$$

Let q (called the upper bound of the symbolic image) be the largest diameter of the images $f(M(i))$, $i = 1, 2, ..., n$. We define the number r as follows. By construction, the cells $M(i)$ are closed sets. If a cell $M(k)$ does not belong to the covering $C(i)$, i.e., $M(k) \cap f(M(i)) = \emptyset$ then the distance

$$r_{ik} = \rho(f(M(i)), M(k)) = \min(\rho(x, y) : x \in f(M(i)), y \in M(k))$$

is positive. Since the number of pairs (i, k) described above is finite then $r = \min r_{ik} > 0$. Thus, r is the smallest distance between the images $f(M(i))$ and the cells $M(k)$ such that $M(k) \cap f(M(i)) = \emptyset$. The value r is called the lower bound of the symbolic image G. Clearly, r depends on the covering C, by varying C one can construct the covering for which r is arbitrarily small. The next propositions describe some properties of the lower bound.

Proposition 9. *If a point $x \in M(j)$ and $\rho(x, f(M(i))) < r$ then the cell $M(j)$ belongs to the covering $C(i)$, i.e., the image $f(M(i))$ intersects the cell $M(j)$.*

Proof. Let $x \in M(j)$. If $\rho(x, f(M(i))) < r$ then $\rho(f(M(i)), M(j)) < r$. By definition of the lower bound, r is the smallest distance between images $f(M(i))$ and cells $M(k)$ which do not intersect. Hence, the cell $M(j)$ has to intersect the image $f(M(i))$. Consequently, the cell $M(j)$ belongs to the covering $C(i)$. $\odot$

Corollary 10. *The set $R_i = \{\cup M(j) : j \in c(i)\}$ contains the r-neighborhood of the image $f(M(i))$:*

$$\{x : \rho(x, f(M(i))) < r\} \subset R_i.$$

Proposition 11. *The lower bound r satisfies the inequality $r \le d$.*

Proof. The number of pairs (i, k) so that $r_{ik} = \rho(f(M(i)), M(k)) > 0$ is finite. Hence there exists a pair (i, m) for which $r = r_{im}$. This means that there exist points $x_j \in f(M(j))$ and $x_m \in M(m)$, $f(M(j)) \cap M(m) = \emptyset$ so that the length of the segment $[x_j, x_m]$ is equal to r. By definition of the lower bound all points of the open segment (x_j, x_m) do not belong to the cell $M(m)$, but belong to some cells of the covering $C(j)$. Since the cells are closed sets, the point x_m belongs to some cell $M(l)$ of $C(j)$. We have $r = \rho(x_m, f(M(j)))$, $x_m \in M(l)$, $M(l) \cap f(M(j)) \ne \emptyset$. Hence, there is a point $x_l \in M(l) \cap f(M(j))$ and the inequality $r = \rho(x_m, f(M(j))) \le \rho(x_m, x_l) \le diam M(l) \le d$ holds. $\odot$

2.3 Pseudo-orbits and Admissible Paths

Definition 12. [6] *For a given $\varepsilon > 0$ an infinite in both directions sequence $\{x_k, \ k \in Z\}$ is called an ε-orbit (a pseudo-trajectory or a pseudo-orbit) of f if for any k*

$$\rho(f(x_k), x_{k+1}) < \varepsilon.$$

A pseudo-orbit $\{x_k\}$ is said to be p-periodic if $x_k = x_{k+p}$ for each $k \in \mathbb{Z}$.

It should be noted that, in the papers [25, 33, 58] the equality sign in the above definition is allowed. In fact this is important if ε is fixed, and it is not important if ε is arbitrary small. In practice, an exact orbit is seldom known, and usually we find nothing more than an ε-orbit for sufficiently small positive ε.

A p-periodic pseudo-orbit (periodic path) will be denoted by its periodic part $\{x_1, ..., x_p\}$ ($\{z_1, ..., z_p\}$).

Example 13. Pseudo-orbit.
On the plane $\mathbb{R}^2$ consider a map of the form

$$f(x, y) = (y, \ 0.05(1 - x^2)y - x).$$

Let us check that the sequence $x_1 = (2, 0)$, $x_2 = (0, -2)$, $x_3 = (-2, 0)$, $x_4 = (0, 2)$, $x_{k+4} = x_k$ forms a 4-periodic ε-orbit for any $\varepsilon > 0.1$, see Fig. 2.3. In fact,

$$
\begin{array}{ll}
f(2, 0) = (0, -2), & \rho(f(x_1), x_2) = 0, \\
f(0, -2) = (-2, -0.1), & \rho(f(x_2), x_3) = 0.1, \\
f(-2, 0) = (0, 2), & \rho(f(x_3), x_4) = 0, \\
f(0, 2) = (2, 0.1), & \rho(f(x_4), x_1) = 0.1.
\end{array}
$$

We can consider the transition from the point $(-2, -0.1)$ to the point $(-2, 0)$ and from the point $(2, 0.1)$ to the point $(2, 0)$ as jumps or corrections of the value 0.1. So we have a 4-periodic ε-orbit for any $\varepsilon > 0.1$.

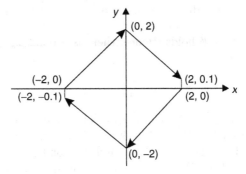

Fig. 2.3. A pseudo-orbit of the mapping $f(x, y) = (y, 0.05(1 - x^2)y - x)$

There is a natural correspondence between admissible paths on the symbolic image G and ε -orbits of the homeomorphism f. Roughly speaking, an admissible path is the trace of an ε-orbit and vice versa. Let us state some relations between the parameters d, q, and r of the symbolic image and ε.

Theorem 14. (Weak shadowing property)

1. *If a sequence $\{z_k\}$ is a path on the symbolic image G and a sequence $\{x_k\}$ is such that $x_k \in M(z_k)$, then the sequence $\{x_k\}$ is an ε-orbit of f for all $\varepsilon > q + d$. In particular, if a sequence $\{z_1, z_2, ..., z_p\}$ of vertices is a p-periodic path and a sequence $\{x_1, x_2, ..., x_p\}$ is such that $x_k \in M(z_k)$, then the sequence $\{x_1, x_2, ..., x_p\}$ is a p-periodic ε-orbit for all $\varepsilon > d + q$.*
2. *If a sequence $\{z_k\}$ is a path on the symbolic image G then there exists a sequence $\{x_k\}$, $x_k \in M(z_k)$ that is an ε-orbit of f for all $\varepsilon > d$. In particular, if a sequence $\{z_1, z_2, ..., z_p\}$ is a p -periodic path on the symbolic image G then there exists a sequence $\{x_1, x_2, ..., x_p\}$, $x_k \in M(z_k)$ which is p-periodic ε-orbit for all $\varepsilon > d$.*
3. *If a sequence $\{x_k\}$ is an ε-orbit of f, $\varepsilon < r$, and $x_k \in M(z_k)$, then the sequence $\{z_k\}$ is an admissible path on the symbolic image G.*
   In particular, if a sequence $\{x_1, x_2, ..., x_p\}$ is a p-periodic ε-orbit, $\varepsilon < r$ and a sequence $\{z_1, z_2, ..., z_p\}$ is such that $x_k \in M(z_k)$, then the sequence $\{z_1, z_2, ..., z_p\}$ is an admissible p-periodic path on the symbolic image G.

Proof. 1. Let $\{z_k\}$ be an admissible path on G. Consequently, there exists the directed edge $z_k \to z_{k+1}$ for every k. This means that the cell $M(z_{k+1})$ belongs to the covering $C(z_k)$. Hence, the image $f(M(z_k))$ intersects the cell $M(z_{k+1})$ and the inequality

$$\rho(f(x_k), x_{k+1}) \leq diam f(M(z_k)) + diam M(z_{k+1}) \leq q + d$$

is fulfilled. Therefore, the sequence $\{x_k\}$ is an ε-orbit of f for all $\varepsilon > q + d$. It should be noted that the point x_k is defined by the inclusion $x_k \in M(z_k)$. Hence the constructed sequence $\{x_k\}$ is not uniquely determined by the path $\{z_k\}$.

2. Let $\{z_k\}$ be an admissible path on G. Consequently, there is the directed edge $z_k \to z_{k+1}$ for every k. This means that the inequality $f(M(z_k)) \cap M(z_{k+1}) \neq \emptyset$ holds. Hence, there is a point $x_k \in M(z_k)$ so that $f(x_k) \in M(z_{k+1})$, i.e.,

$$x_k \in M(z_k) \bigcap f^{-1}(M(z_{k+1})). \tag{2.3}$$

We say that the pair $x_k \to f(x_k)$ corresponds to the directed edge $z_k \to z_{k+1}$. Let us fix x_k for each $k \in N$, and check that the sequence $\{x_k\}$ is an ε-orbit for every $\varepsilon > d$. In fact, the image $f(x_k)$ and the point x_{k+1} belong to the cell $M(z_{k+1})$ by construction. Hence, the following inequalities hold

$$\rho(f(x_k), x_{k+1}) \leq diam M(z_{k+1}) \leq d < \varepsilon.$$

Note that in the case considered the sequence $\{x_k\}$ is not unique, although the point x_k is determined by the inclusion (2.3). It should be emphasized that in the previous case x_k may be arbitrary point in $M(z_k)$.

3. Let the hypotheses of the statement 3 hold. Fix an integer $k \in N$. Since $\rho(f(x_k), x_{k+1}) < r$ and $x_k \in M(z_k)$ then $\rho(f(M(z_k)), M(z_{k+1})) < r$. As r is the smallest distance between $f(M(i))$ and $M(k)$ so that $M(k) \cap f(M(i)) = \emptyset$, the cell $M(z_{k+1})$ has to intersect $f(M(z_k))$. Thus there exists the directed edge $z_k \to z_{k+1}$ for every $k \in N$, and the sequence $\{z_k\}$ is an admissible path on the symbolic image G.

$$\odot$$

An admissible path $\{i_n\}$ on symbolic image can be considered as a coding of a trajectory or an orbit. If there is an orbit $\{x_n = f^n(x_0)\}$ such that $x_n \in M(i_n)$ then the path is called the true coding else we have the false coding.

2.4 Transition Matrix

The directed graph G is uniquely determined by its $n \times n$ (adjacency) matrix of transitions $\Pi = (\pi_{ij})$, where $\pi_{ij} = 1$ if and only if there is the directed edge $i \to j$, otherwise $\pi_{ij} = 0$. It should be remarked that we can consider the matrix of transitions independently of the symbolic image by putting

$$\pi_{ij} = 1 \; if \; f(M(i)) \cap M(j) \neq \emptyset$$
$$\pi_{ij} = 0 \; if \; f(M(i)) \cap M(j) = \emptyset.$$

Let

$$\Pi^2 = \left(\pi_{ij}^2\right)$$

be the square the transition matrix, where $\pi_{ij}^2 = \sum_{k=1}^{n} \pi_{ik} \pi_{kj}$, and upper script 2 stands for an index (not for power). Clearly, $\pi_{ik} \pi_{kj} = 1$ if and only if $\pi_{ik} = 1$ and $\pi_{kj} = 1$, otherwise $\pi_{ik} \pi_{kj} = 0$. So $\pi_{ik} \pi_{kj} = 1$ if and only if there exists the path $i \to k \to j$ from i to j through k. Then the sum $\sum_{k=1}^{n} \pi_{ik} \pi_{kj} = \pi_{ij}^2$ is the number of all admissible paths of length 2 from i to j. In the similar way one can verify that the entry π_{ij}^p of is the number of all admissible paths of length p. In particular, π_{ii}^p is the number of p-periodic paths through the vertex i. Thus, the trace of the matrix Π^p

$$tr\Pi^p = \sum_{i=1}^{n} \pi_{ii}^p$$

is the number of all p-periodic paths.

Definition 15. *A vertex of the symbolic image is called recurrent if there is a periodic path passing through it. The set of recurrent vertices is denoted by RV. Two recurrent vertices i and j are called equivalent if there is a periodic path containing i and j.*

The recurrent vertices $\{i\}$ are uniquely determined by the nonzero diagonal entries $\pi_{ii}^m \neq 0$ of the powers Π^m of the transitions matrix for $m \leq n$, where n is the number of the covering cells. According to Definition 15, the set of recurrent vertices RV decomposes into classes $\{H_k\}$ of equivalent recurrent vertices. In the graph theory the classes H_k are called a strongly connected components of the graph G. Each periodic path ω determines a unique class $H_k = H(\omega)$.

2.5 Subdivision Process

We will apply a process of adaptive subdivision to coverings and will construct a sequence of symbolic images. At first, let us consider a main step of the process a subdivision of covering. Let $C = \{M(i)\}$ be a covering of M and G be the symbolic image with respect to C. Suppose a new covering NC is a subdivision of C. It is convenient to designate cells of the new covering as $m(i, k)$. This means that each cell $M(i)$ is subdivided by cells $m(i, k)$, $k = 1, 2, ...$, which form a subdivision of the cell $M(i)$, i.e.,

$$\bigcup_k m(i, k) = M(i).$$

Denote by NG the symbolic image to the new covering NC. The vertices of the new symbolic image are declared as (i, k). Of course it is possible that some cells will not be subdivided, i.e., $m(i, 1) = M(i)$, and the vertex i in G and the vertex $(i, 1)$ in NG. The subdivision just described generates a natural mapping s from NG onto G which takes the vertices (i, k) onto the vertex i. From $f(m(i, k)) \cap m(j, l) \neq \emptyset$ it follows that $f(M(i)) \cap M(j) \neq \emptyset$, so the directed edge $(i, k) \to (j, l)$ is mapped onto the directed edge $i \to j$. Hence, the mapping s takes the directed graph NG onto the directed graph G. It is convenient to express this property by a diagram. As before, let us denote the vertices of G and the new graph NG by V and NV respectively. Each graph G and NG can be considered as multi-valued mappings $G : V \to V$, $NG : NV \to NV$. Thus we have the diagram

$$\begin{array}{ccc} NV & \xrightarrow{NG} & NV \\ \downarrow s & & \downarrow s \\ V & \xrightarrow{G} & V \end{array} \qquad (2.4)$$

that commutes. The commutativity has the same meaning as before, i.e.

$$s(NG(i, k)) \subset G(s(i, k)). \qquad (2.5)$$

From this it follows that every admissible path on the graph NG is transformed by s on some admissible path on the graph G. In particular, the image of a periodic path is a periodic path and the image of a recurrent vertex is a recurrent vertex. Moreover, the image of a class NH of equivalent recurrent vertices on NG is a class H of equivalent recurrent vertices on G.

Let P be a space of admissible paths on G, NP be a space of the admissible paths on NG, and $P_0 = s(NP)$ be the image of the space NP, $P_0 \subset P$.

Proposition 16. *1. For any sequence $\xi = \{i_n\} \in P_0$ there exists a sequence $\{x_n\}, x_n \in M(i_n)$, which is an ε-trajectory of f for any $\varepsilon > d^* = \mathrm{diam}(NC)$.*
2. Let a sequence $\xi = \{i_n\} \in P \setminus P_0$ and $\delta < r_2$, where r_2 is the low bound of the symbolic image NG. Then there is no any δ-trajectory $\{x_n\}$, such that $x_n \in M(i_n)$.

Proof. 1. Let $\omega = \{i_n\}$ be a sequence from the space P_0. It means that $\{i_n\}$ is an admissible path on G and there exists an admissible path $\gamma = \{(i_n, j_n)\}$ on NG such that $\omega = s(\gamma)$. By Theorem 14 for the path γ there exists an ε-trajectory $\{x_n\}$ for any $\varepsilon > d^* = \mathrm{diam}(NC)$ and $x_n \in m(i_n, j_n)$. As $m(i_n, j_n) \subset M(i_n)$, the $\{x_n\}$ is the required ε-trajectory.

2. Let $\omega = \{i_n\} \in P \setminus P_0$. Suppose, on the contrary, that there exists a sequence $\{x_n\}, x_n \in M(i_n)$ which is a δ-trajectory for $\delta < r_2$. By Theorem 14, there is an admissible path $\gamma = \{(i_n, j_n)\}$ on the symbolic image G_2 such that $m(i_n, j_n) \ni x_n, \forall n \in \mathbb{Z}$. This means that there exists such a sequence $\gamma \in NP$, that $\omega = s(\gamma)$, i.e. $\omega \in P_0$. We obtained a contradiction. Thus, there is no any $\omega = \{i_n\}$ from the space $P \setminus P_0$ which could be matched to an δ-trajectory x_n for $\delta < r_2$ such that $x_n \in M(i_n)$.

$\odot$

Corollary 17. *If $\xi \in P \setminus P_0$ then the admissible path ξ is a false coding.*

2.6 Sequence of Symbolic Images

Let $\{C_t, t \in N\}$ be the sequence of coverings of the manifold M by cells which are consecutive subdivisions. Let us denote by $M(z^t)$ cells of the covering C_t, where z^t is the cell index, and by d_t the maximal diameter of cells from the covering C_t. Let $\{G_t\}$ be the sequence of symbolic images of a continuous mapping $f : M \to M$ corresponding to the sequence of coverings C_t. We have two sequences of mappings $\{s_t\}$ and $\{G_t\}$ and the diagram

$$
\begin{array}{cccccc}
V_1 & \xleftarrow{s_1} & V_2 & \xleftarrow{s_2} & V_3 & \xleftarrow{s_3} \dots \\
G_1 \downarrow & & G_2 \downarrow & & G_3 \downarrow & \\
V_1 & \xleftarrow{s_1} & V_2 & \xleftarrow{s_2} & V_3 & \xleftarrow{s_3} \dots
\end{array}
\tag{2.6}
$$

where $s_t(z^{t+1}) = z^t$ if $M(z^{t+1}) \subset M(z^t)$, that commutes.

Let us consider the extreme case of the covering refinement when each cell consists of a unique point, i.e. $M(x) = \{x\}$. In this case the set of vertices coincides with the set of points of M endowed with discrete topology. Hence, the set of vertices has the continuum cardinality. The set of edges is a collection of pairs $(x, f(x))$. Thus, we can assume that this symbolic image coincides with the initial mapping $f : M \to M$. Let $C_t = \{M(1), M(2), \dots\}$ be a finite covering and G_t be the corresponding symbolic image. The mapping s which relates to a point the index of a cell the point belongs to, yields the diagram

$$
\begin{array}{ccc}
V & \xleftarrow{\; s \;} & M \\
G_t \downarrow & & f \downarrow \\
V & \xleftarrow{\; s \;} & M,
\end{array}
\tag{2.7}
$$

where $s(x) = i$ if $x \in M(i)$, that commutes. Notice that in this case the mapping s is multivalued on cells boundaries. Since diagram (2.7) commutes then an orbit of a system is mapped on an admissible path of the symbolic image. Thus, for the sequence of symbolic images $\{G_t\}$ which corresponds to the sequence of covering subdivisions $\{C_t\}$ we obtain the diagram

$$
\begin{array}{ccccccccc}
V_1 & \xleftarrow{\; s_1 \;} & V_2 & \xleftarrow{\; s_2 \;} & V_3 & \xleftarrow{\; s_3 \;} & \dots & \xleftarrow{\; s \;} & M \\
G_1 \downarrow & & G_2 \downarrow & & G_3 \downarrow & & & & f \downarrow \\
V_1 & \xleftarrow{\; s_1 \;} & V_2 & \xleftarrow{\; s_2 \;} & V_3 & \xleftarrow{\; s_3 \;} & \dots & \xleftarrow{\; s \;} & M
\end{array}
\tag{2.8}
$$

that commutes.

Each s_t is a mapping of directed graphs and maps an admissible path on an admissible path. Let $\xi^t = \{z^t(k), k \in Z\}$ be an admissible path on the symbolic image G_t. We denote by P_t the space of admissible paths on the symbolic image G_t. The mapping $s_t : G_{t+1} \to G_t$ generates the mapping in the spaces of paths, $s_t(P_{t+1}) \subset P_t$, however, $s_t(P_{t+1}) \neq P_t$, in general. If we fix a path ξ^t on each symbolic image G_t then we obtain the sequence of paths $\{\xi^t \in P_t\}$. Each orbit $T(x_0) = \{x_k = f^k(x_0), k \in Z\}$ generates the admissible path $\xi^t = \{z^t(k), x_k \in M(z^t(k))\}$ on G_t and for paths of this kind we have $\xi^t = s_t(\xi^{t+1})$. The path just described is the coding of the orbit $T(x_0)$ corresponding to the covering C_t. Let Cod_t be the set of codings of all true orbits of f corresponding to the covering C_t. Clearly, the set of the orbit codings is contained in the set of admissible paths, i.e. $Cod_t \subset P_t$.

Theorem 18. (Strong shadowing property)

Let $\{C_l\}$ be a sequence of coverings which are consecutive subdivisions with diameters $d_l = d(C_l) \to 0$ as $l \to \infty$. Suppose that there exists a sequence of admissible paths $\{\omega_l \in P_l\}$, $\omega_l = \{i_k^l, k \in Z\}$, such that $\omega_l = s_l(\omega_{l+1})$. Then there exists the unique trajectory $T = \{x_{k+1} = f(x_k)\}$ such that $x_k \in M(i_k^l)$ for any l.

Proof. Fix $k \in Z$ and consider a sequence of the cells $\{M(i_k^l), l = 1, 2, \dots\}$. Since $s_l(i_k^{l+1}) = i_k^l$ then

$$
M(i_k^1) \supset M(i_k^2) \supset \dots \supset M(i_k^l) \supset M(i_k^{l+1}) \supset \dots.
\tag{2.9}
$$

As the cells are closed and their maximal diameters d_l tend to 0 then there is the only point

$$x_k = \lim_{l \to \infty} M(i_k^l) = \bigcap_l M(i_k^l).$$

As the path $\omega_l = \{i_k^l, \ k \in \mathbb{Z}\}$ is admissible then

$$f(M(i_k^l)) \cap M(i_{k+1}^l) \neq \emptyset$$

for any l.

As (2.9) holds and $d_l \to 0$, the closed sets $\{f(M(i_k^l)), \ l = 1, 2, ...\}$ are embedded and their maximal diameters q_l tend to 0. Then there exists the unique point

$$x_{k+1}^* = \lim_{l \to \infty} f(M(i_k^l)).$$

Similarly, the sequence of the closed sets $\{f(M(i_k^l)) \cap M(i_{k+1}^l), \ l = 1, 2, ...\}$ has the limit point

$$\lim_{l \to \infty} f(M(i_k^l)) \cap M(i_{k+1}^l) = x_{k+1}^{**}.$$

Since

$$x_{k+1}^{**} \in f(M(i_k^l)) \cap M(i_{k+1}^l) \subset M(i_{k+1}^l)$$

for any l then x_{k+1}^{**} is in $M(i_{k+1}^l)$, and as a consequence, in

$$\bigcap_l M(i_{k+1}^l) = x_{k+1}.$$

Hence $x_{k+1}^{**} = x_{k+1}$. As

$$x_{k+1}^{**} \in f(M(i_k^l)) \cap M(i_{k+1}^l) \subset f(M(i_k^l))$$

for any l then x_{k+1}^{**} is in $f(M(i_k^l))$, and as a consequence, in

$$\bigcap_l f(M(i_k^l)) = x_{k+1}^*,$$

i.e. $x_{k+1}^{**} = x_{k+1}^*$. Since $f(x_k) \in f(M(i_k^l))$ for any l, then $f(x_k)$ is in

$$\bigcap_l f(M(i_k^l)) = x_{k+1}^*,$$

i.e. $f(x_k) = x_{k+1}$.

$$\odot$$

Remark. According to this theorem for any sequence of paths $\{\omega_l\}$ there exists the only trajectory T. The converse does not hold: a trajectory may generate more than one sequence of the kind $\{\omega_l\}$. For example, a fixed point of the investigated map which is on the boundary of a cell generates infinitely many sequences of the described type.

3

Periodic Trajectories

The purpose of this chapter is localization and construction of periodic trajectories of a fixed period in a given compact domain. By localization we mean an algorithm which gives a sequence of neighborhoods for a desired set. The sequence is monotone decreasing, i.e., the neighborhoods are imbedded one inside the other, and converges to the desired set. The set desired is the set of p-periodic trajectories. By investigating the symbolic image one can separate the cells through which p-periodic trajectories may pass from those through which periodic trajectories do not pass. The union of these cells is a closed neighborhood of the desired set. Then we apply a method of adaptive subdivision for cells and construct a sequence of symbolic images which generates a sequence of embedded neighborhoods. It turns out that if the maximal diameter d of the divided cells tends to 0, the constructed sequence of neighborhoods converges to a set of p-periodic trajectories. On this way an algorithm for localization of periodic trajectories with a fixed period is obtained. Moreover, by Proposition 3 we can find the periodic ε-trajectories in each step of the algorithm. In the next chapter we apply the Newton method to find a sufficient condition for the existence of a true p-periodic trajectory near an ε-trajectory.

3.1 Periodic ε-Trajectories

Recall that a point $x \in M$ is periodic with period $p > 0$ if $f^p(x) = x$. The last equality leads $f^{kp}(x) = x$ for any $k \in N = \{1, 2, ...\}$, so if x is p-periodic then x is kp-periodic for any positive integer k. The set of all periodic points, called periodic set, is denoted by Per, and the set of periodic points of period p, called p-periodic, is denoted by $Per(p)$. Because Per and $Per(p)$ are unions of trajectories, they are invariant. Since a limit of p-periodic trajectories is a p-periodic trajectory, the set $Per(p)$ is closed. However, the set of all periodic trajectories Per may not be closed, because a limit of periodic trajectories may be non-periodic if least periods tend to infinity. It is clear that

$$\bigcup_{p \in \mathbb{N}} Per(p) = Per.$$

Definition 19. *Let $\varepsilon > 0$ and an infinite sequence $\{x_k, k \in \mathbb{Z}\}$ be an ε-trajectory of f. If the sequence $\{x_k\}$ is periodic then it is called a p-periodic ε-trajectory and the points x_k are called (p, ε)-periodic. We say that a point is ε-periodic if it is (p, ε)-period with some $p > 0$.*

Denote the sets of ε-periodic points and (p, ε)-periodic points by $Q(\varepsilon)$ and $Q(p, \varepsilon)$, respectively. From the above definition it follows that

$$Per(p) \subset Q(p, \varepsilon),$$

$$\bigcup_{p \in \mathbb{N}} Q(p, \varepsilon) = Q(\varepsilon),$$

$$Q(p, \varepsilon) \subset Q(np, \varepsilon),$$

for each $n \in \mathbb{N}$. The following proposition describes the properties of $Q(p, \varepsilon)$.

Proposition 20. *1. The sets $Q(p, \varepsilon)$ and $Q(\varepsilon)$ are open.*
2. If $\varepsilon_1 > \varepsilon_2 > 0$ then $Q(p, \varepsilon_2) \subset Q(p, \varepsilon_1)$ and $Q(\varepsilon_2) \subset Q(\varepsilon_1)$.
3.

$$\lim_{\varepsilon \to 0} Q(p, \varepsilon) = \bigcap_{\varepsilon > 0} Q(p, \varepsilon) = Per(p). \tag{3.1}$$

Proof. 1. Let a point x is in $Q(p, \varepsilon)$. There exists p-periodic ε-orbit $\{x = x_1, x_2, ..., x_p\}$ through x. Put $\max \rho(f(x_k), x_{k+1}) = r < \varepsilon$. Since the mapping f is uniformly continuous on the compact M then for $\varepsilon_1 = \epsilon - r$ there exists $\delta > 0$ such that from $\rho(y, z) < \delta$ it follows $\rho(f(y), f(z)) < \varepsilon_1$. Without loss of generality we can assume that $\delta < \varepsilon - r$. Let us show that δ-neighborhood of x is in $Q(p, \varepsilon)$. For a point y : $\rho(y, x) < \delta$ consider the p-periodic sequence $\{y = x_1^*, x_2, ..., x_p\}$. We have

$$\rho(f(y), x_2) \leq \rho(f(y), f(x)) + \rho(f(x), x_2) < \varepsilon - r + r = \varepsilon,$$

$$\rho(f(x_p), y) \leq \rho(f(x_p), x) + \rho(x, y) < r + \delta < \varepsilon.$$

So the p-periodic sequence $\{y = x_1^*, x_2, ..., x_p\}$ is an ε-trajectory, and $y \in Q(p, \varepsilon)$. Thus the set $Q(p, \varepsilon)$ is open. In a similar manner, the set $Q(\varepsilon)$ is open.

2. Each ε_2-trajectory is an ε_1-trajectory provided $\varepsilon_1 > \varepsilon_2$. The inclusions stated follow directly from the given fact.

3. It is evident that $Per(p) \subset \bigcap_{\varepsilon > 0} Q(p, \varepsilon)$. Let us show the opposite inclusion. Suppose that $x \in \bigcap_{\varepsilon > 0} Q(p, \varepsilon)$, i.e., for each $\varepsilon > 0$ there exists a p-periodic orbit through x $\{x = x_1, x_2(\varepsilon), ..., x_p(\varepsilon)\}$ such that $\rho(f(x_k(\varepsilon)), x_{k+1}(x\varepsilon)) < \varepsilon$. Since M is a compact there is a subsequence $\varepsilon_l \to 0$ as $l \to \infty$ such that $\lim_{l \to \infty} x_k(\varepsilon_l) = x_k^*$ exists for each k. Let us show that the p-periodic sequence $\{x = x_1, x_2^*, ..., x_p^*\}$ is an orbit. We have

$\rho(f(x_1), x_2^*) \le \rho(f(x_1), x_2(\varepsilon_l)) + \rho(x_2(\varepsilon_l), x_2^*) \to 0$, i.e., $f(x_1) = x_2^*$. In the similar manner, we obtain $f(x_k^*) = x_{k+1}^*$. Thus, the point x is p-periodic and $\bigcap_{\varepsilon>0} Q(p, \varepsilon) \subset Per(p)$.

$\odot$

Thus, the open sets $\{Q(p, \varepsilon), \varepsilon > 0\}$ forms a fundamental family of neighborhoods of p-periodic set $Per(p)$. (Recall that a sequence $\{U_k, \ k \in \mathbb{N}\}$ forms a fundamental family of neighborhoods of a set A if $A \subset U_k, \ k \in \mathbb{N}$, and for each neighborhood U of A there exists $U_k \subset U$.)

Definition 21. *A vertex of the symbolic image is called p-periodic if a periodic path of the period p passes through it.*

The p-periodic vertices $\{i\}$ are uniquely determined by nonzero diagonal entries $\{\pi_{ii}^p \ne 0\}$ of powers Π^p of the transition matrix, (see Subsection 2.4). Denote by $P(p, d)$ the union of cells $M(i)$ for which the vertices are p-periodic:

$$P(p, d) = \left\{ \bigcup M(i) : i \text{ is } p - periodic \right\},$$

where by $d = d(C)$ is the diameter of the covering C. Notice that the set $P(p, d)$ depends on the covering C. However, in what follows we need only to consider the dependence of P on the largest diameter d. Let us denote by $T(p, d)$ the union of the cells $M(k)$ for which the vertices k are not p-periodic:

$$T(p, d) = \left\{ \bigcup M(k) : k \text{ is not } p - periodic \right\}.$$

Theorem 22. *1. The set $P(p, d)$ is a closed neighborhood of the p-periodic set $Per(p)$. Moreover, $P(p, d)$ is a subset of (p, ε)-periodic points set for any $\varepsilon > q + d$, i.e.,*

$$P(p, d) \subset Q(p, \varepsilon), \ \varepsilon > q + d.$$

2. For any neighborhood V of $Per(p)$ there exists $d_0 > 0$ such that

$$Per(p) \subset P(p, d) \subset V, \ d < d_0,$$

i.e., the neighborhood $P(p, d)$ is small provided the largest diameter d is small enough.

3. The p-periodic set $Per(p)$ coincides with intersection of the sets $P(p, d)$ for all positive d:

$$Per(p) = \bigcap_{d>0} P(p, d). \tag{3.2}$$

4. The points of $T(p, d)$ are not p-periodic. Moreover, if $\varepsilon < r$ there is no p-periodic ε-trajectory passing through x from $T(p, d)$, i.e.,

$$Q(p, \varepsilon) \bigcap T(p, d) = \emptyset, \ \varepsilon < r.$$

Proof. 1. Let ε_1 and ε_2 be such that $\varepsilon_1 < r < q + d < \varepsilon_2$. At first we prove that

$$Q(p, \varepsilon_1) \subset P(p, d) \subset Q(p, \varepsilon_2). \tag{3.3}$$

In fact, if a point x belongs to $Q(p, \varepsilon_1)$ then there exists a p-periodic ε_1-trajectory $\{x_1, \ldots, x_p\}$ passing through $x = x_1$. Consider the sequence of cells $\{M(z_i)\}$ with $x_i \in M(z_i)$. Because $\varepsilon_1 < r$, according to the Theorem 1, item 3, the sequence $\{z_1, \ldots, z_p\}$ is a periodic path through the vertex z_1. Thus, the vertex z_1 is p-periodic. Hence, the cell $M(z_1)$ containing x is in $P(p, d)$. From this it follows that $Q(p, \varepsilon_1) \subset P(p, d)$.

Consider a point x belonging to $P(p, d)$. There exists a cell $M(z)$ such that $x \in M(z)$. The vertex z is recurrent with period p. In other words, on the symbolic image G there exists a periodic path $\{z_1, \ldots, z_p\}$, $z_1 = z$. Let us construct a periodic sequence $\{x_1, \ldots, x_p\}$, so that $x_1 = x$ and $x_i \in M(z_i)$. By Theorem 1, item 1, the sequence $\{x_1, ..., x_p\}$ is a periodic ε_2-trajectory with period p for any $\varepsilon_2 > q + d$. Hence, the point $x = x_1$ lies in $Q(p, \varepsilon_2)$. Since x is a point from $M(z)$ we have the inclusion $P(p, d) \subset Q(p, \varepsilon_2)$. Thus (3.3) holds. From the inclusions $Per(p) \subset Q(p, \varepsilon_1) \subset P(p, d)$ it follows that $P(p, d)$ is a closed neighborhood of the p-periodic set $Per(p)$ and $P(p, d) \subset Q(p, \varepsilon)$, $\varepsilon > q + d$.

2. Let V be an arbitrary neighborhood of Q. Since f is a continuous mapping and M is compact, the largest diameter q of the images $f(M(i))$ tends to 0 as the largest diameter of cells $d \to 0$. Set $\varepsilon_2 = \frac{3}{2}(q + d)$. We have $\varepsilon_2 \to 0$ as $d \to 0$. Because $\{Q(\varepsilon), \varepsilon > 0\}$ is a fundamental system of neighborhoods for Q, there is $\varepsilon^* > 0$ so that $Q(\varepsilon^*) \subset V$. Shoose d_0 so that $\varepsilon_2 = \frac{3}{2}(q(d_0) + d_0) \le \varepsilon^*$. For such $d < d_0$ we have $P(d) \subset Q(\varepsilon_2) \subset Q(\varepsilon^*) \subset V$ by Proposition 20 and inclusion (3.3).

3. Since f is a continuous mapping and M is compact, the largest diameter q of the images $f(M(i))$ tends to 0 as the largest diameter of cells $d \to 0$. Set $\varepsilon_1 = (1/2)r$, $\varepsilon_2 = \frac{3}{2}(q + d)$. By Proposition 11, $r \le d$ and we have $\varepsilon_1 \to 0$ and $\varepsilon_2 \to 0$ as $d \to 0$. Because $\{Q(p, \varepsilon), \varepsilon > 0\}$ is a fundamental system of neighborhoods of $Per(p)$ we have

$$Per(p) = \bigcap_{\varepsilon > 0} Q(p, \varepsilon) = \lim_{\varepsilon \to 0} Q(p, \varepsilon) = \lim_{\varepsilon_1 \to 0} Q(p, \varepsilon_1) = \lim_{\varepsilon_2 \to 0} Q(p, \varepsilon_2).$$

The last equalities and inclusion (3.3) imply the equality (3.2).

4. We prove the statement by contradiction. Let $x \in M(k)$, where k is not p-periodic. Let $\{x_1, \ldots, x_p\}$ be a p-periodic ε-trajectory passing through $x = x_1$ and $\varepsilon < r$. Consider a sequence $\{z_1, \ldots, z_p\}$ such that $z_1 = k$, $x_i \in M(z_i)$. As $\varepsilon < r$, by Theorem 1, the sequence $\{z_1, ..., z_p\}$ is a p-periodic path on the symbolic image G. Because $z_1 = k$, the vertex k is p-periodic. The obtained contradiction completes the proof of the theorem.

⊙

By construction, the set $T(p, d)$ is closed and the pair $\{P(p, d), \ T(p, d)\}$ forms a closed covering of M. Hence, $P(p, d) \backslash T(p, d)$ is an open neighborhood of the p-periodic set $Per(p)$. Theorem 22 implies the following inclusions:

$$Per(p) \subset Q(p, \varepsilon_1) \subset M \setminus T(p, d) = P(p, d) \setminus T(p, d) \subset P(p, d) \subset Q(p, \varepsilon_2),$$

where $\varepsilon_1 < r < q + d < \varepsilon_2$. Thus, the set $P(p, d)$ contains the neighborhood $Q(p, \varepsilon_1)$ and is contained in $Q(p, \varepsilon_2)$. However, in general, the set $P(p, d_1)$ is not embedded in $P(p, d_2)$ even though $d_1 < d_2$. Theorem 22 makes possible to get a neighborhood of the p-periodic set $Per(p)$ without any preliminary information on dynamical system.

3.2 Localization Algorithm

Let us describe the algorithm that localizes the p-periodic set. We apply the process of adoptive subdivision. In this context the adaptive subdivision means that some cells are excluded, while others are subdivided.

Algorithm localizing the p-periodic set.

1. Starting with an initial covering C, the symbolic image G of the map f is found. Notice that cells of the initial covering may have an arbitrarily diameter d_0.
2. The p-periodic vertices $\{i_k\}$ on the graph G are recognized. Using the p-periodic vertices, a closed neighborhood $P = \{\bigcup M(i_k) : i_k \text{ is } p\text{-periodic}\}$ of the p-periodic set $Per(p)$ is found.
3. The cells $\{M(i_k)\}$ corresponding to the p-periodic vertices are subdivided. For example, the largest diameter of the cells may be divided by 2. Denote by d the maximal diameter of new cells after the subdivision. A new covering is defined.
4. The symbolic image G is constructed for the new covering. Notice that the new symbolic is constructed only for the set $P = \{\bigcup M(i_k) : i_k \text{ is } p\text{-}periodic\}$. In other words, cells corresponding to non p-periodic vertices are excluded from construction process in the new covering and the new symbolic image.
5. Then one goes back to the second step.

By repeating the process of adaptive subdivision we obtain (by Theorem 22) a sequence of neighborhoods P_0, P_1, $P_2, \ldots$ for the p-periodic set $Per(p)$ and a sequence of largest diameters d_0, d_1, $d_2, \ldots$. The following theorem substantiates the algorithm outlined above.

Theorem 23. *The sequence of sets* $P_0, P_1, P_2, \ldots$ *offers the following properties:*

1. The neighborhoods P_k *are imbedded, i.e.*

$$P_0 \supset P_1 \supset P_2 \supset \ldots \supset Per(p),$$

2. If the largest diameters $d_k \to 0$ as $k \to \infty$ then

$$\lim_{k \to \infty} P_k = \bigcap_k P_k = Per(p).$$

Proof. 1. Let $C = \{M(i)\}$ be a covering of M, and G be the symbolic image for the covering C. Suppose a new covering NC is produced by an adaptive subdivision of C. Denote by NG the symbolic image for NC. Cells of the covering NC are denoted by $m(i, k)$ so that $\bigcup_k m(i, k) = M(i)$ and the vertices of NG are designated as (i, k). Since $f(m(i, k)) \subset f(M(i))$ and $m(j, l) \subset M(j)$, we have $f(M(i)) \bigcap M(j) \neq \emptyset$, provided $f(m(i, k)) \bigcap m(j, l) \neq \emptyset$. Hence, the existence of the edge $(i, k) \to (j, l)$ leads to the existence of the edge $i \to j$. Define a mapping h from the vertices of NG to the vertices of G so that $h(i, k) = i$ The mapping $h : NG \to G$ takes a directed edge $(i, k) \to (j, l)$ on the directed edge $i \to j$. This implies that every path on the graph NG is transformed on some path of G. In particular, p-periodic paths on the graph NG are mapped onto p-periodic paths of the graph G. Hence, p -periodic vertices of NG are mapped into p-periodic vertices of G. Set

$$NP = \left\{ \bigcup m(i, k) : (i, k) \text{ is } p\text{-periodic on } NG \right\}.$$

From the previous it follows that NP is a subset of $P : NP \subset P$. Thus, the described algorithm of localization gives the sequence of imbedded neighborhoods $P_0 \supset P_1 \supset P_2 \supset \ldots$ of the p-periodic set.

2. Let $d_k \to 0$ as $k \to \infty$. Inclusions (3.3) hold for an arbitrary symbolic image. Hence, these inclusions hold for the neighborhoods P_k. Thus, $P_k \subset Q(p, \varepsilon_k)$ provided $q_k + d_k < \varepsilon_k$, where d_k and q_k are the largest diameters of cells and their images, respectively. Set $\varepsilon_k = (3/2)(q_k + d_k)$. Since X is a continuous mapping and M is compact, $q_k \to 0$ as $d_k \to 0$. Thus, we have $\varepsilon_k \to 0$ as $k \to \infty$. Because $Per(p) \subset P_k \subset Q(p, \varepsilon_k)$ and

$$\lim_{k \to \infty} Q(p, \varepsilon_k) = \bigcap_k Q(p, \varepsilon_k) = Per(p),$$

we obtain

$$Per(p) \subset \bigcap_k P_k \subset \bigcap_k Q(p, \varepsilon_k) = Per(p), \ i.e.,$$

$$\lim_{k \to \infty} P_k = \bigcap_k P_k = Per(p).$$

$\odot$

Example 24. Periodic orbits of the Ikeda mapping.
Let us consider a mapping of the form

$$T : z \to R + C_2 z e^{i(C_1 - C_3/(1+|z|^2))},$$

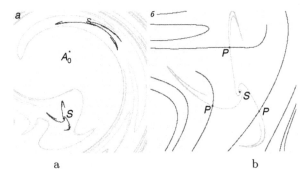

Fig. 3.1. Periodic orbits of the Ikeda mapping

where $z = xi + y$ is complex, d, C_1, C_2, C_3 are real constants (mapping parameters). The Ikeda map occurs in the modeling of optical recording media (crystals) [60]. Numerical simulations of the dynamical behavior of the map T have been carried out with $C_1 = 0.4, C_2 = 0.9, C_3 = 6.0$, and R $= 0.6$. By applying the method described above we detected that the dynamical system has the sink A_0 (0.3397, 0.2809), the 2-periodic hyperbolic orbit H (1.0094, −0.1100), (−0.2110, −0.4211), and the 2-periodic sink S (0.5997, 0.6757), (0.2188, −0.7184) (see Fig. 3.1,a). Near the sink S there is the 6-periodic hyperbolic orbit P (0.1869, −0.5785), (0.3556, 0.7053), (0.2818, −0.7800), (0.6249, 0.6969), (0.1343, −0.7635), (0.8751, 0.4730). The displacement of the point (0.2188, −0.7184) of the orbit S and three points (0.1869, −0.5785), (0.2818, −0.7800), (0.1343, −0.7635) of the orbit P is shown in the Fig. 3.1,b.

4

Newton's Method

The algorithm of localization of the set of p-periodic trajectories does not allow to determine in details the position of a single trajectory. Nevertheless, it enables to construct a p-periodic pseudo-trajectory. Actually, given a p-periodic path $\{z_1, \ldots, z_p\}$ on a symbolic image G, we construct a sequence $\{x_1, \ldots, x_p\}$ with an arbitrary chosen point x_i in the cell $M(z_i)$. According to results of the previous chapter this sequence is a periodic ε-trajectory for any $\varepsilon > d + q(d)$, where d and $q(d)$ are parameters of G. Since for a continuous map f the value $d + q(d) \to 0$ as $d \to 0$, we are able to construct p-periodic trajectories for small values $\varepsilon > 0$. Notice that we cannot be sure that on any step of the algorithm there is a true trajectory $\{y_1, \ldots, y_p\}$ that shadows the constructed ε-trajectory $\{x_1, \ldots, x_p\}$. To solve this problem we apply Newton's method [67] which not only ensures the existence of a true trajectory $\{y_1, \ldots, y_p\}$ but also allows to estimate the distance between the true trajectory and an ε-trajectory. This permits to construct a sequence of periodic pseudo trajectories converging to a true periodic trajectory.

In this chapter we establish a *shadowing theorem* which gives sufficient conditions for the existence of a true p-periodic trajectory near an approximate periodic one, being these conditions are formulated as the conditions of convergence of Newton's method. Therefore we obtain an algorithm for construction of successive approximations of a true trajectory.

4.1 Basic Results

Theorem 25. [67] *Let V, W be open sets in a Banach space H, $0 \in W$. Suppose that a mapping $F : V \to W$ is differentiable and F' meets the Lipschitz condition with constant L. Let the operator $F'(x_0)$ be invertible at a point $x_0 \in V$ and $KRL < 1/4$, where $K = \|(F'(x_0))^{-1}\|$, $R = \|(F'(x_0))^{-1}F(x_0)\|$. Assume that the ball $\{x : \|x - x_0\| \le 2R\}$ lies in V. Then the equation $F(x) = 0$ has unique solution $x^* \in V$ and*

$$\|x_n - x^*\| \le \frac{R}{2^{n-1}}, \tag{4.1}$$

where

$$x_{n+1} = x_n - (F'(x_0))^{-1} F(x_n), \ \ n = 0, 1, 2, \dots . \tag{4.2}$$

Let x_0 be an approximate solution of the equation $F(x) = 0$, i.e. $\|F(x_0)\|$ be small enough. If $\det F'(x_0) \ne 0$ then there exists the inverse matrix $(F'(x_0))^{-1}$. So we can compute $K = \|(F'(x_0))^{-1}\|$ and $R = \|(F'(x_0))^{-1} F(x_0)\|$. Easily seen that R is sufficiently small if $\|F(x_0)\|$ is small, i.e. x_0 is a good approximation to the solution. The Lipschitz constant L can be estimated by means of the second derivative of F. If $KRL < 1/4$ then firstly, we can assert that there exists the true solution x^*, and secondly, for the approximations constructed by formula (4.2) the inequality (4.1) holds.

Let f be a diffeomorphism defined on a manifold M and $\{x_1, x_2, \dots, x_p\}$ be a p-periodic ε-trajectory of f. As M is a manifold, there are neighbourhoods $V(x_i) \equiv V_i$ which we identify with balls of radii a_i. Set $D = \bigcup_i V_i, i = 1, \dots, p$.

Theorem 26. *Suppose that the following conditions are fulfilled:*

1. *f' meets Lipschitz condition with constant L in neighborhoods V_i, $i = 1, 2, \dots, p$;*
2. *the mapping $C = A_p A_{p-1} \dots A_1 - I$ is invertible, where $A_i = f'(x_i)$, $i = 1, 2, \dots, p$;*
3. *for every $i = 1, 2, \dots, p$,*

$$a_i \ge 2K \frac{a^p - 1}{a - 1} \varepsilon \equiv \sigma,$$

where $K = \max_i \|(A_{i-1}A_{i-2}\dots A_1 A_p \dots A_i - I)^{-1}\|$, $a = \max_D \|f'(x)\|$;
4. *$LK^2(\frac{a^p-1}{a-1})^2 \varepsilon < \frac{1}{4}$.*

Then there exists the unique periodic trajectory $\{y_1, y_1, \dots, y_p\}$ of f such that $\|x_i - y_i\| < \sigma$, $i = 1, 2, \dots, p$.

Proof. For convenience we identify each p-periodic ε-trajectory $\{x_k, \ x_k = x_{k+p}\}$ with a finite sequence $\{x_1, x_2, \dots, x_p\}$. Let $H = \oplus_{i=1}^p R^n$ be the Banach space of all finite sequences $v = \{v_1, v_2, \dots, v_p\}$, $v_i \in R^n$ endowed with the norm $\|v\| = max_i |v_i|$. Consider the set $D = \oplus_{i=1}^p V_i \subset H$ and the map F on D defined by

$$F(v) = \{f(x_1 + v_1) - (x_2 + v_2), \dots, f(x_p + v_p) - (x_1 + v_1)\}.$$

Apply Newton's method for the equation $F(v) = 0$ with the initial point $v_0 = 0$. The operator $F'(0)$ has the form

$$F'(0) = \begin{pmatrix} A_1 & -I & 0 & \dots & 0 \\ 0 & A_2 & -I & \dots & 0 \\ \dots & \dots & \dots & \dots & \dots \\ -I & 0 & 0 & \dots & A_p \end{pmatrix}. \tag{4.3}$$

At first we prove that $F'(0)$ is invertible. It follows from (4.3) that

$$u_{i+1} = A_i u_i - w_i, \ i = 1, \ldots, p-1; u_1 = A_p u_p - w_p. \tag{4.4}$$

Using the first equality (4.4) for $i = p-1, \ldots, 1$ we obtain

$$(A_{i-1} A_{i-2} \ldots A_1 A_p A_{p-1} \ldots A_i - I) u_i = t_i,$$

where $t_i = (A_{i-1} \ldots A_1 A_p \ldots A_{i+1} w_i + \ldots + A_{i-1} \ldots A_2 w_1 + \ldots + w_{i-1})$. By condition the operator $C = A_p A_{p-1} \ldots A_1 - I$ is invertible. From this it follows that the operators

$$C_i = (A_{i-1} A_{i-2} \ldots A_1 A_p A_{p-1} \ldots A_i - I) = A_{i-1} A_{i-2} \ldots A_1 C A_1^{-1} A_2^{-1} \ldots A_{i-1}^{-1}$$

$i = 1, 2, \ldots, p$ are invertible. Then

$$u_i = C_i^{-1} t_i, \tag{4.5}$$

and hence, $F'(0)$ is invertible. By (4.5) we obtain

$$\|u\| \le K(a^{p-1} + a^{p-2} + \ldots + 1) \|w\|. \tag{4.6}$$

As a is an estimation of the derivative norm, we can consider $a \ne 1$. From (4.6) it follows that

$$\|(F'(0))^{-1}\| \le K \frac{a^p - 1}{a - 1}.$$

For a periodic ε-trajectory $\{x_1, x_2, \ldots, x_p\}$ we have $F(0) = \{f(x_1) - x_2, \ldots, f(x_p) - x_1\}$, and

$$\|(F'(0))^{-1} F(0)\| \le K \frac{a^p - 1}{a - 1} \varepsilon.$$

Since the periodic trajectory $\{x_1, x_2, \ldots, x_p\}$ can be considered as a point of the Banach space H, we apply Theorem 25 and complete the proof.
$\odot$

Theorem 26 allows to formulate the following algorithm of construction of p-periodic trajectory.

1. Construct a p-periodic ε-trajectory $\{x_1, x_2, \ldots, x_p\}$ by methods of symbolic dynamics.
2. Verify the hypotheses of Theorem 26. If they are fulfilled then there exists a p-periodic trajectory $\{y_1, y_1, \ldots, y_p\}$ near the trajectory $\{x_1, x_2, \ldots, x_p\}$. Otherwise we should return to the step 1 and construct a periodic ε-trajectory using subdivision method.
3. Construct a sequence of p-periodic ε-trajectories $x^k = \{x_1 + v_1^k, x_2 + v_2^k, \ldots, x_p + v_p^k\}$ by the formula

$$v^{k+1} = v^k - (F'(0))^{-1} F(v^k),$$

where $F(v) = \{f(x_1 + v_1) - (x_2 + v_2), \ldots, f(x_p + v_p) - (x_1 + v_1)\}$.

According to Theorem 26 the sequence

$$\{x^k = \{x_1 + v_1^k, ..., x_p + v_p^k\}\}$$

converges to the true p-periodic trajectory $x^* = \{y_1, y_2, \ldots, y_p\}$ and

$$\left\|x^k - x^*\right\| \leq \frac{R}{2^{k-1}},$$

where $R = \left\|(F'(0))^{-1}F(0)\right\|$.

Thus, using symbolic dynamics methods and Newton's algorithm we can construct an approximation to a true periodic trajectory. The last inequality allows to control the distance between ε- and the true trajectories.

4.2 Component of Periodic ε-Trajectories

Let us consider conditions ensuring the existence of periodic trajectories.

Definition 27. *Let E be a nonempty subset of $Q(p, \varepsilon)$. The set E is said to be a component of $Q(p, \varepsilon)$, if for any $x \in E$ every p-periodic ε-trajectory passing through x lies in E and there is no any subset $E_1 \subset E$ with such a property.*

It is easy to check that a component E is open.

Proposition 28. *Let $\overline{E}$ be the closure of E and $\{x_1, x_2, ..., x_p\}$ be a p-periodic ε-trajectory, where $x_i \in \overline{E}$, $i = 1, 2, ..., p$. Then $x_i \in E$, $i = 1, 2, ..., p$, i.e. a p-periodic ε-trajectory cannot reach the boundary of a component.*

Proof. Suppose on the contrary that there is $x_j \in \overline{E} \setminus E$ for some j. Then there exists $x \in E$ near x_{j-1} such that the sequence $\{x_1, ..., x_{j-2}, x, x_j, ..., x_p\}$ is a p-periodic ε-trajectory. According to Definition 27 if $x \in E$ then $x_j \in E$. This leads us to a contradiction.

$$\odot$$

Theorem 29. *Let E be a component of $Q(p, \varepsilon)$ and for any p-periodic ε-trajectory $\{x_1, x_2, ..., x_p\}$, $x_i \in E$ the operator with the matrix $A_p A_{p-1}...A_1 - I$, $A_i = f'(x_i)$, is invertible. Then there exists a true p-periodic trajectory of f in E.*

Proof. Let Σ be the space of sequences $\{x_1, x_2, ..., x_p\}$, such that $x_i \in \overline{E}$, $\rho(f(x_i), x_{i+1}) \leq \varepsilon$, $i = 1, 2, ..., p$, and $x_{p+1} = x_1$. The space Σ is a compact with respect to the metric

$$d(\{x_1, ..., x_p\}, \{y_1, ..., y_p\}) = \max_i \rho(x_i, y_i).$$

Consider the function

$$F(\{x_i\}) = \max_i \rho(f(x_i), x_{i+1})$$

defined on Σ. As F is continuous and Σ is a compact then, there exists a sequence $\{y_i\} \in \Sigma$ such that

$$F(\{y_i\}) = \min_\Sigma F(\{x_i\}).$$

According to Proposition 28 $y_i \in E$, $i = 1, 2, \ldots, p$. To prove that the sequence $\{y_i\}$ is a true periodic trajectory it is sufficient to verify that $F(\{y_i\}) = 0$. Suppose on the contrary that $F(\{y_i\}) = \alpha > 0$. Since $y_i \in E$ and E is open the function F is well defined for all $\{x_i\}$ from some neighborhood of $\{y_i\}$ in Σ. In order to get a contradiction we find a sequence $\{u_i\}$ such that $F(\{y_i + u_i\}) < \alpha$ and $\{y_i + u_i\} \in \Sigma$. Let V_i be a neighborhood of y_i, which we identify with a ball in R^n equipped with the norm $| \, . \, |$. Setting $w_i = f(y_i) - y_{i+1}$ and $B_i(u_i) = f(y_i + u_i) - f(y_i)$ we have

$$f(y_i + u_i) - (y_{i+1} + u_{i+1}) = B_i(u_i) + w_i - u_{i+1}.$$

We prove that for some sufficiently small $\delta > 0$ there exists a solution $\{u_i\}$ of the equation

$$B_i(u_i) - u_{i+1} = -\delta w_i. \tag{4.7}$$

Consider the map B of the form

$$B(\{u_i\}) = \{B_1(u_1) - u_2, B_2(u_2) - u_3, \ldots, B_p(u_p) - u_1\}.$$

Obviously, $B(0) = 0$ and

$$det B'(0) = det \begin{pmatrix} A_1 & -I & 0 & \ldots & 0 \\ 0 & A_2 & -I & \ldots & 0 \\ \ldots & \ldots & \ldots & \ldots & \ldots \\ -I & 0 & 0 & \ldots & A_p \end{pmatrix} = det(A_p \ldots A_1 - I) \neq 0.$$

By the Implicit Function Theorem, equation (4.7) has a solution for each sufficiently small $\delta > 0$.

Then

$$f(y_i + u_i) - (y_{i+1} + u_{i+1}) = (1 - \delta)w_i.$$

In this case we have

$$F(\{y_i + u_i\}) = (1 - \delta)F(\{y_i\}) < \alpha,$$

that leads us to a contradiction. Hence, $F(\{y_i\}) = 0$ and $\{y_i\}$ is a true periodic trajectory.

$\odot$

4.3 Component of Periodic Vertices

Let G be a symbolic image of a diffeomorphism f which corresponds to a covering C. Denote by $G(p)$ the set of all p-periodic vertices of the graph G.

Definition 30. *A nonempty set $\Lambda \subset G(p)$ is said to be a component of $G(p)$, if for every vertex $i \in \Lambda$ any p-periodic path passing through i lies in Λ and there is no nonempty subsets of Λ with this a property.*

Denote by $R(p)$ the set $\{\bigcup M(i) : i \in \Lambda\}$ and let a_i be the centers of the cells $M(i)$.

Definition 31. *A component Λ is called isolated if the sets $\{\bigcup M(i) : i \in \Lambda\}$ and $\{\bigcup M(k) : k \in G(p) \setminus \Lambda\}$ do not intersect.*

Let C be a covering of M by the balls $M(i) = V(a_i, d/2)$ with radii $d/2$ and the centers a_i. Consider a covering C^* of M by the balls $M^*(i) = V(a_i, d)$. Let G^* be the symbolic image associated with the covering C^* and $G^*(p)$ be the set of all p-periodic recurrent vertices in G^*.

Proposition 32. *Let Λ^* be an isolated component of $G^*(p)$, $R^*(p) = \{\bigcup M^*(i) : i \in \Lambda^*\}$ and E be a component of $Q(p, d/2)$. Then either $E \subset R^*(p)$ or $E \cap R^*(p) = \emptyset$.*

Proof. Consider a set $E_1 = E \cap R^*(p)$. Let $E_1 \neq \emptyset$. It is sufficient to prove that $E_1 = E$, i.e., each p-periodic $d/2$-trajectory $\{x_1, ..., x_p\}$ lies in E_1 provided $x_1 \in E_1$.

Consider a periodic $d/2$-trajectory $\{x_1, ..., x_p\}$, $x_1 \in E_1$. Each x_k, $k = 1, ..., p$, lies in E because E is a component of $Q(p, d/2)$. Let us prove that $x_k \in R^*(p)$. As C is a covering of M there are the vertices $i_1, ..., i_p \in G$ such that $\rho(x_k, a_{i_k}) \leq d/2$, where a_{i_k} is the centre of the ball $M^*(i_k) = V(a_{i_k}, d/2)$.

We have to verify that $f(x_k) \in M^*(i_{k+1})$, $k = 1, ..., p$, $i_{p+1} = i_1$. As

$$\rho(f(x_k), a_{i_{k+1}}) \leq \rho(f(x_k), x_{k+1}) + \rho(x_{k+1}, a_{i_{k+1}}) < d/2 + d/2 = d,$$

for each k we have

$$f(M^*(i_k)) \bigcap M^*(i_{k+1}) \neq \emptyset.$$

Hence, the p-periodic sequence $\{i_1, ..., i_p\}$ is an admissible path on G^*. By construction, i_1 is a p-periodic vertex, $x_1 \in R^*(p) \bigcap M^*(i_1) \neq \emptyset$ and as Λ^* is isolated, we obtain that $i_1 \in \Lambda^*$. Since Λ^* is a component of $G^*(p)$, then that $i_k \in \Lambda^*$, $k = 1, ..., p$. Thus, $x_k \in M^*(i_k) \subset R^*(p)$ and as a consequence, $R^*(p) \bigcap E = E$.   ⊙

By Proposition 32, in order to establish that $R^*(p) \subset Q(p, d/2)$ it is sufficient to verify that

$$R^*(p) \bigcap Q(p, d/2) \neq \emptyset. \tag{4.8}$$

The last relation means the following. Let G^* and C^* be as above. Subdivide the cells $M^*(i_k)$ in $R^*(p)$ so that their maximal diameter $d_1 < d/2$. As above let $m(i,j)$ be a subdivisions of the cell $M(i)$, i.e. $\bigcup_j m(i,j) = M(i)$. If the obtained covering has a p-periodic cell $m(i,j)$ lying in $R^*(p)$, then by Proposition 32 there exists a p-periodic $d/2$-trajectory passing through $m(i,j)$. This gives (4.8). The following theorem establishes sufficient conditions of existence of a true periodic trajectory in a component of periodic ε-trajectories.

Theorem 33. *Let C^*, G^*, Λ^*, $R^*(p)$ be as above. Assume that $R^*(p) \bigcap Q(p, d/2) \neq \emptyset$ and for any p-periodic path $\{z_1, z_2, ..., z_p\}$ in Λ^* the operator $B_p B_{p-1}...B_1 - I$, $B_i = f'(a_i)$, is invertible, a_i are centers of the cells $M^*(z_i)$, and*

$$\|(B_p B_{p-1}...B_1 - I)^{-1}\|(a + \alpha(d/2))^p - a^p) < 1,$$

$\alpha(\, . \,)$ is the module of continuity of f' on $R^*(p)$ and a as in Theorem 26. Then there exists a true p-periodic trajectory $\{y_1, y_2, ..., y_p\}$ of f, which lies in $R^*(p)$.

Proof. By Proposition (32) there is a component E of $Q(p, d/2)$ contained in $R^*(p)$. According to Theorem 29, it is sufficient to verify that the operator $A_p A_{p-1}...A_1 - I$ is invertible for each periodic $d/2$-trajectory $\{x_1, x_2, ..., x_p\}$ in E, where $A_i = f'(x_i)$. Let a_{z_i} be the center of the cell $M^*(z_i)$ such that $\rho(x_i, a_{z_i}) \leq d/2$. Then $|A_i - B_i| \leq \alpha(d/2)$, where $B_i = f'(a_{z_i})$. Hence,

$$\left| \prod_{i=1}^{p} A_i - \prod_{i=1}^{p} B_i \right| \leq (a + \alpha(d/2))^p - a^p.$$

To prove that the operator $\prod_{i=1}^{p} A_i - I$ is invertible it is sufficient to prove that

$$\left| \left(\prod_{i=1}^{p} A_i - I \right) u \right| \geq \mu|u|$$

for any $u \in R^n$. It is easy to see that

$$\left| \left(\prod_{i=1}^{p} A_i - I \right) u \right| \geq \left| \left(\prod_{i=1}^{p} B_i - I \right) u \right| - \left| \left(\prod_{i=1}^{p} B_i - \prod_{i=1}^{p} A_i \right) u \right|.$$

Since the operator $\prod_{i=1}^{p} B_i - I$ is invertible, we have

$$\left| \left(\prod_{i=1}^{p} A_i - I \right) u \right| \geq \left| \left(\prod_{i=1}^{p} B_i - I \right)^{-1} \right|^{-1} |u| - ((a + \alpha(d/2))^p - a^p)|u| = \mu|u|,$$

where $\mu = |(\prod_{i=1}^{p} B_i - I)^{-1}|^{-1}(1 - |(\prod_{i=1}^{p} B_i - I)^{-1}|((a+\alpha(d/2))^p - a^p))$. As $\mu > 0$ the operator $\prod_{i=1}^{p} A_i - I$ is invertible. This completes the proof. $\odot$

An application of the Newton's method is available in the Appendix A.

5

Invariant Sets

The present chapter is devoted to the localization of invariant sets of various types which are maximal (up to inclusion) in a given compact. The algorithm for localization is based on the symbolic image construction and is a modification of the algorithm for searching strongly connected components. To localize a positive invariant set of a system we associate the set of so-called non-leaving vertices and construct neighborhood of the positive invariant set. Using the adaptive subdivision method one can construct a sequence of imbedded sets converging to the desired set. The results of this chapter were obtained by S. Kobyakov, D. Matiassevitch and G. Osipenko.

5.1 Definitions and Examples

Let $f : M \to M$ be a homeomorphism generating a discrete dynamical system on a manifold M.

Definition 34. *The set I^+ is called positive invariant if for any point from I^+ its image lies in I^+, i.e. the inclusion $x \in I^+$ implies $f(x) \in I^+$. Similarly, the set I^- is called negative invariant if $f^{-1}(x) \in I^-$ for any $x \in I^-$. The set I is called invariant if $f(x) \in I$ and $f^{-1}(x) \in I$ for any $x \in I$.*

From the definition it follows that a positive (negative) invariant set is the union of positive (negative) half-orbits. The invariant set I consists of entire orbits.

Let K be a given compact in M. We denote the maximal up to inclusion positive invariant set in K by I^+, the maximal negative invariant set by I^- and by I the intersection of I^+ and I^-. It is easy to see that I^+ consists of positive half-orbits lying in the compact considered. The set I consists of orbits entirely lying in K.

Invariant sets of continuous dynamical systems. Let us consider a continuous dynamical system described by a flow $F(t, x)$, $x \in M$, and $t \in R$.

Definition 35. *The set* I^+ *is called positive invariant if for any point* x_0 *from* I^+ *and* $t \geq 0$ *the image* $F(t, x_0)$ *lies in* I^+, *Similarly, the set* I^- *is called negative invariant if for any point* x_0 *from* I^- *and* $t \leq 0$ *the image* $F(t, x_0)$ *lies in* I^-. *The set* I *is called invariant if for any point* x_0 *from* I *and* $t \in R$ *the image* $F(t, x_0)$ *lies in* I.

Let K be a given compact in M. The maximal invariant sets are defined the same way as for discrete systems. For the given continuous system F we defined the shift operator $f(x) = F(1, x)$ which generates the discrete dynamical system. It is necessary to note that the invariant sets constructed for the continuous system F may be different from the ones for the discrete system f (see the Example 36). It can be proved that the invariant sets of f contain the invariant sets of F.

Example 36. Invariant sets of linear system.
In the plane consider the linear system of differential equations

$$\dot{x} = -x,$$
$$\dot{y} = y.$$

Clearly, the system has a hyperbolic saddle equilibrium at the origin $(0, 0)$. Let K be a square-shaped region centered at the origin. Let f be a time-one shift along trajectories of the system. Then (see Fig. 5.1) I^+ is a line segment $\{y = 0\} \bigcap K$ on the x-axis, I^- is a line segment $\{x = 0\} \bigcap K$ on the y-axis, I is the equilibrium point, and $I^- \cup I^+$ is the union of two line segments on the axes.

The main results of this chapter are algorithms for the localization of the sets I^+, I^-, I, and $I^+ \cup I^-$. Localization problems as described above are widely met in applications. As an example, if one needs to find initial data of trajectories that do not leave a given domain then the positive invariant set I^+ ought to be constructed.

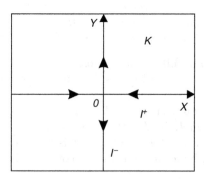

Fig. 5.1. Positive and negative invariant sets

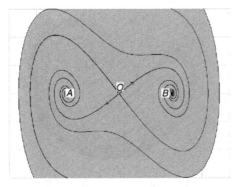

Fig. 5.2. Invariant sets I^+ and $I^- = I$ for the damped Duffing equation

Example 37. Duffing equation.
Consider the damped Duffing equation

$$\dot{x} = y,$$
$$\dot{y} = x - 0.27x^3 - 0.48y$$

in the domain $K = [-4.3, 4.3] \times [-3, 3]$. In Fig. 5.2, the positive invariant set I^+ maximal in K is colored in grey.

The domain K contains three equilibrium points A, O, and B. The equilibriums A and B are focuses (sinks), while O is a hyperbolic saddle. The invariant sets I^- and I coincide and consist of the equilibriums indicated and the unstable separatrices of the hyperbolic saddle O, which approach the equilibriums A and B by spirals.

The invariant sets I, I^+, and I^- may be extremely complicated and even have a fractional dimension. The last relates generally to the existence of chaotic domains in dynamical systems.

Example 38. Modified Ikeda mapping.
Consider the modified Ikeda mapping:

$$T : (x, y) \rightarrow (r + a(x \cos \tau - y \sin \tau), b(x \sin \tau + y \cos \tau)),$$

where $r = 2, a = -0.9$, and $b = 0.9, \tau = C_1 - C_3/(1 + x^2 + y^2) \, C_1 = 0.4, C_3 = 6$. We notice that for these parameters, the mapping T is orientation reversing. In the domain $K = [-2, 4] \times [-3, 2]$ the invariant sets I and I^- coincide and are depicted in Fig. 5.3. Locally, this invariant set is homeomorphic to a product of a segment and the Cantor set and has a fractal dimension between 1 and 2.

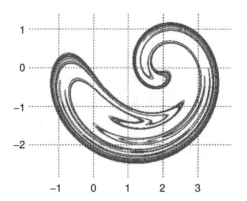

Fig. 5.3. Invariant set of the orientation reversing Ikeda mapping

5.2 Symbolic Image and Invariant Sets

Let $C = \{M(1), \ldots, M(n)\}$ be a covering of a compact K by closed sets (cells) $\{M(i)\}$. We set $d = \max\{\text{diam} M(i)\}$. Let G be the symbolic image of the mapping f with respect to the covering C. A cell $M(i)$ is called an outgoing (from K) cell if $f(M(i)) \bigcap K = \emptyset$. A cell $M(i)$ is called a leaving cell if all the admissible paths from the corresponding vertex of the symbolic image lead to vertices corresponding to outgoing cells. Vertices of the symbolic image will be also called "outgoing" and "leaving". It is obvious that a vertex is "non-leaving" if there exists an infinite path going through the vertex. We denote the set of non-leaving vertices of the graph G by V^+.

Theorem 39. *Let* $M_d^+ = \{\bigcup M(i) : i \in V^+\}$ *be the union of all non-leaving cells of the covering* C, $d = \max \text{diam } M(i)$. *Then*

1) M_d^+ *is a closed neighborhood of the positive invariant set* I^+ *maximal in* K, *i.e.*
$$I^+ \subset M_d^+,$$

2) if C_2 *is a covering which is the subdivision of a covering* C_1, *then*
$$I^+ \subset M_{d_2}^+ \subset M_{d_1}^+,$$

3) if C_k *be a sequence of coverings such that* C_{k+1} *is the subdivision of* C_k, *and the maximal diameter* $d_k \to 0$ *as* $k \to \infty$ *then*
$$\lim_{k \to \infty} M_k^+ = \bigcap_k M_k^+ = I^+,$$

where M_k^+ *is the union of all non-leaving cells of* C_k.

Proof. 1. Suppose the contrary, i.e. there exist $x \in I^+$, but $x \notin M^+$. Then, by definition of M^+, the point x is in a leaving cell $M(i) \subset K$. There exist such

$n \in \mathbb{N}$ and an outgoing cell $M(j)$ so that $f^n(x) \in M(j)$. Hence $f^{n+1}(x) \notin K$ which contradicts the assumption $x \in I^+ \subset K$.

2. Let C_2 be a subdivision of C_1. We denote the symbolic images of f with respect to C_1 and C_2 by G_1 and G_2. Let M_1^+ and M_2^+ be the unions of all non-leaving cells of the coverings C_1 and C_2 respectively. Consider the cells $M^2(i) \in M_2^+$ and $M^1(j)$ so that $M^2(i) \subseteq M^1(j) \in C_1$. It is sufficient to proof that $M^1(j)$ is non-leaving. It follows from the definition of M_2^+ that there exists an infinite path $i_1, i_2, \ldots$ on G_2 with the initial vertex $i_1 = i$. An infinite path $j_1, j_2, \ldots$ on G_1 could be obtained as follows. Let j_p be a vertex which corresponds to the cell $M^1(j_p) \in C_1$ so that $M^1(j_p) \supseteq M^2(i_p)$, whereby $j_1 = j$ and $M^2(i_p) \in C_2$. The constructed path is admissible on G_1. So the cell $M^1(j)$ is non-leaving.

3. From the previous part we have the equality

$$\lim_{k \to \infty} M_k^+ = \bigcap_k M_k^+.$$

It follows from the part 1) that $\bigcap_k M_k^+ \supseteq I^+$. Now we will prove that

$$\bigcap_k M_k^+ \subset I^+.$$

Suppose the contrary i.e. that there exists $x \in \bigcap_k M_k^+$, but $x \notin I^+$. It implies that there exists an integer $m > 0$ so that $f^m(x) \notin K$, $f^l(x) \in K$, $l = 1, 2, \ldots, m - 1$. We denote $dist(f^m(x), K)$ by r_1. Consider a fine subdivision C_1 so that the maximal diameter of the images of its cells is less then r_1. We denote M_1 as a cell of C_1 containing $f^{(m-1)}(x)$. Obviously, M_1 is an outgoing cell. Let C_2 be a fine subdivision of C_1 so that the image of $M_2 \in C_2$, where by $M_2 \ni f^{(m-2)}(x)$ lies in M_1. The cell M_2 is the leaving cell. Now we repeat such a procedure $m - 2$ more times. On the step number i we choose a subdivision C_i so that the image of the cell containing $f^{(m-i)}(x)$ lies in the cell C_{i-1}. So we will obtain a chain of a cells where $x \in M_m$, $f(M_{i+1}) \subseteq M_i$. The cell M_1 is outgoing, hence M_m is a leaving cell of the covering C_m. Hence $x \notin M_m^+$ which contradicts the assumption $x \in \bigcap_k M_k^+$.

$$\odot$$

The set M_d^+ can be described in more detail by using ε-invariant sets. Recall that the sequence of points $\{x_i, \ i = 1, 2, \ldots\}$ is called a positive ε-half-orbit if $\rho(f(x_i), x_{i+1}) < \varepsilon$, $i = 1, 2, \ldots$, where $\rho(.,.)$ is the distance on M.

Definition 40. *The set I_ε^+ of all positive ε-half-orbits lying in K is called the maximal positive ε-invariant set in K.*

It is easy to see that the positive invariant set is contained in the positive ε-invariant set: $I^+ \subset I_\varepsilon^+$.

Theorem 41. *1. The mapping $\varepsilon \to I_\varepsilon^+$ is increasing with respect to inclusion, i.e. if $\varepsilon_1 < \varepsilon_2$ then*

$$I^+ \subset I_{\varepsilon_1}^+ \subset I_{\varepsilon_2}^+.$$

2. If $\varepsilon \to 0$, $\varepsilon > 0$ then $I_\varepsilon^+ \to I^+$, i.e.

$$I^+ = \bigcap_{\varepsilon > 0} I_\varepsilon^+.$$

Proof. The first statement follows from the fact that if $\varepsilon_1 < \varepsilon_2$ then each ε_1-half-orbit is ε_2-half-orbit.

To prove the second statement show that if there exists a positive ε-half-orbit in K passing through x_0 for every $\varepsilon > 0$ then x_0 lies in the maximal positive invariant set. Denote points of an ε-half-orbit in K passing through x_0 by $x_j(\varepsilon)$, $j \in \mathbb{N}$ in such a way that $x_0(\varepsilon) = x_0$. Consider a sequence $\{\varepsilon\}$, $\varepsilon \to 0$. From the inequality $\rho(x_1(\varepsilon), f(x_0)) < \varepsilon$ it follows that there exists

$$x_1 = \lim_{\varepsilon \to 0} x_1(\varepsilon) = f(x_0).$$

Since each $x_1(\varepsilon)$ lies in the compact K, the image $f(x_0)$ is in K as well. Consider the sequence $\{x_2(\varepsilon)\}$. We have

$$\rho(x_2(\varepsilon), f(x_1)) \leq \rho(x_2(\varepsilon), f(x_1(\varepsilon))) + \rho(f(x_1(\varepsilon)), f(x_1))$$

$$< \varepsilon + \rho(f(x_1(\varepsilon)), f(x_1)).$$

So from $x_1 = \lim_{\varepsilon \to 0} x_1(\varepsilon)$ is follows that there exists

$$x_2 = \lim_{\varepsilon \to 0} x_2(\varepsilon) = f(x_1),$$

where $x_2 \in K$. The same way we can prove that

$$x_k = \lim_{\varepsilon \to 0} x_k(\varepsilon) = f(x_{k-1})$$

where $x_k \in K$ for any positive integer k. The obtained sequence $\{x_k\}$ lies in K and one is a positive half-orbit through x_0. Hence x_0 lies in the positive invariant set I^+ maximal in K. Thus we have the inclusion $\bigcup_{\varepsilon > 0} \subset I^+$ that together with the first statement gives the desired equality.

⊙

Notice that the monotonicity of the sequence of sets I_ε^+ allows to determine $\lim_{\varepsilon \to 0} I_\varepsilon^+$ as $\bigcap_{\varepsilon > 0} I_\varepsilon^+$. Let q be the maximal diameter of images of cells and r be the lowest bound of the symbolic image.

Theorem 42. *Let M_d^+ be the union of all non-leaving cells of a covering C. Then*

$$I_r^+ \subset M_d^+ \subset I_{q+d}^+.$$

Proof. Let us prove the first inclusion. Consider a point $x_1 \in I_r^+$. It follows from the definition of I_r^+ that there exists an r-half-orbit $\{x_i\}$ through x_1 so that $x_i \in K$ and $dist(f(x_i), x_{i+1}) \leq r$ for every i. We denote by $M(j_i)$ the cell of M_d^+ that contains x_i. By this choice and according to the Theorem 14 on weak shadowing property, the sequence of vertices $\{j_i\}$ is an admissible path on the symbolic image. Therefore there are no outgoing cells among $M(j_i)$, hence $M(j_1)$ is non-leaving, and it follows that $x_1 \in M_d^+$.

To prove the second inclusion, suppose that $x_1 \in M_d^+$. We will find a positive $(q + d)$-orbit starting from x_1 and lying in K. By definition of M_d^+ there exists an infinite positive path $\{j_i\}$ on the symbolic image so that $x_1 \in M(j_1)$ and the vertices of the path are non-leaving. Let us fix a point x_i in $M(j_i)$. According to the Theorem 14 on weak shadowing property, the obtained sequence $\{x_i\}$ is a $(q + d)$-half-orbit and by our construction this sequence is in K. So the point x_1 lies in I_{q+d}^+.

$\odot$

Notice that, by the continuity of f, $d_k \to 0$ implies $q_k, r_k \to 0$. Then from Theorem 41 and Theorem 42 it follows that the third assertion of Theorem 39, namely $\lim_{d_k \to 0} M_{d_k}^+ = I^+$, remains valid without assumption that the sequence C_k is formed by subdivisions of coverings. However, without this assumption the sequence $\{M_{d_k}^+\}$ is not monotone with respect to inclusion as $d_k \to 0$.

Thus, the positive invariant set can be localized via the sets M_d^+. In a similar manner, we can construct sequences of sets localizing I^-, $I^+ \bigcup I^-$, and $I^+ \bigcap I^-$ in K. We restrict ourselves to stating results for the negative invariant set I^- and for $I = I^+ \bigcap I^-$. Denote by G^{-1} the graph obtained from the symbolic image by reversing the direction of all edges of the symbolic image on the opposite one. Let V^- be the set of non-leaving vertices of the graph G^{-1}.

Theorem 43. *Let $M_d^- = \{\bigcup M(i) : i \in V^-\}$ be the union of all non-leaving cells of the reversed symbolic image G^{-1}. Then*

1) M_d^- is a closed neighborhood of the negative invariant set I^- maximal in K, i.e.

$$I^- \subset M_d^-,$$

2) if C_2 is a subdivision of a covering C_1, then

$$M_{d_2}^- \subset M_{d_1}^-,$$

3) if C_k is a sequence of subdivisions and $d_k \to 0$, then

$$\lim_{d_k \to 0} M_{d_k}^- = \bigcap_k M_{d_k}^- = I^-.$$

The set V_0 of vertices is called invariant if for each vertex of V_0 there exists an infinite (in both directions) path going through the vertex and contained in V_0. It is clear that $V^i = V^+ \cap V^-$ is the maximal (with respect to inclusion) invariant set of vertices. The invariant set I maximal in a compact K satisfies the following properties.

Theorem 44. *Let $M_d^i = \{\bigcup M(j) : j \in V^i\}$. Then*

1) M_d^i is a closed neighborhood of the invariant set I maximal in K, i.e.

$$I \subset M_d^i,$$

2) if C_2 is the subdivision of a covering C_1, then

$$M_{d_2}^i \subset M_{d_1}^i,$$

3) if C_k is a sequence of subdivisions and $d_k \to 0$, then

$$\lim_{d_k \to 0} M_{d_k}^i = \bigcap_k M_{d_k}^i = I.$$

Theorem 44 gives a possibility to get the procedure for localization of the maximal invariant set. The algorithm is based on the adaptive subdivision method: on each step of the procedure some cells are excluded from consideration while the others are subdivided.

Algorithm

Step 1. An initial covering C of a compact K is determined. For C a symbolic image G of a dynamical system is constructed.

Step 2. All non-leaving vertices of G are detected. The neighborhood $U = \{\bigcup M(i) : i \text{ is a non-leaving vertex}\}$ of the positive invariant set is obtained.

Step 3. Cells corresponding to non-leaving vertices are subdivided, while cells corresponding to leaving vertices are excluded.

Step 4. For the collection of cells obtained the new symbolic image is constructed.

Step 5. Return to Step 2.

By Theorem 44, the sequence U_k of imbedded sets converges to the desired set $I^+ = \lim_{k \to \infty} U_k$.

5.3 Construction of Non-leaving Vertices

Let a compact $K \subset M$ be given. For a covering C of K denote the symbolic image by G. Let V and E be the sets of vertices and edges of G, respectively. The aim of this section is to describe a procedure for determining the non-leaving vertices of the symbolic image G, i.e. a method realizing Step 2 of the algorithm outlined above.

Before we proceed, recall some notions of graph theory. A class of equivalent recurrent vertices is called a strongly connected component. An outgoing vertex is a vertex without any out-come edges. A leaving vertex is a vertex such that all admissible finite paths starting from it end at outgoing vertices. In other words, there is no infinite path starting from a leaving vertex. Thus the maximal set of non-leaving vertices is the union of all strongly connected components of and the set of vertices for which at least one such component can be reached from. The algorithm for detecting non-leaving vertices is a modification of the well-known efficient Tarjan method [141] for detecting strongly connected components. The following statements form the basis of the procedure.

Proposition 45. *1. A periodic path consists of non-leaving vertices.*

2. If a path $\{i_k\}$ goes through a non-leaving vertex i_m then all preceding vertices i_k, $k < m$ are non-leaving.

3. If all edges issuing from a vertex i_m end with leaving vertices then i_m is a leaving vertex.

The proposed algorithm involves the construction of paths described by Proposition 45. In doing so, traveling through a path we arrive either at a non-leaving vertex (thus, all vertices traveled are non-leaving) or arrive at a leaving vertex. In the last case we need to return to the previous vertex and continue the path choosing another edge. This procedure allows to divide vertices into two groups: non-leaving and leaving ones.

Algorithm complexity.
The algorithm is organized in such a way that each vertex is just twice tested: when we arrive at a vertex not considered earlier and when we return to a vertex. Thus, the total number of algorithm steps does not exceed the double number of vertices and is not less than the number of vertices. This implies that the temporal algorithm complexity is estimated as $O(\max(|E|, |V|))$, where $|E|$ and $|V|$ stand for the number of edges and vertices, respectively. It is evident that any other algorithm cannot be radically better with respect to the temporal complexity since each vertex and each edge ought to be tested at least once.

Remark. The operating time depends heavily on the design of graph storing. Suppose that the design of graph storing maintains a fast access to vertices and edges. To localize the negative invariant set we need only to reverse graph edges. Then we can apply the above-described algorithm or, what is the same, perform a search for non-leaving vertices by incoming (instead of out-coming) edges. For instance, let edges be stored as a file of out-coming edges. To form an additional list of incoming edges we increase the operating time by $O(|E|)$.

The algorithm for construction of the maximal set of non-leaving vertices can be complemented by the Tarjan algorithm in such a way that the resulting algorithm allows to isolate strongly connected components and the set of vertices incoming to such components but not belonging to them.

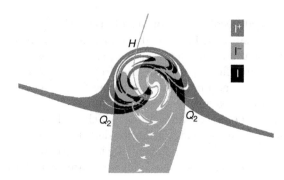

Fig. 5.4. Localization of the invariant sets I^+ and I^- for the modified Ikeda mapping

To illustrate the software as an example we present below localization of the union of positive and negative invariant sets (maximal in a given domain) of the modified Ikeda mapping.

Example 46. Invariant sets of the modified Ikeda mapping.
Consider the modified Ikeda mapping:

$$T : (x, y) \rightarrow (r + a(x \cos \tau - y \sin \tau), b(x \sin \tau + y \cos \tau)),$$

with $r = 1$, $a = -0.9$, $b = 1.2$, $\tau = C_1 - C_3/(1 + x^2 + y^2)$, $C_1 = 0.4$, and $C_3 = 6$.
    The Fig. 5.4 depicts the hyperbolic fixed point $H(-0.0950, 2.1937)$ and the sets I^+ and I^- in $K = [-10, 10] \times [-10, 10]$ constructed by symbolic dynamics methods. The intersection $I = I^+ \bigcap I^-$ contains orbits homoclinic to the 2-periodic orbit $Q_2(-1.5584, -1.9046), (3.0088, -1.2438)$ generating a chaotic behavior.

5.4 A Set-oriented Method

Now we consider a method for the localization of invariant sets which does not require construction of the symbolic image. We operate only with cells and their images. The method has advantages and disadvantages that we discuss later.
    Let $C = \{M(i)\}$ be a covering of the compact K. Recall that a cell M^* is outgoing from K if $f(M^*) \bigcap K = \emptyset$. It is clear that the positive invariant set I^+ does not intersect outgoing cells. So, if these cells are removed, we get that

$$K^* = \left\{ \bigcup M(j) \ : \ f(M(j)) \bigcap K \neq \emptyset \right\}$$

is a neighborhood of I^+. Considering the compact K^* as an initial one, we can repeatedly apply the same technique. It is not difficult to see that by the

described way all leaving cells are excluded in finite steps. In this case we have
the equality

$$K^* = \left\{ \bigcup M(j) \; : \; f(M(j)) \bigcap K^* \neq \emptyset \right\}.$$

The next step is a subdivision of the covering obtained and then the removing
technique is repeated. The described method was suggested by M. Dellnitz
and A. Hohmann [31], at 1996.

So algorithm is as follows.

1. For an initial covering C set $R = K$.
2. Form the lists $C_1 = C$ and $C_2 = \emptyset$.
3. For a cell $M^* \in C_1$ find the intersection $f(M^*) \bigcap R = \Lambda$.
4. If $\Lambda = \emptyset$ then the cell M^* is eliminated from consideration, i.e. M^* is
   removed from C and C_1, while C_2 remains unchanged. Set

$$R = \left\{ \bigcup M(j) \; : \; M(j) \in C \right\}.$$

5. If $\Lambda \neq \emptyset$ then the cell M^* is moved from C_1 in C_2, while C remains
   unchanged.
6. If $C_1 \neq \emptyset$ then we go to step 3.
7. If $C_1 = \emptyset$ and the cell M^* was eliminated from the list C_1 (i.e., the step 4
   was carried out and the covering was changed) then we go to step 2 with
   a new covering C.
8. If $C_1 = \emptyset$ and we do not remove a cell M^*, i.e. step 4 is dropped then the
   covering is subdivided, a new covering C is formed, and we go to step 2.

By this algorithm we obtained the sequences of the coverings $\{C^p\}$ and the
sets $\{R_p = \{\bigcup M(j) \; : \; M(j) \in C^p\}\}$.

Theorem 47. *The sequence $\{R_p\}$ offers the following properties:*

*1) R_p is a closed neighborhood of the positive invariant set I^+ maximal in
K, i.e.*

$$I^+ \subset R_p,$$

2) the sets R_p are embedded:

$$R_{p+1} \subset R_p,$$

3) if the maximal diameter of the coverings $d_p \to 0$ as $p \to \infty$ then

$$\lim_{p \to \infty} R_p = \bigcap_p R_p = I^+.$$

A proof is similar to the proof of Theorem 39, and we left it to the reader.

Let us compare now the set oriented algorithm just described with the
algorithm based on the symbolic image construction. The set oriented algorithm has two iterative loops: in the first loop the covering is fixed while

subdivision is the main step in the second one. The subdivision is the main step for the construction of symbolic images sequence. So, the investigation of a fixed symbolic image corresponds to the work of the set oriented algorithm between two subdivisions. The advantage of the set oriented method is that the algorithm operates only with cells and their images, it is not necessary to compute and store links between cells. Hence, memory is not needed to store adjacent lists. The disadvantage of the method is that the image of a cell has be calculated whenever needed. This can happen many times that increases the operating time.

6

Chain Recurrent Set

Our purpose is the localization of a chain recurrent set, which includes all the types of returning trajectories: periodic, almost-periodic, recurrent, homoclinic, nonwandering and so on. All needed estimations can be obtained by the traditional numerical methods. Indeed, we apply the same technique as we use for localization of p-periodic set, but in this case a period is not fixed, i.e., all ε-periodic points are considered. By investigating the symbolic image we can separate the cells through which ε-periodic trajectories may pass from those through which ε-periodic trajectories do not pass. This allows us to construct a neighborhood of the chain recurrent set. By using the method of adaptive subdivision we obtain a sequence of embedded neighborhoods for the chain recurrent set. The constructed sequence converges to the chain recurrent set.

6.1 Definitions and Examples

Let $f : M \to M$ be a homeomorphism on a smooth compact manifold M. Recall that the set of all ε-periodic points is denoted by $Q(\varepsilon)$, i.e., for any point $x \in Q(\varepsilon)$ there is a periodic ε-trajectory through x. By Proposition 20 the ε-periodic set has the following properties:

$$1) \text{ the sets } Q(\varepsilon), \ \varepsilon > 0 \text{ are open,}$$

$$2) \ Q(\varepsilon_2) \subset Q(\varepsilon_1), \ if \ \varepsilon_2 < \varepsilon_1. \tag{6.1}$$

Definition 48. [28] *A point x is called chain recurrent if x is ε-periodic for each positive ε, i.e., there exists a periodic ε-trajectory passing through x. A chain recurrent set, denoted Q, is the set of all the chain recurrent points.*

The concept of "recurrence" in the dynamical systems theory has a long history. We consider the main notions.

Recurrent point. A point x_0 is recurrent if for any $\varepsilon > 0$ and n_0 there is $n > n_0$ so that the distance $\rho(f^n(x_0), x_0) < \varepsilon$.

Nonwandering point. A point x_0 is nonwandering if for any $\varepsilon > 0$ and n_0 there is $n > n_0$ so that $f^n(V(\varepsilon, x_0)) \cap V(\varepsilon, x_0) \neq \emptyset$, where $V(\varepsilon, x_0) = \{x : \rho(x, x_0) < \varepsilon\}$ is an ε-neighborhood of x_0. It is clear that a periodic point is recurrent, a recurrent point is nonwandering, and a nonwandering point is chain recurrent.

Weak nonwandering point [129]. A point x_0 is weak nonwandering if for any $\varepsilon > 0$ there is a perturbed system g so that the C^0-distance $\rho(f, g) = \sup \rho(f(x), g(x)) < \varepsilon$ and x_0 is nonwandering for g.

π-point [129], [131]. A point x_0 is called π-point (or generates a periodic trajectory) if for any $\varepsilon > 0$ there is a perturbed system g so that $\rho(f, g) < \varepsilon$ and x_0 is periodic for g. V.A. Pliss [115] has considered C^1 π-point. A point x_0 generates a periodic trajectory in C^1-topology if for any $\varepsilon > 0$ there is a C^1-perturbed system g such that $\rho(f, g) < \varepsilon$, $\rho(\frac{\partial f}{\partial x}, \frac{\partial g}{\partial x}) < \varepsilon$, and x_0 is periodic for g. A.N. Sharkovsky [129] and M. Shub [131] showed that the set of weak nonwandering points and the set of π-points coincide with the chain recurrent set. So we have the inclusions

$$Per \subset \Omega_0 \subset \Omega \subset Q, \tag{6.2}$$

were Ω_0 is the recurrent set and Ω is a nonwandering set and Q is the chain recurrent set. Let us show that the inclusions 6.2 are embeddings.

Example 49. Trajectory on the torus.

Let us consider a flow on the torus T^2. Consider a unit square and identify its opposite sides, in this way we get a model for the torus T^2. Let (x, y) be coordinates on $\mathbb{R}^2$ which are equivalent by $mod 1$, i.e., a point (x, y) is equivalent with a point $(x+n, y+m)$ where n and m are integers. For example (see Fig. 6.1), the point A is equivalent with the point B, and the point C is equivalent with the point D. The system

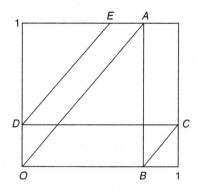

Fig. 6.1. Trajectory on the torus

$$x' = a,$$
$$y' = b, \tag{6.3}$$

where a and b are constants, defines a flow on the torus. The system (6.3) can be rewritten as the equation $\frac{dy}{dx} = \frac{b}{a}$, which has the solution $y = y_0 + \frac{b}{a}(x - x_0)$. The trajectories are parallel straight lines on the plane (x, y). On the torus these lines have to be identified. Thus the trajectory of the point $O = (0,0)$ is

$$T(O) = \{[OA], A \equiv B, [BC], C \equiv D, [DE], ...\}.$$

If $\frac{a}{b} = \frac{n}{m}$, where n and m are integers, i.e., $\frac{a}{b}$ is rational, then each trajectory $y = y_0 + \frac{n}{m}(x - x_0)$ is closed. Indeed if $(x - x_0) = m$ and $y - y_0 = n$ then $x = x_0(mod 1)$ and $y = y_0(mod 1)$, i.e., the points (x_0, y_0) and $(x_0 + m, y_0 + n)$ are equivalent on the torus and we have only periodic trajectories. If $\frac{a}{b}$ is irrational, then each trajectory does not return to its initial point, and it is everywhere dense on the torus. We get a so-called irrational winding of the torus. In the last case there are no periodic trajectories and each point is recurrent. Thus $Per = \emptyset$ and the recurrent set Ω_0 coincides with the torus T^2.

Example 50. A recurrent set does not coincide with a nonwandering set.
  Let us consider the system of an undamped pendulum

$$x' = y,$$
$$y' = -\sin x.$$

It has the hyperbolic points (saddles) $(\pm\pi, 0)$. They are connected by their separatrices, see Fig. 6.2,a. Every trajectory inside a cell bounded by the separatrices is periodic. Each point on the separatrix is nonwandering because it is a limit of periodic points. However these points are not recurrent. So, in this case we have $Per = \Omega_0 \subset \Omega = Q$, $\Omega_0 \neq \Omega$.

Example 51. A Nonwandering set does not coincide with a chain recurrent set.
  To prove the strict inclusion $\Omega \subset Q$, we consider a dynamical system on the circle S^1 with a single fixed point O. The coordinate on S^1 has the

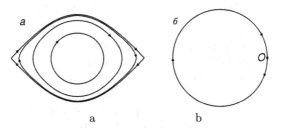

Fig. 6.2. The recurrent, nonwandering, and chain recurrent sets

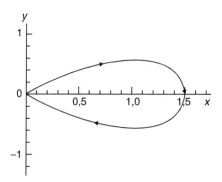

Fig. 6.3. A homoclinic trajectory

angle φ, $0 \le \varphi < 2\pi$. The equation of motion is defined by $\varphi' = \cos\varphi - 1$. For all points $\varphi \neq 0$ we have $\varphi' < 0$. This means that the angle decreases. Because a trajectory can not go through the equilibrium O, the nonwandering set consists of the fixed point O, see Fig. 6.2,b. However, because a pseudo trajectory allows jumps, the chain recurrent set Q is the circle S^1. So in this case we have $O = Per = \Omega_0 = \Omega \subset Q = S^1$, $\Omega \neq Q$.

A trajectory $T(x)$ is called homoclinic if the α-and ω-limit sets $\alpha(x) = \omega(x)$ coincide with a periodic trajectory T_0. In particular, T_0 may be a fixed point (equilibrium) x_0. One may additionally suppose that the periodic trajectory T_0 is hyperbolic, i.e., the differential $Df(x_0)$ has the eigenvalues with non zero real parts. From the definition it follows that a homoclinic trajectories are in chain recurrent sets. The trajectory $T \neq O$ of the previous example is homoclinic.

Example 52. Homoclinic trajectory.
The system of differential equations

$$x' = y,$$
$$y' = x - x^2$$

has two equilibriums $(0,0)$, $(1,0)$ and an energy integral $H(x,y) = \frac{1}{2}(y^2 - x^2) + \frac{1}{3}x^3$. The invariant curve $H(x,y) = 0$ contains the equilibrium $(0,0)$ and the homoclinic trajectory which has the beginning and the end on this equilibrium.

Sometimes the chain recurrent set is connected with a given domain D. This means that in Definition 48 the ε-periodic trajectories have to be in D. It is known [19], that the chain recurrent set Q is invariant, closed, and contains periodic, homoclinic, nonwandering and other singular trajectories. It should be remarked that if a chain recurrent point is not periodic and dim $M > 1$ then there exists as small as one likes perturbation of f in C^0-topology for which this point is periodic [113,129,131]. One may say that a chain recurrent point may become periodic under a small C^0-perturbation of the map f, or the chain recurrent point generates periodic orbits. Since there is a periodic ε-trajectory through each point $x \in Q(\varepsilon)$, it immediately follows from the definition of the chain recurrent set that

$$Q = \lim_{\varepsilon \to 0} Q(\varepsilon) = \bigcap_{\varepsilon > 0} Q(\varepsilon).$$

Thus the family of open sets $\{Q(\varepsilon), \ \varepsilon > 0\}$ forms a fundamental system of neighborhoods for the chain recurrent set.

6.2 Neighborhood of Chain Recurrent Set

Let us consider a symbolic image of the mapping f with respect to a covering C. Recall that a vertex of the symbolic image is called recurrent if a periodic path passes through it.

Denote by $P(d)$ the union of the cells $M(i)$ for which the vertex i is recurrent, i.e.,

$$P(d) = \{\cup M(i) : i \ is \ recurrent\}, \tag{6.4}$$

where by d is the largest diameter of the cells $M(i)$. Like before, we consider the dependence of P on the largest diameter d. Denote by $T(d)$ the union of the cells $M(k)$ for which the vertex k is nonrecurrent, i.e.,

$$T(d) = \{\cup M(k) : \ k \ is \ nonrecurrent\}.$$

The following theorem describes the properties of the sets $P(d)$ and $T(d)$.

Theorem 53. *1. The set $P(d)$ is a closed neighborhood of the chain recurrent set. Moreover, $P(d)$ is a subset of the set of ε-periodic points for each $\varepsilon > q + d$, i.e.,*

$$Q \subset P(d) \subset Q(\varepsilon), \ \varepsilon > q + d. \tag{6.5}$$

2. For any neighborhood V of Q there exists a $d > 0$ so that $P(d) \subset V$, i.e., if the largest diameter d is small enough then the neighborhood $P(d)$ is also reasonably small.

3. The chain recurrent set Q coincides with the intersection of the sets $P(d)$ for all positive d:

$$Q = \bigcap_{d > 0} P(d). \tag{6.6}$$

4. The points of T are not chain recurrent, i.e., $T \cap Q = \emptyset$. Moreover, if $\varepsilon < r$ then there is no periodic ε-trajectory passing through $x \in T$, i.e.,

$$Q(\varepsilon) \cap T = \emptyset, \ \varepsilon < r.$$

Proof. 1. Suppose $\varepsilon_1, \varepsilon_2$, are given so that $\varepsilon_1 < r < q + d < \varepsilon_2$. At first we prove that

$$Q(\varepsilon_1) \subset P(d) \subset Q(\varepsilon_2). \tag{6.7}$$

If a point x belongs to $Q(\varepsilon_1)$, then there exists a periodic ε_1-trajectory $\{x_1, ..., x_p\}$ passing through $x = x_1$. Consider a finite sequence of cells $\{M(z_i) : i = 1, ..., p\}$ such that $x_i \in M(z_i)$. Because $\varepsilon_1 < r$, according to Theorem 14, item 3, the sequence $\{z_i\}$ is an admissible periodic path passing through the vertex z_1. Thus the vertex z_1 is recurrent and the cell $M(z_1)$ is in $P(d)$. From this it follows that $Q(\varepsilon_1) \subset P(d)$.

Consider a point x belonging to $P(d)$. There exists a cell $M(z)$ so that $x \in M(z)$ and the vertex z is recurrent. In other words, on the symbolic image G there exists a periodic path $\{z_1, z_2, ..., z_p\}$, $z_1 = z$. Let us construct a periodic sequence $\{x_1, ..., x_p\}$, such that $x_1 = x$ and $x_i \in M(z_i)$. By Theorem 14, item 1, the sequence $\{x_i\}$ is a periodic ε_2-trajectory for any $\varepsilon_2 > q + d$. Hence the point $x = x_i$ lies in $Q(\varepsilon_2)$, i.e. $P(d) \subset Q(\varepsilon_2)$. Thus (6.5) and (6.7) hold. From the inclusions $Q \subset Q(\varepsilon_1) \subset P(d)$ it follows that $P(d)$ is a closed neighborhood of the chain recurrent set.

2. Let V be arbitrary neighborhood of Q. Since f is a continuous mapping and M is compact, the largest diameter q of the images $f(M(i))$ tends to 0 as the largest diameter of cells $d \to 0$. Set $\varepsilon_2 = \frac{3}{2}(q + d)$. We have $\varepsilon_2 \to 0$ as $d \to 0$. Because $\{Q(\varepsilon), \varepsilon > 0\}$ is a fundamental system of neighborhoods of Q, there is $\varepsilon^* > 0$ such that $Q(\varepsilon^*) \subset V$. Take d such that $\varepsilon_2 = \frac{3}{2}(q(d) + d) \le \varepsilon^*$. Since $Q(\varepsilon)$ depends monotonically on ε, the inclusion (6.7) means that we have $P(d) \subset Q(\varepsilon_2) \subset Q(\varepsilon^*) \subset V$.

3. The family of open sets $\{Q(\varepsilon), \varepsilon > 0\}$ forms a fundamental system of neighborhoods for chain recurrent sets Q, $Q = \bigcap_{\varepsilon > 0} Q(\varepsilon)$. From the inclusions (6.7) we get

$$Q \subset \bigcap_{d > 0} P(d) \subset \bigcap_{\varepsilon_2} Q(\varepsilon_2) = Q,$$

where $\varepsilon_2 = \frac{3}{2}(q(d) + d)$ and $\varepsilon_2 \to 0$ as $d \to 0$. It should be noted, that generally speaking $P(d_1)$ is not contained in $P(d_2)$ even though $d_1 < d_2$. As we will see later, there is an algorithm for the construction of a sequence of imbedded neighborhoods $P_1 \supset P_2 \supset ... \to Q$.

4. We prove this proposition by contradiction. Let $x \in M(k)$ where by k is a non recurrent vertex. Let $\{x_1, ..., x_p\}$ be a periodic ε -trajectory passing through $x = x_1$ and $\varepsilon < r$. Consider a sequence $\{z_1, ..., z_p\}$ so that $z_1 = k$ and $x_i \in M(z_i)$. As $\varepsilon < r$, according to Theorem 14, item 3, the sequence $\{z_i\}$ is a periodic path on the symbolic image G. Because $z_1 = k$, the vertex k is recurrent. The resulting contradiction completes the proof of proposition and the theorem.

$$\odot$$

Theorem 53 leads us to the following inclusions

$$Q \subset Q(\varepsilon_1) \subset M \backslash T = P(d) \backslash T \subset P(d) \subset Q(\varepsilon_2), \tag{6.8}$$

where $\varepsilon_1 < r < q + d < \varepsilon_2$. By definition, the set T is closed and $T \cap Q = \emptyset$. Hence, the set $P(d) \backslash T$ is an open neighborhood of the chain recurrent set.

6.3 Algorithm for Localization

Theorem 53 permits us to construct a neighborhood of the chain recurrent set without any preliminary information about a dynamic system. Now we apply the process of adaptive subdivision in order to localize the chain recurrent set.

Algorithm localizing the chain recurrent set.

1. Starting with an initial covering C, the symbolic image G of the map f is found.
2. The recurrent vertices $\{i_k\}$ of the graph G are recognized. Using the recurrent vertices, a closed neighborhood $P = \{ \cup M(i_k) : i_k \text{ is recurrent}\}$ of the chain recurrent set is found.
3. The cells corresponding to the recurrent vertices $\{M(i_k) : i_k \text{ is recurrent}\}$ are subdivided while cells corresponding to nonrecurrent vertices are excluded. Denote by d the largest diameter of the new cells achieved by subdivision. Thus the new covering is defined.
4. The symbolic image G is constructed for the new covering. The cells corresponding to non recurrent vertices do not participate in the construction of the new covering and the new symbolic image.
5. Then one goes back to the second step.

By repeating this process of adaptive subdivision we obtain a sequence of neighborhoods $P_0, P_1, P_2, \ldots$ for the chain recurrent set Q and a sequence of the largest diameters $d_0, d_1, d_2, \ldots$ for the cells. The following theorem substantiates the described algorithm for localization of the chain recurrent set.

Theorem 54. *The sequence of sets* $P_0, P_1, P_2, \ldots$ *offers the following properties:*

1) the neighborhoods P_k *are imbedded into each other, i.e.,*

$$P_0 \supset P_1 \supset P_2 \supset \ldots \supset Q,$$

2) if the largest diameter $d_k \to 0$ *as k tends to infinity then*

$$\lim_{k \to \infty} P_k = \bigcap_k P_k = Q. \tag{6.9}$$

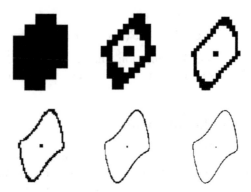

Fig. 6.4. Localization of the chain recurrent set of the Van der-Pol system

Proof. 1. Let $C = \{M(i)\}$ be a covering of M and G be the symbolic image for the covering C. Suppose a new covering NC is produced by applying an adaptive subdivision on C. Denote by NG the symbolic image for NC. For convenience the cells of the new covering NC are designated as $m(i,k)$ so that $\bigcup_k m(i,k) = M(i)$, i.e., the cells $\{m(i,k) : k = 1,2,...\}$ of the new covering form a partition of the cell $M(i)$. In this case the vertices of NG are designated as (i,k). Since $f(m(i,k)) \subset f(M(i))$ and $m(j,l) \subset M(j)$, the intersection $f(M(i)) \cap M(j) \neq \emptyset$ provided us $f(m(i,k)) \cap m(j,l) \neq \emptyset$. That means that there exists a mapping from the graph NG to the graph G so that the vertex (i,k) is mapped to the vertex i and the directed edge $(i,k) \to (j,l)$ is mapped to the directed edge $i \to j$. Hence, each admissible path on the new graph NG is transformed to some admissible path on the graph G. Particularly, the periodic paths on the graph NG are mapped into the periodic paths of the graph G. Thus the recurrent vertices of the new symbolic image NG are mapped on the recurrent vertices of G. Set

$$P = \{\cup M(i) : i \text{ is recurrent on } G\},$$

$$NP = \{\cup m(i,k) : (i,k) \text{ is recurrent on } NG\}.$$

It follows from the above that NP is a subset of $P : NP \subset P$. Applying this result to each step of the partitioning process, we see that the algorithm returns the sequence of imbedded neighborhoods $P_0 \supset P_1 \supset P_2 \supset ...$ of the chain recurrent set.

2. Let $d_k \to 0$ as $k \to \infty$. The inclusions (6.7) are obtained for an arbitrary symbolic image. Hence they hold for the neighborhoods P_k. It follows that $P_k \subset Q(\varepsilon)$ provided $q_k + d_k < \varepsilon$, where q_k is the largest diameter of the images $f(M(i))$. Set $\varepsilon_k = (3\backslash 2)(q_k + d_k)$. Since X is a continuous mapping and M is compact, $q_k \to 0$ as $d_k \to 0$. Thus we have $\varepsilon_k \to 0$ as $k \to \infty$. Because the family $\{Q(\varepsilon), \varepsilon > 0\}$ is a fundamental system of neighborhood of Q, $Q \subset P_k \subset Q(\varepsilon_k)$ and $\lim_{k\to\infty} Q(\varepsilon_k) = \bigcap_k Q(\varepsilon_k) = Q$ we obtain $Q \subset \bigcap_k P_k \subset \bigcap_k Q(\varepsilon_k) = Q$, i.e.,

$$\lim_{k\to\infty} P_k = \bigcap_k P_k = Q.$$

⊙

Example 55. Van-der-Pol system.

Let us consider the Van-der-Pol system

$x' = y,$

$y' = \varepsilon(1 - x^2)y - x$

where by $\varepsilon = 1.5$. It is well known that the chain recurrent set of the Van der-Pol system consists of an equilibrium point (0,0) and a periodic orbit. The system has been studied numerically in the square $M = [-3,5;3,5] \times [-3,5;3,5]$. The initial covering consists of 49 cells, which are 1×1 squares. The subsequent subdivisions cells are into 4 equal squares. The picture presents the neighborhoods $P_1, P_2, P_3, P_4, P_5, P_6$ of the chain recurrent set. According to Theorem 54 these neighborhoods are embedded one inside the other and tend to the equilibrium point and the periodic orbit. Notice that the initial covering does not separate the equilibrium point and the periodic orbit. The separation appears at the first subdivision of P_0. As can be seen on the picture, the neighborhoods of the periodic orbit and the equilibrium point are small enough, will be eventually obtained at the sixth step of the algorithm. A computer program realizing the algorithm described above has been implemented in St. Petersburg Technical University, 1991 by A. Moiseev.

7

Attractors

Our purpose is it to construct attractors and their domain of attraction without any preliminary information about the system. The investigation is based on the methods of symbolic dynamics and all needed estimations can be obtained by traditional numerical methods. When investigating the symbolic image, one can obtain neighborhoods of the attractors and estimate their domains of attraction. This allows us to construct an estimation of an attractor-repellor pair. By applying a subdivision on the covering, a sequence of embedded approximations will be constructed so that it converges to the desired attractor-repellor pair. It should be emphasized that our investigation was stimulated by the basic ideas of A. M. Laypunov about stability [75].

7.1 Definitions and Examples

We consider a discrete dynamical system generated by a mapping $f : M \to M$. Let $\rho(x, A) = \inf(\rho(x, y) : y \in A)$ be the distance between a point x and a set A. Denote by $V(\varepsilon, A) = \{x : \rho(x, A) < \varepsilon\}$, $\varepsilon > 0$ an ε-neighborhood of A. The ω-limit set of a subset B is defined as

$$\omega(B) = \bigcap_{n>0} cl f^n(T^+(B)),$$

where $cl\cdot$ means the closure and $T^+(B) = \{x = f^k(x_0), \ x_0 \in B, \ k > 0\}$. The α-limit set of a subset B is defined as

$$\alpha(B) = \bigcap_{n<0} cl f^n(T^-(B)),$$

where $T^-(B) = \{x = f^k(x_0), \ x_0 \in B, \ k < 0\}$. Recall that a set Λ is invariant if for each point $x_0 \in \Lambda$ the trajectory $T(x_0) = \{x = f^k(x_0), \ x_0, k \in Z\}$ passing through x_0 is in Λ.

Let us consider the first concepts of attraction iniciated by Lyapunov at 19-th century.

Definition 56. *1. An invariant set Λ is called stable (by Lyapunov) if for each $\varepsilon > 0$ there exists $\delta > 0$ such that if $x \in V(\delta, \Lambda)$, the positive semi-trajectory $T^+(x) \subset V(\varepsilon, \Lambda)$.*

2. A stable by Lyapunov set Λ is called asymptotically stable if there exists a neighborhood V of Λ such that for each $x \in V$

$$\lim_{n \to \infty} \rho(f^n(x), \Lambda) = 0.$$

3. Closed asymptotically stable set Λ is attractor.
4. The set

$$W^s(\Lambda) = \{x : \lim_{n \to \infty} \rho(f^n(x), \Lambda) = 0\}$$

is called the domain of attraction or the basin of Λ.

Note that the existence of a neighborhood $V(\Lambda)$ such that for each point $x \in V$ its ω-limit be in Λ, it is not suficient in order to Λ be an attractor, see Example 59. There is other (equivalent) definition of attractor.

Definition 57. [12] *A set Λ is an attractor if it has a neighborhood U so that its ω-limit set $\omega(U) = \Lambda$.*

We have to indicate that sometimes the sets described by Definition 56 or 57 are called the attracting sets and additional conditions are involved in the definition of attractor. For example, Λ is assumed to have a dense trajectory [48], or the domain of attraction W^s is supposed to coincide with the phase space. The last supposition is produced to systems of partial differential equations. However we will use Definition 56 or 57.

Example 58. Consider the system of differential equations

$$x' = x - x^3,$$
$$y' = -y.$$

The system has three attractors (see Fig. 7.1): the equilibriums $A_1 = (-1, 0)$, $A_2 = (1, 0)$ and the interval $A_3 = [-1, 1] \times \{0\}$. The equilibrium $(-1, 0)$ has the domain of attraction $W^s(A_1) = (-\infty, 0) \times (-\infty, \infty)$, the equilibrium $(1, 0)$ has the domain of attraction $W^s(A_2) = (0, \infty) \times (-\infty, \infty)$, and the attractor A_3 has the domain of attraction $W^s(A_3) = \mathbf{R}^2$.

As it follows from the definition of attraction domain that $W^s(\Lambda) = \{x : \omega(x) \subset \Lambda\}$. The following example shows that the condition "$\omega(x) \subset \Lambda$ for any x from a neighborhood $V(\Lambda)$" is not sufficient in order to the set Λ be attractor.

Example 59. The elliptic domain.
Consider a plane system with one equilibrium point x_0 with domain Λ filled with homoclinic trajectories, i.e. the trajectories $T(x)$ for which their α- and ω-limit sets are the origin O (see Fig. 7.2). As Λ is the union of the trajectories,

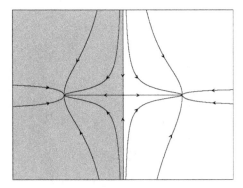

Fig. 7.1. Attractors of the system $x' = x - x^3$, $y' = -y$

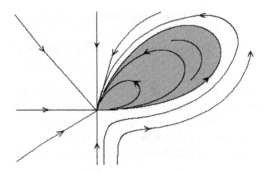

Fig. 7.2. The elliptic domain

Λ is invariant set, which is named the elliptic domain. The equilibrium point O is the ω-limit set for any point $x \in R^2$, i.e. $\omega(x) = O$. However, 0 is not attractor, in this case the elliptic domain Λ is attractor. This example shows that in general $\omega(\Lambda) \neq \cup_{x \in \Lambda} \omega(x)$, we can only guarantee the inclusion $\cup_{x \in \Lambda} \omega(x) \subset \omega(\Lambda)$.

In modern practice it is very important to know position of an attractor and its domain of attraction. The reason is that a basin is an open set, every point from the basin comes in a small neighborhood of the attractor for a finite number of iterations. Thus for the Example 58, starting at the right half-plane a trajectory is brought near the equilibrium point (1,0) within a finite iteration. The following example shows that domain of attraction may be rather complicated.

Example 60. Basin of the 2-periodic attractor *triple hook*.
The Ikeda mapping in the real notation takes the form

$$I : (x, y) \rightarrow (R + C_2(x cos\tau - y sin\tau), C_2(x sin\tau + y cos\tau)),$$

where by $\tau = C_1 - C_3/(1 + x2 + y2)$. Numerical simulations of the dynamical behavior of the map I have been carried out with $C_1 = 0.4, C_2 = 0.9, C_3 = 6.0$.

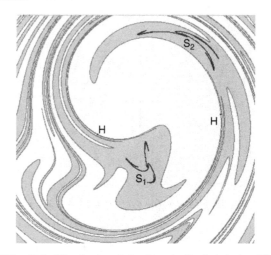

Fig. 7.3. The basin of the 2-periodic "triple hook"

The parameter R takes the values 0.6. The system has the 2-periodic attractor $A = S_1 + S_2$ so that each component has the form of a triple hook, see Fig. 7.3. The 2-periodicity means that $I(S_1) = S_2$ and $I(S_2) = S_1$. The structure of the component S_1 can be seen in Example 24. As the Fig. 7.3 shows the domain D of attraction is more complicate than in previous examples. The basin D is 2-periodic as well. The boundary ∂D is formed by the unstable manifold of the hyperbolic two-periodic orbit H $(1.0094, -0.1100), (-0.2110, -0.4211)$.

Strange attractor.

There are many attractors with so nontrivial structure that they are called "strange attractors". By this is meant that the attractor is chaotic, i.e., it has a dense orbit and a nontrivial fractal structure that is similar to Cantor set. Henon [53] explored numerically the diffeomorphism

$$f(x,y) = (1 - ax^2 + y, bx).$$

Henon's computation represents numerical evidence of the existence of a strange attractor for the case a=1.4 and b=0.3, see Fig. 7.4. The Henon attractor is locally the product of a curve and a Cantor set [48].

The domain of attraction is invariant set and a neighborhood of Λ [12] is invariant as well. It is clear that an invariant set for the homeomorphism f is as well invariant for the inverse mapping f^{-1}.

Definition 61. *An invariant set Λ is called repellor if there exists a neighborhood U such that its α-limit set $\alpha(U) = \Lambda$.*

An invariant set Λ is repellor for f if Λ is attractor for f^{-1}. The set $\Lambda^* = M \setminus W^s(\Lambda)$ is a repellor [12] which is called dual repellor of Λ and the pair $\{\Lambda, \Lambda^*\}$

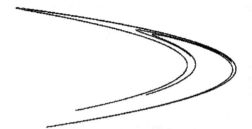

Fig. 7.4. The Henon attractor

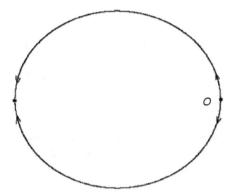

Fig. 7.5. Attractor-repellor pair

is called attractor-repellor pair. Thus the dynamics between an attractor-repellor pair is very simple: each trajectory starts at the repellor and ends at the attractor.

Example 62. Attractor-repellor pair.
Consider the equation $\varphi' = \sin \varphi$ on a circle S^1, see Fig. 7.5. It has two equilibrium points $\varphi = 0$ and $\varphi = \pi$, being $\varphi = \pi$ is attractor and $\varphi = 0$ is its dual repellor.

Proposition 63. [12] *An invariant set Λ is attractor if and only if there exists a neighborhood U of Λ such that*

$$f(cl\ U) \subset U,\ \Lambda = \bigcap_{n>0} f^n(U),\ W^s(\Lambda) = \bigcup_{n<0} f^n(U).$$

The set U is called fundamental neighborhood of Λ. One can say that each trajectory through $x \in W^s(\Lambda) \setminus \Lambda$ starts at the repellor Λ^* and ends at the attractor Λ. One of our aims is to construct the attractor, repellor and domain of attraction without any preliminary information about the dynamical system. As might be expected, the properties of the attractor and its domain of attraction hold not only for true trajectories, but they also can be applied to ε-trajectories.

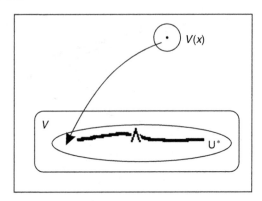

Fig. 7.6. An attractor and pseudo trajectories

Proposition 64. *Let Λ be an attractor with a neighborhood V and $x \in W^s(\Lambda)$. Then*

1) *there exists a neighborhood U^* of Λ, $U^* \subset V$ and $\varepsilon_1 > 0$ such that every positive ε_1-semi-trajectory passing through U^* remains in U^*,*
2) *there exists ε_2 such that every ε_2-trajectory passing through an ε_2-neighborhood $V(\varepsilon_2, x)$ reaches U^*.*

Proof. Let U be a fundamental neighborhood of Λ and V be a neighborhood of Λ. According to Proposition 63, for every V there exists $k > 0$ such that $f^k(U) \subset V$ and $cl f^{k+1}(U) \subset f^k(U)$. Set $U^* = f^k(U)$, see Fig. 7.6. The distance

$$\rho(cl\ f(U^*), M \setminus U^*) = \min(\rho(x, y), \ x \in cl\ f(U^*), \ y \in M \setminus U^*)$$

is positive, because U^* is an open set and $cl\ f(U^*) \subset U^*$. Set

$$\varepsilon_1 = \rho(cl\ f(U^*), M \setminus U^*).$$

In this case any ε_1-neighborhood of $f(U^*)$ lies in U^*. We have to show that every positive ε_1-semi-trajectory passing through U^* remains in U^*. Let x_k, x_{k+1} be a pair of consecutive points on an ε_1-semi-trajectory and let $x_k \in U^*$. Since $\rho(f(x_k), x_{k+1}) < \varepsilon_1$, the point x_{k+1} is in an ε_1-neighborhood of $f(U^*)$. Hence $x_{k+1} \in U^*$.

Let x be a point in $W^s(\Lambda)$. We prove by contradiction that there is an $\varepsilon_2 > 0$ such that every ε_2-trajectory passing through $V(\varepsilon_2, x)$ reaches U^*. Conversely, suppose that for each $\varepsilon > 0$ there exists a positive ε-semi-trajectory $w(\varepsilon)$ passing through $V(\varepsilon, x)$ which misses U^*. Let $\varepsilon_n \to 0$ and $\{w(\varepsilon_n)\}$ be the sequence of positive ε_n-semi-trajectories described above. Since the sequence $\{w(\varepsilon_n)\}$ lies in the compact M, there is a converging subsequence $w(\varepsilon_{n_k}) \to w$. The trajectory w passes through x and is outside U^*. We come to a contradiction, because the trajectory w has to end in $\Lambda \subset U^*$.

Corollary 65. *It is obviously that $\varepsilon = \min(\varepsilon_1, \varepsilon_2)$ satisfies the both conclusions of the proposition, i.e., each ε-trajectory passing through $V(\varepsilon, x)$ reaches U^* and remains there.*

Proposition 66. *Let V_1 be an arbitrary small neighborhood of an attractor Λ and let V_2 be an arbitrary large neighborhood such that*

$$\Lambda \subset V_1 \subset V_2 \subset clV_2 \subset W^s(\Lambda).$$

Then there exist $\varepsilon > 0$ and the neighborhoods U_1, U_2 of Λ,

$$\Lambda \subset U_1 \subset V_1 \subset V_2 \subset U_2 \subset clU_2 \subset W^s(\Lambda)$$

such that (see Fig. 7.7)

1) any ε-trajectory passing through $U_2 \setminus U_1$ starts outside U_2 and ends in U_1;
2) any positive ε-semi-trajectory passing through U_1 remains there;
3) any negative ε-semi-trajectory passing through $M \setminus clU_2$ remains there.

Proof. According to Proposition 64, there exist an ε_1 and a neighborhood U_1 for the neighborhood V_1 so that $U_1 \subset V_1$ and every positive ε_1-semi-trajectory passing through U_1 remains in U_1. Moreover, for a point $x \in W^s(\Lambda)$, there is $\varepsilon_{11} > 0$ such that any ε_{11}-trajectory passing through $V(\varepsilon_{11}, x)$ ends in U_1.

Recall that the set $\Lambda^* = M \setminus W^s(\Lambda)$ is repellor for f and attractor for f^{-1}. The set $M \setminus clV_2$ is a neighborhood of Λ^* because $clV_2 \subset W^s(\Lambda)$. In view of the symmetry between attractor and repellor, there are an ε_2 and a neighborhood $U^* \subset (M \setminus clV_2)$ of Λ^* such that every negative ε_2-semi-trajectory passing through U^* remains in U^*. Set $U_2 = M \setminus clU^*$. Thus any negative ε_2-semi-trajectory passing through $U^* = M \setminus clU_2$ remains there. Moreover, for any point $x \in W^s(\Lambda)$, there is $\varepsilon_{22} > 0$ such that each ε_{22}-trajectory through $V(\varepsilon_{22}, x)$ starts outside U_2.

Consider a compact $K = cl(U_2 \setminus U_1)$. For any point $x \in K$ there are ε_{11} and ε_{22} as described above. Set $\varepsilon(x) = \min\{\varepsilon_{11}, \varepsilon_{22}\}$. According to the

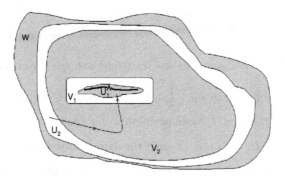

Fig. 7.7. Pseudo trajectories, an attractor and its basin

construction process every $\varepsilon(x)$ -trajectory passing through $V(\varepsilon(x), x)$ starts outside U_2 and ends in U_1. The family of the neighborhoods $\{V(\varepsilon(x), x);\ x \in K\}$ forms an open covering of the compact K. There exists a finite covering of K by the neighborhoods $\{V(\varepsilon_m, x_m) :\ \varepsilon_m = \varepsilon(x_m),\ m = 3, 4, ..., r\}$. We set $\varepsilon = \min(\varepsilon_m,\ m = 1, 2, 3, ..., r)$. Since $\varepsilon \le \varepsilon_1$, any positive ε-semi-trajectory passing through U_1 remains there. Since $\varepsilon \le \varepsilon_2$, any negative ε-semi-trajectory passing through $M \setminus clU_2 = U^*$ remains there. As $\varepsilon \le \varepsilon_m,\ m = 3, ..., r$, any ε-trajectory passing through $U_2 \setminus U_1$ starts outside U_2 and ends in U_1.

$$\odot$$

7.2 Attractor on Symbolic Image

Consider a symbolic image G of a homeomorphism f. A set of vertices $L \subset Ver$ generates subgraph $G(L)$ which contains the vertices L and the edges $i \to j$ if and only if the vertices i and j belong to L. We say that the set L is invariant if for each vertex $i \in L$ there exist the edges $j \to i$ and $i \to k$ in $G(L)$. In order to construct an invariant set we consider an infinite in both direction admissible path $\omega = \{..., i_{-1}, i_0, i_1, ...\}$. The set of vertices $Ver(\omega)$ of the path ω forms an invariant set because for each $i_k \in \omega$ there are edges $i_{k-1} \to i_k$ and $i_k \to i_{k+1}$. The same way, the family of paths $S = \{\omega\}$ gives the invariant set of vertices $Ver(S)$. We can say that the set L is invariant if for each vertex $i \in L$ there is an admissible path through i which lies in L. According to Theorem 14, each trajectory $\{x_k\}$ of f generates an admissible path $\{z_k :\ x_k \in M(z_k)\}$ on the symbolic image G. Hence an invariant set $\Lambda \subset M$ generates an invariant set of vertices of the form

$$L(\Lambda) = \{z : M(z) \cap \Lambda \ne \emptyset\}.$$

In particular the set Ver is invariant. Let L be an invariant set of vertices on the symbolic image G. The set of vertices

$$En(L) = \{j \in L :\ there\ exists\ an\ edge\ i \to j,\ i \notin L\}$$

is called entrance of L. The set of vertices

$$Ex(L) = \{i \in L,\ there\ exists\ an\ edge\ i \to j,\ j \notin L\}$$

is called exit of L.

Definition 67. • We say that an invariant set $L \subset Ver$ is attractor if $Ex(L) = \emptyset$.
• We say that an invariant set $L \subset Ver$ is repellor if $En(L) = \emptyset$.

Let L be an attractor. A basin or domain of attraction is the set of vertices

$$D(L) = \{j :\ each\ path\ through\ j\ ends\ in\ L\},$$

i.e., for each path $\{..., j, ..., i_k, ...\}$ there exists a number k^* such that the vertices i_k with $k > k^*$ belong to L.

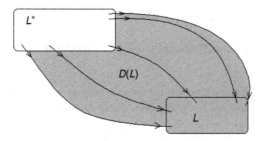

Fig. 7.8. Attractor-repellor pair on a symbolic image

Proposition 68. *Let $L \subset Ver$ be an attractor. Then (see Fig. 7.8)*

1) the vertices from $D(L) \setminus L$ are non-recurrent,
2) the set of vertices $L^* = Ver \setminus D(L)$ is repellor.

Proof. 1. If $i \in D(L) \setminus L$ is recurrent then there is a periodic path $\omega = \{i = i_0, ..., i_m = i\}$. As $i \notin L$ the path ω does not end in L. It means that i does not belong to $D(L)$. We come to a contradiction. Hence the vertices from $D(L) \setminus L$ are non-recurrent.

2. We consider only connected graphs. This means that any two vertices i and j might be connected by a non-directed path. If $L = Ver$, then $DL = L$ and $L^* = \emptyset$. Suppose that $L \neq Ver$. At first we prove that $L^* \neq \emptyset$. If $L = D(L)$ then $L^* = Ver(G) \setminus L \neq \emptyset$. Let $D(L) \setminus L \neq \emptyset$. As the number of vertices is finite, the beginning of each path through $j \in D(L) \setminus L$ has a periodic part, i.e., the path is of the form $\{..., i_k, ..., i_l, ..., j, ...\}$ with $i_k = i_l$. Since the exit of L is empty, no vertex of a closed path $\{i_k, ..., i_l = i_k\}$ can belong to L. As the vertices from $D(L) \setminus L$ are non-recurrent, $i_k \notin D(L)$ and $i_k \in L^*$, i.e., $L^* \neq \emptyset$.

Now we prove by contradiction that L^* is invariant. Let $i \in L^*$. On the contrary, suppose there is no edge $i \to j$ for any $j \in L^*$. Then every edge $i \to k$ ends in $D(L)$. In this case each path passing through i ends in L, i.e., $i \in D(L)$. Hence $i \notin L^*$, a contradiction. The same way we can prove that there is an edge $j \to i$, $j \in L^*$ for each $i \in L^*$ for all $i \in L^*$.

Next we establish that the exit $Ex(L^*) \neq \emptyset$. As $L \neq Ver$, we have $En(L) \neq \emptyset$, i.e., there is an edge $i \to j$, $j \in L$, $i \notin L$. It means that either $i \in L^*$ or $i \in D(L) \setminus L$. In the first case, $i \in Ex(L^*) \neq \emptyset$. In the second case, since the set $D(L) \setminus L$ consists of non-recurrent vertices, each path through i starts in L^* and ends in L. Hence there is an edge $k \to l$, $k \in L^*$, $l \in D(L)$, i.e., $k \in Ex(L^*) \neq \emptyset$.

At last we establish that the entrance $En(L^*) = \emptyset$. Let $j \in En(L^*)$. This means that there is an edge $i \to j$, $i \in D(L)$, $j \in L^*$. As each path through i ends in L, each path through j ends in L as well, i.e., $j \in D(L)$ and $j \notin L^*$, a contradiction.

$\odot$

If $L \subset Ver$ is an attractor, the repellor $L^* = Ver \setminus D(L)$ is called dual for the attractor L.

The following proposition describes the structure of an attractor on a symbolic image.

Proposition 69. *Each attractor L consists only of some classes of equivalent recurrent vertices and all paths between these classes.*

Proof. Let L be an attractor; $i, j \in L$; $\omega = \{i, \dots i_k, \dots, j\}$ be a path between i and j. Since the exit $Ex(L)$ is empty, each vertex i_k from ω belongs to L. In particular if i is a recurrent vertex, the class of recurrent vertices equivalent to i lies in L. In a similar manner we obtain that all paths between these classes lie in L as well.

We have to show there are no other types of vertices in L. Let $i \in L$ be a non-recurrent vertex and ω be a path through i. Since $Ex(L) = \emptyset$, ω ends in L. Suppose there is no path through i between the recurrent vertices from L. In this case every path ω passes between L and L^* through possible non-recurrent vertices. As there exists only a finite number of non-recurrent vertices, we can find one, say k, which is either i itself or precedes i on some ω so that for every edge $j \to i$ the relation $j \notin L$ holds. It means that L is not invariant, a contradiction.

$\odot$

7.3 Attractors of a System and its Symbolic Image

As one would expect, there is a natural correlation between the attractors of a dynamical system and the attractors of its symbolic image.

Theorem 70. *If L and $D(L)$ are attractor and its domain of attraction on a symbolic image G, then there are an attractor Λ of the homeomorphism f and its basin $W^s(\Lambda)$ such that*

1)
$$U = int \left\{ \bigcup M(i), \ i \in L \right\},$$

where int· denotes the interior, is the fundamental neighborhood of Λ;
2) $W = \{\bigcup M(j), \ j \in D(L)\}$ is an approximation of the basin $W^s(\Lambda)$ such that
$$W \subset W^s(\Lambda).$$

Proof. 1. At first we establish the inclusion: $f(clU) \subset U$. As each cell $M(i)$ is closed, $clU \subset \{\bigcup M(i), \ i \in L\}$. By definition of attractor, $Ex(L) = \emptyset$. Hence, if $i \in L$ and there is an edge $i \to j$ then $j \in L$, i.e., the set $c(i) = \{j : \ M(j) \cap f(M(i)) \neq \emptyset\}$ lies in L. According to Corollary 10 we have the inclusion

$$f(M(i)) \subset int \left(\bigcup_{j \in c(i)} M(j) \right).$$

Hence, the following inclusions hold

$$f(clU) \subset f \left(\bigcup_{i \in L} M(i) \right) = \bigcup_{i \in L} f(M(i)) \subset \qquad (7.1)$$

$$\bigcup_{i \in L} int \left(\bigcup_{j \in c(i)} M(j) \right) \subset int \bigcup_{j \in L} M(j) = U$$

By Proposition 63 the set $\Lambda = \bigcap_{k>0} f^k(U)$ is an attractor of f, where U is the fundamental neighborhood.

2. Let $x \in \{\bigcup M(i) :\ i \in D\}$ where D is the domain of attraction for L. Consider the positive semi-trajectory $T^+(x) = \{f^k(x) :\ k \in Z^+\}$. According to Theorem 14, $T^+(x)$ generates an admissible path $\omega = \{i_k :\ f^k(x) \in M(i_k),\ k \in Z^+\}$ on the symbolic image G with $i_0 \in D$. Being D the domain of attraction, the path ω has to end in L. This infers the existence of an integer k_0 such that $i_k \in L$ for all $k > k_0$ and as a consequence the inclusion $f^k(x) \in M(i_k) \subset \{\bigcup M(j),\ j \in L\}$. Considering that (7.1) $f(\bigcup_{i \in L} f(M(i)) \subset U$ and U is a fundamental neighborhood of Λ, so we have $f^k(x) \to \Lambda$ as $k \to \infty$, i.e., $x \in W^s(\Lambda)$.

$\odot$

The following theorem shows that both an attractor of a dynamical system and its domain of attraction can be constructed as precisely as one likes by a symbolic image, being its diameter small enough.

Theorem 71. *Let $\Lambda \subset M$ be an attractor, V_1 be its arbitrarily small neighborhood, and V_2 be an arbitrarily large neighborhood such that*

$$\Lambda \subset V_1 \subset V_2 \subset clV_2 \subset W^s(\Lambda).$$

Then there exists $d_0 > 0$ such that each symbolic image G with the maximal diameter of covering cells $d < d_0$ has an attractor L and a domain of attraction $D(L)$ so that (see Fig. 7.9)

$$\Lambda \subset \left\{ \bigcup M(i),\ i \in L \right\} \subset V_1 \subset V_2 \subset \left\{ \bigcup M(j),\ j \in D(L) \right\} \subset W^s(\Lambda).$$

Proof. According to Proposition 66, there are $\varepsilon_0 > 0$ and the neighborhoods U_1, U_2 of Λ,

$$\Lambda \subset U_1 \subset V_1 \subset V_2 \subset U_2 \subset clU_2 \subset W^s(\Lambda)$$

such that 1) every ε_0-trajectory passing through $U_2 \setminus U_1$ starts outside U_2 and ends inside U_1; 2) every positive ε_0-semi-trajectory passing through U_1

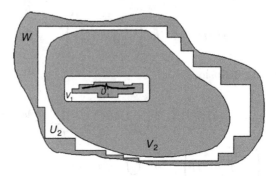

Fig. 7.9. An approximation of an attractor and its basin

remains there; 3) each negative ε_0-semi-trajectory passing through $M \setminus cl U_2$ remains there. Recall that $q(d)$ is the maximal diameter of images of cells and d is the maximal diameter of the covering cells. Since M is compact and f is continuous, the mapping f is uniformly continuous. Let $\alpha(d)$ be the modulus of continuity of the mappings f and f^{-1}. Hence $\alpha(d) \to 0$ as $d \to 0$. As $q(d)$ is the maximal diameter of the images of cells, we have $q(d) \leq \alpha(d)$. Choose d_0 so that $\alpha(d_0) + d_0 = \varepsilon_0$. Fix some positive $d < d_0$ and consider a symbolic image G with diameter d. According to Theorem 14, any path $\{z_k\}$ on the symbolic image G generates an ε-trajectory $\{x_k : x_k \in M(z_k)\}$ so that $q(d) + d < \varepsilon < \alpha(d_0) + d_0 = \varepsilon_0$. Set

$$L_1 = \{i : M(i) \subset U_1\},$$

$$L_2 = \{j : M(j) \cap cl(U_2 \setminus U_1) \neq \emptyset\},$$

$$L_3 = \{k : M(k) \subset M \setminus U_2\}.$$

According to Proposition 64, the neighborhood U_1 is constructed as an iteration of the fundamental neighborhood. Hence $cl\, f(U_1) \subset U_1$. Moreover, in the proofs of Propositions 64 and 66 the number ε_0 is obtained as the distance $\rho(cl\, f(U_1), M \setminus U_1) \geq \varepsilon_0$. As $\Lambda \subset f(U_1)$, the set of vertices $L_0 = \{i : M(i) \cap \Lambda \neq \emptyset\}$ is in L_1. Being Λ an invariant set and every trajectory of f generates a path on G, the set L_0 is invariant in G. Let L be the maximal invariant set of vertices in L_1. Obviously, $L_0 \subseteq L$. We have to show that L is attractor. Consider a path $\omega = \{z_k\}$ through $i \in L$ and suppose that it does not lie in L. That means the path ω passes through a vertex j so that $M(j) \cap (M \setminus U_1) \neq \emptyset$. According to Theorem 14, a sequence $\{x_k : x_k \in M(z_k), z_k \in \omega\}$ is an ε-trajectory, where $q(d) + d < \varepsilon < \varepsilon_0$. The sequence $\{x_k\}$ can be chosen so that $x_i \in U_1$, $x_j \in M \setminus U_1$. It is no loss of generality to set $\min\{k : x_k \in M(z_k) \subset U_1, z_k \in \omega\} = 0$. According to Proposition 66, the positive ε-trajectory $\{x_k, k > 0\}$ remains in U_1. Moreover, since $x_{-1} \in M \setminus U_1$, the negative ε-trajectory $\{x_k, k < 0\}$ has to start outside U_2 and pass through $U_2 \setminus U_1$. That means the path ω starts off L, passes through L_2 and ends in L. In this case, the exit $Ex(L) = \emptyset$, i.e., L is an attractor. The same way we can prove that any path through L_2 ends in L. Denote by D the domain of attraction for L. The above argument leads us to the conclusion that the domain of attraction $D \supset (L_1 \cup L_2)$ and $U_2 \subset \{\bigcup M(j), \ j \in D\} \subset W^s(\Lambda)$.

$$\odot$$

Thus the construction of an attractor of a dynamical system and its domain of attraction is reduced by Theorem 71 to the same problem on symbolic image.

7.4 Transition Matrix and Attractors

Let us introduce a quasi-order relation between the vertices of a symbolic image. We assume that $i \prec j$ if and only if there exists an admissible path of the form

$$i = i_0, i_1, i_2, ..., i_m = j.$$

Hence, a vertex i is recurrent if and only if $i \prec i$, and recurrent vertices i, j are equivalent if and only if $i \prec j \prec i$.

Proposition 72. [4] *The vertices of a symbolic image G can be renumbered so that*

- *the equivalent recurrent vertices are numbered with consecutive integers;*
- *the new numbers i, j of other vertices are chosen so that $i < j$ if $i \prec j \not\prec i$.*

In other words, the transition matrix has the form

$$\Pi = \begin{pmatrix} (\Pi_1) \cdots & \cdots & \cdots & \cdots \\ & \ddots & & \\ 0 & (\Pi_k) \cdots & \cdots \\ & \ddots & \ddots & \\ 0 & 0 & (\Pi_s) \end{pmatrix} \qquad (7.2)$$

where the elements under the diagonal blocks are zeros, each diagonal block Π_k corresponds to either a class of equivalent recurrent vertices H_k or a non-recurrent vertex. In the last case Π_k coincides with a single zero. The renumbering described in Proposition 72 is not uniquely defined. Indeed, if there are no admissible paths from i to j and from j to i, i.e., $i \not\prec j$, $j \not\prec i$, then the order between i and j is not fixed by Proposition 72. In this case the order between the vertices i, j can be arbitrary chosen. It follows from Propositions 69 and 72 that for every attractor $L = \{i\}$, its domain of attraction $D(L) = \{j\}$, and its corresponding repellor $L^* = \{k\}$, there is a renumbering such that

$$k < j < i, \ where \ j \in D(L) \setminus L.$$

In fact, there are the relations

$$j \nprec k, \ i \nprec j,$$

as $i \in L$, $j \in D(L) \setminus L$ and $k \in L^*$. There conceivably are the relations:

$$k \prec j, \ j \prec i.$$

This leads us to a renumbering of the form: $k < j < i$. But according to Proposition 68, the vertices from $D(L) \setminus L$ are non-recurrent. Thus, the transition matrix takes the form

$$\Pi = \begin{pmatrix} (\Pi_1) & \cdots & \cdots & \cdots & \cdots \\ & 0 & \cdots & \cdots & \cdots \\ 0 & & \ddots & & \\ & \ddots & & 0 & \cdots \\ 0 & & 0 & & (\Pi_2) \end{pmatrix}, \tag{7.3}$$

where the blocks Π_1, Π_2 correspond the repellor L^* and the attractor L, respectively. Thus we can find the attractors of a symbolic image by using the representations of the transition matrix in the form (7.3). It is important to bear in mind that the transition matrix of the form (7.2) is not uniquely defined.

7.5 The Construction of the Attractor-Repellor Pair

In order to construct an attractor, its domain of attraction and its repellor, we apply the process of adaptive subdivision.

Let $C = \{M(i)\}$ be a covering of M and G be the symbolic image for C. A new covering NC is a subdivision of C, i.e., each cell $M(i)$ is subdivided so that new cells $m(i,k)$, $k = 1, 2, \dots$, form a subdivision of the cell $M(i)$

$$\bigcup_k m(i,k) = M(i).$$

Denote by NG the symbolic image for the new covering NC. There is the natural projection $h : G \to NG$, $h(i,k) = i$ which maps the directed graph NG onto the directed graph G.

Consider a symbolic image G with an attractor L, the domain of attraction $D(L)$ and the corresponding dual repellor L^*. Suppose that the cells $M(i)$, $i \in L$ are subdivided and that the other cells remain as before. The new symbolic image has the same repellor L^* for which the dual attractor NL exists. It is not difficult in understanding that the dual attractor is a maximal invariant set in $h^{-1}(L)$. So the new attractor NL is determined by the initial attractor L and may be constructed as the maximal invariant set in $h^{-1}(L)$.

The repellor L^* is an attractor for the graph G with the inverse orientation. For such a graph L^* has different domains of attraction on G and NG — $D(L^*) = Ver(G) \setminus L$ and $ND(L^*) = Ver(NG) \setminus NL$ respectively. The restrictions of the symbolic images G and NG on $Ver(G) \setminus L$ coincide because only the cells $M(i)$, $i \in L$ are subdivided. This leads us to the conclusion that $D(L^*) \subset ND(L^*)$ and the image $h(NL)$ is in L. Hence, the new attractor NL lies in the set of vertices $\{(i, k) : i \in L\}$. So we get the inclusion

$$\left\{\bigcup M(i),\ i \in L\right\} \supset \left\{\bigcup m(i,k),\ (i,k) \in NL\right\}.$$

The domain of attraction for the new attractor NL is $D(NL) = Ver(NG) \setminus L^*$. Hence, we have

$$\left\{\bigcup M(j),\ j \in D(L)\right\} \subset \left\{\bigcup m(j,l),\ (j,l) \in D(NL)\right\}.$$

Now consider the second subdivision such that the cells $M(j)$, $j \in L^*$ are exposed to subdivision. The same way we obtain the inclusions

$$A_1 = \left\{\bigcup M(i),\ i \in L\right\} \supset \left\{\bigcup m(i,k),\ (i,k) \in NL\right\} = A_2,$$

$$W_1 = \left\{\bigcup M(j),\ j \in D(L)\right\} \subset \left\{\bigcup m(e,l),\ (e,l) \in D(NL)\right\} = W_2,$$

$$R_1 = \left\{\bigcup M(k),\ k \in L^*\right\} \supset \left\{\bigcup m(k,q),\ (k,q) \in NL^*\right\} = R_2,$$

where NL^* is a new repellor. The repellor NL^* is the maximal invariant set in $h^{-1}(L^*)$ We can consider each subdivision as three successive subdivisions: a) a subdivision of the cells $M(i)$, $i \in L$; b) a subdivision of the cells $M(k)$, $k \in L^*$, and c) a subdivision of the cells $M(j)$, $j \in D(L) \setminus L$. Note that the equalities $A_1 = A_2$, $R_1 = R_2$ and $W_1 = W_2$ hold under the subdivision of cells $M(j)$, $j \in D(L) \setminus L$. In fact, in this case the attractor L and the repellor L^* do not change. Hence, the new domain of attraction is $ND(L) = \{(i,k) : i \in D(L)\}$ and $ND(L) \setminus L = \{(j,k) : j \in D(L) \setminus L\}$. If $\bigcup_k m(j,k) = M(j)$, we have $W_1 = W_2$.

Consider a covering C and the corresponding symbolic image G. Suppose we pick some attractor L on G and choose a sequence of subdivisions so that the maximal diameter d of covering cells tends to zero. We get decreasing sequences of enclosed sets $A_1, A_2, \ldots, R_1, R_2, \ldots$ and an increasing sequence of enclosed sets $W_1, W_2, \ldots$ as well. It follows from Theorems 70 and 71 that there exists an attractor Λ, its domain of attraction $D(\Lambda)$ and the corresponding repellor Λ^* so that

$$\lim_{s \to \infty} A_s = \Lambda,\ \lim_{s \to \infty} W_s = W^u(\Lambda),\ \lim_{s \to \infty} R_s = \Lambda^*. \tag{7.4}$$

Moreover, it follows from Theorem 71 that each attractor of the homeomorphism f can be constructed by the described process. Thus we come to the following result.

Algorithm for the construction of an attractor, its domain of attraction and its dual repellor.

1) Let C be a covering of M by cells whose maximal diameter d is small. Construct the symbolic image G for the given covering.
2) Localize an attractor L, its domain of attraction $D(L)$ and its repellor L^*.
3) Obtain the sets

$$A = \left\{ \bigcup M(i), \ i \in L \right\},$$

$$W = \left\{ \bigcup M(j), \ j \in D(L) \right\},$$

$$R = \left\{ \bigcup M(k), \ k \in L^* \right\}.$$

Let $d = \max\{diam M(i), \ diam M(k) : \ i \in L, \ k \in L^*\}$.
4) Subdivide the cells corresponding to the attractor L and the repellor L^* and construct a new covering.
5) Construct the symbolic image for the new covering.
6) Return to the second step.

Repeating this process we obtain the sequences of sets $A_1, \ A_2, \ldots;$ $W_1, W_2, \ldots;$ $R_1, R_2, \ldots,$ and the sequence of numbers $d_1, d_2, \ldots.$ Here the adaptive subdivision means that the cells corresponding to the attractor and the repellor are subdivided, but the cells corresponding to the basin are not subdivided. So, we have proved the following

Theorem 73. *1. The described algorithm gives the sequences of embedded sets*

$$A_1 \supset A_2 \supset \cdots$$

$$W_1 \subset W_2 \subset \cdots$$

$$R_1 \supset R_2 \supset \cdots.$$

2. If $d_s \to 0$ when s becomes infinite then

$\lim_{s \to \infty} A_s = \Lambda$ *is attractor,*
$\lim_{s \to \infty} W_s = W^s(\Lambda)$ *is its domain of attraction,*
$\lim_{s \to \infty} R_s = \Lambda^*$ *is the dual repeller.*

3. Any attractor Λ can be constructed by such an algorithm.

Consider a limitation of the proposed algorithm. If an attractor Λ is fixed then according to Theorem 71 the maximal diameter d for an initial covering is defined by Λ. So the choice of an initial covering requires some previous information. Actually, the attractor Λ is defined by the choice of the attractor L on the initial symbolic image G. Subsequent subdivisions localize the attractor Λ. In the next chapter we consider the concept of filtration which helps to affect the initial choice.

Example 74. Let us consider a system defined on the domain $M = [-5, 2] \times [-2, 2]$. We divide M into three parts: $M_1 = [-5, -3.5] \times [-2, 2]$, $M_2 = [-2.5, -2] \times [-2, 2]$, and $M_3 = [-2, -2] \times [-2, 2]$. At M_1 we put the system

$$x' = -x(x + 4)(x + 3),$$
$$y' = -y.$$

At the domain M_3 we put the perturbed Duffing system

$$x' = y,$$
$$y' = x - x^3 - 0.25y.$$

At M_2 we put a smooth vector field connecting the described systems (see Fig. 7.10). Near the boundary ∂M we put a vector field so that the trajectories passing through ∂M are inside M. The system has two equilibriums $(-4, 0)$ and $(-3, 0)$ at M_1. The point $(-4, 0)$ is an attractor, the point $(-3, 0)$ is hyperbolic. The Duffing system has the two attractors $(\pm 1, 0)$ and the hyperbolic point $(0, 0)$. Except three point-attractors $(-4, 0)$ and $(\pm 1, 0)$ the unstable separatrices $W^u(O)$ of the origin $O(0, 0)$ form an attractor. Moreover, the union of the unstable separatrices $W^u(-4, 0)$ and $W^u(O)$ is an attractor as well. So the system has 5 attractors. The biggest one is shown by the thick line at the Fig. 7.10. Using the localizing algorithm we can construct each of them. Each attractor is determined by the choice of an initial estimation of attractor A_1. So, assuming $A_1 = M$ we construct the biggest one.

Now we consider an example of the numerical construction of an attractor and its basin. The described above algorithm for the construction was implemented

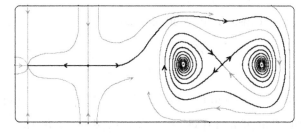

Fig. 7.10. The maximal attractor in the given domain

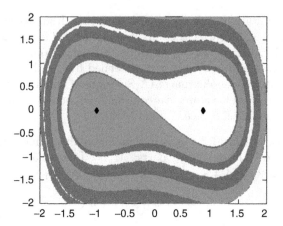

Fig. 7.11. The estimations of the attractors and its basins

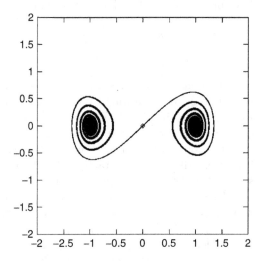

Fig. 7.12. The estimations of the attractor maximal in K

by Danny Fundinger during his visit St. Petersburg Polytechnic University at 2003.

Example 75. Let us consider the Duffing system

$$x' = y,$$
$$y' = x - x^3 - \varepsilon y$$

for the parameter $\varepsilon = 0.15$ in the domain $K = [-2, 2] \times [-1.5, 1.5]$.

The last means that if a trajectory leaves the domain K then we assume that it goes to infinity. The system in K has three attractors: equilibrium points $(\pm 1, 0)$ and attractor A_{max} maximal by inclusion. The attractor A_{max} is formed by the equilibrium $O(0,0)$ and its unstable separatrices which are spirals around the points $(-1, 0)$ and $(+1, 0)$, respectively, see Fig. 7.12. The estimations of the given attractors and its domains of attraction were obtained. The result of calculation is shown in Fig. 7.11. To localize the attractors $(\pm 1, 0)$ and their basins 2653 and 113424 cells with the size 0.004688×0.004688 respectively were used. In addition an estimation of the attractor maximal in K was obtained with the size of cells 0.002344×0.002344. The result is illustrated in Fig. 7.12.

8

Filtration

A dynamical system has generally a lot of attractors and repellors. The dynamics of an attractor-repellor pair is so structured that a trajectory outside the attractor and repellor starts at the repellor and ends at the attractor. If the system has many attractors, the dynamics that results is not so trivial. In this case the concept of filtration is very useful because filtration generates sequence of extending attractors. Filtration is a discrete analog of global Lyapunov function. The main filtration property is persistence with respect to perturbation of system. The proposed investigation is based on the ideas of C. Conley [28] about chain recurrent set and on results of Z. Nitecki and M. Shub [93] about Morse decomposition.

8.1 Definition and Properties

Let $f : M \to M$ be a homeomorphism on a manifold M.

Definition 76. [93] *A filtration for the homeomorphism f is a finite sequence $F = \{U_0, U_1, \ldots, U_m\}$ of open sets so that $\emptyset = U_0 \subset U_1 \subset \cdots \subset U_m = M$ and for each $k = 0, 1, \ldots, m$, $f(clU_k) \subset U_k$.*

The second condition is a property of the fundamental neighborhood of an attractor, see Proposition 63.

Example 77. Let us consider the system

$$x' = x - x^3,$$
$$y' = -y$$

on the plane. The system has three equilibrium points $(-1,0)$, $(0,0)$, $(1,0)$. The equilibrium $(0,0)$ is hyperbolic point, the equilibriums $(\pm 1, 0)$ are attractors. Let U_1 be a fundamental neighborhood of $(-1,0)$ and V_1 be a fundamental neighborhood of $(1,0)$. There is an open set U_3 that contains

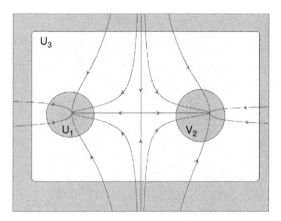

Fig. 8.1. The filtration $\emptyset \subset U_1 \subset U_1 + V_1 \subset U_3 \subset \mathbf{R}^2$

the neighborhoods U_1, V_1, the interval $[-1, 1] \times 0$ and each trajectory goes through the boundary inside U_3 (see Fig. 8.1). In this case we have the filtration $\emptyset = U_0 \subset U_1 \subset U_2 \subset U_3 \subset U_4 = \mathbf{R}^2$, where $U_2 = U_1 + V_1$. Evidently, we can replace U_1 by V_1 due to the symmetry and construct new filtration. So the filtration is not uniquely determined by system.

The next proposition describes the structure of attractors generated by a filtration.

Proposition 78. [19] *For a given filtration F the maximal invariant subset in U_k, $k = 0, 1, \ldots, m$*

$$A_k = \left\{ \bigcap f^n(U_k) : \ n \in Z^+ \right\}$$

is an attractor for f and $\emptyset = A_0 \subset A_1 \subset \cdots \subset A_m = M$.

It follows from the proposition that each U_k is the fundamental neighborhood of the attractor A_k. The intersection of an attractor and a repellor is called Morse set. The maximal invariant set I_k in $U_k \backslash U_{k-1}$ is the intersection of the attractor A_k and the repellor R_{k-1} dual for A_{k-1}:

$$I_k = A_k \cap R_{k-1}.$$

The described finite collection $\{I_k\}$ is called Morse decomposition.

Example 79. The above system

$$x' = x - x^3,$$
$$y' = -y.$$

has the filtration $\emptyset = U_0 \subset U_1 \subset U_1 + V_1 \subset U_3 \subset U_4 = \mathbf{R}^2$. The attractor A_1 consists of the single point $(-1, 0)$, the attractor A_2 consists of two points $(-1, 0)$ and $(+1, 0)$, the attractor A_3 is the interval $[-1, 1] \times \{0\}$. We have the inclusions $\emptyset = A_0 \subset A_1 \subset A_2 \subset A_3 \subset A_4 = \mathbf{R}^2$.

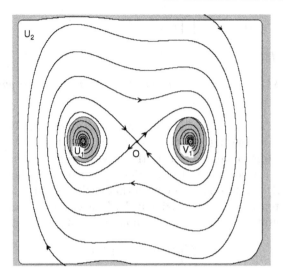

Fig. 8.2. Filtration of Duffing system

Example 80. Consider the perturbed Duffing system

$$x' = y,$$
$$y' = x - x^3 - \varepsilon y,$$

where $\varepsilon > 0$. The topology of trajectories is similar to the topology of Example 77 but behavior of trajectories is more complicated, see Fig. 8.2. The system has three equilibrium points $(-1,0)$, $(0,0)$, $(1,0)$. The equilibrium $(0,0)$ is hyperbolic point, the equilibriums $(\pm 1, 0)$ are attractors. Let U_1 be a fundamental neighborhood of $(-1,0)$ and V_1 be a fundamental neighborhood of $(1,0)$. The system has the filtration of the form $\emptyset = U_0 \subset U_1 \subset U_1 + V_1 \subset U_3 \subset U_4 = \mathbf{R}^2$. The attractor A_1 consists of the single point $(-1,0)$, the attractor A_2 consists of the points $(-1,0)$ and $(+1,0)$, the attractor A_3 consists of the equilibrium $O(0,0)$ and its unstable separatrices which are spirals around the points $(-1,0)$ and $(+1,0)$, respectively. We have the inclusions $\emptyset = A_0 \subset A_1 \subset A_2 \subset A_3 \subset A_4 = \mathbf{R}^2$. The phase portrait of the system for $\varepsilon = 0.15$ is shown in Fig. 8.2, being the trajectories represented on the picture are the separatrices of the equilibrium point O.

The next proposition shows that a filtration is preserved under perturbation.

Proposition 81. [19] *If a sequence F is a filtration for f then there is such a neighborhood V of f in C^0-topology that the sequence F is a filtration for any map $g \in V$.*

Let $F = \{\emptyset = U_0 \subset U_1 \subset U_2 \subset .... \subset U_m = M\}$ be a filtration. By Proposition 78 the maximal invariant subset in $U_k \setminus U_{k-1}$ is a Morse set

$$I_k(F) = \left\{\bigcap f^n(U_k \setminus U_{k-1}) : \ n \in \mathbb{Z}\right\}.$$

Set $I(F) = \{\bigcup I_k(F) : \ k = 1, \ldots m\}$.

Proposition 82. *The chain recurrent set Q of a homeomorphism f lies in $I(F)$.*

Proof. Consider an attractor Λ, its domain of attraction $W^s(\Lambda)$ and the repellor Λ^* corresponding to Λ. First we prove that any point x from $W^s(\Lambda) \setminus \Lambda$ is nonrecurrent. In fact, there is a neighborhood V such that $x \in W^s(\Lambda) \setminus clV$. According to Proposition 64, for the neighborhood V and a point $x \in W^s(\Lambda) \setminus \Lambda$, there is $\varepsilon > 0$ such that each positive ε-trajectory passing through x reaches V and remains in V. Hence, there is not any periodic ε-trajectory passing through x, and the point x is not recurrent. It means that

$$Q \bigcap (W^s(\Lambda) \setminus \Lambda) = \emptyset.$$

Consider a filtration $F = \{U_0, U_1, \ldots, U_m\}$. Fix a chain recurrent point x. It follows from the above that if the point x does not lie in the attractor $A_k = \bigcap_{n>0} f_n(U_k)$ then $x \notin W^s(A_k)$. Since the attractors $A_0, A_1, \ldots, A_m$ form a sequence of extending sets, there exists an attractor A_l such that $x \in A_l$ and $x \notin A_{l-1}$. Being U_k is a fundamental neighborhood of A_k,

$$A_l = \bigcap_n f^n(U_l), \ W^s(A_{l-1}) = \bigcup_n f^n(U_{l-1}).$$

We have $x \in \bigcap_n f^n(U_l)$ and $x \notin W^s(A_{l-1}) = \bigcup_n f^n(U_{l-1})$. It follows from the equality $(A \setminus B) \cap (C \setminus D) = (A \cap C) \setminus B \cup D$ that

$$K_l(F) = \bigcap_n f^n(U_l \setminus U_{l-1}) = \bigcap_n f^n(U_l) \setminus \bigcup_n f^n(U_{l-1}) = A_l \setminus W^s(A_{l-1}) \ni x.$$

Thus $Q \subset I(F) = \bigcup_k I_k(F)$.

$\odot$

The following example shows that a filtration is not unique.

Example 83. Let us consider the system

$$x' = \sin x,$$
$$y' = -y$$

in the domain $D = [-12, 12] \times [-5, 5]$. It has the equilibrium points $(k\pi, 0)$, $k \in \mathbb{Z}$. The trajectories passing through the boundary ∂D come into D, see Fig. 8.3. The equilibrium points $(0, 0)$, $(\pm 2\pi, 0)$ are hyperbolic. The equilibrium points $(\pm \pi, 0)$, $(\pm 3\pi, 0)$ are attractors. The set $U_1 = [-13/4\pi, -11/4\pi] \times [-1, 1]$ is a fundamental neighborhood of the attractor $(-3\pi, 0)$. The sets $V_1 = [-5/4\pi, -3/4\pi] \times [-1, 1]$, $V_3 = [3/4\pi, 5/4\pi] \times [-1, 1]$ and $V_5 =$

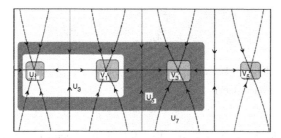

Fig. 8.3. The filtration of the system $x' = \sin x$, $y' = -y$

$[11/4\pi, 13/4\pi] \times [-1,1]$ are fundamental neighborhoods of the attractors $(-\pi, 0)$, $(\pi, 0)$ and $(3\pi, 0)$, respectively. The system has the filtration $\emptyset \subset U_1 \subset U_2 \subset U_3 \subset U_4 \subset U_5 \subset U_6 \subset U_7 \subset \mathbb{R}^2$, where $U_3 = [-7/3\pi, -1/2\pi] \times [-2, 2]$, $U_5 = [-11/3\pi, 3/2\pi] \times [-3, 3]$, $U_2 = U_1 + V_1$, $U_4 = U_3 + V_3$, $U_6 = U_5 + V_5$. For the given filtration $I_1 = (-3\pi, 0)$, $I_2 = (-\pi, 0)$, $I_3 = (-2\pi, 0)$, $I_4 = (\pi, 0)$, $I_5 = (0, 0)$, $I_6 = (3\pi, 0)$, $I_7 = (2\pi, 0)$. It is clear that there are few filtrations. For example, there is a filtration of the form $\emptyset \subset V_1 \subset U_2 \subset U_3 \subset U_4 \subset U_5 \subset U_6 \subset U_7 \subset \mathbb{R}^2$.

A filtration F is said to be fine if $I(F)$ coincides with a chain recurrent set Q. It should be noted that in [93] a filtration F is called fine if $I(F)$ coincides with a nonwandering set.

Example 84. Let us consider the system of Example 77

$$x' = x - x^3,$$
$$y' = -y.$$

The described filtration $\emptyset = U_0 \subset U_1 \subset U_1 + V_1 \subset U_3 \subset U_4 = \mathbb{R}^2$ has $I_1 = (-1, 0)$, $I_2 = (+1, 0)$, $I_3 = (0, 0)$. This filtration is fine, because the chain recurrent set Q consists of three equilibriums $(0, 0)$ and $(\pm 1, 0)$, i.e., $Q = I_1 + I_2 + I_3$.

If f does not admit any fine filtration, it is common practice to consider a sequence $\{F_l : l = 1, 2, ...\}$ of filtrations for which the sequence $I(F_l)$ tends to Q as $l \to \infty$.

Definition 85. • *A filtration* $F^* = \{U_0^*, \ldots, U_p^*\}$ *refines a filtration* $F = \{U_0, \ldots, U_q\}$ *if for each* $\alpha = 1, \ldots p$ *there exists* $\beta(\alpha)$, $1 \leq \beta(\alpha) \leq q$ *such that*

$$U_\alpha^* \setminus U_{\alpha-1}^* \subset U_{\beta(\alpha)} \setminus U_{\beta(\alpha)-1}.$$

• *A sequence* F_1, $F_2, \ldots$ *of filtrations for* f *is said to be fine if* F_{k+1} *refines* F_k *and*

$$\bigcap_k I(F_k) = Q.$$

Fig. 8.4. Filtration of the system $x' = x^2 \sin \frac{\pi}{x}$

Example 86. The equation

$$x' = x^2 \sin \frac{\pi}{x},$$

where $x \in \mathbb{R}$, does not admit any fine filtration, because any neighborhood of the equilibrium 0 contains the infinite number of equilibriums which form separated components of the chain recurrent set. The equation has the equilibriums at the points $x = \frac{1}{n}$, $n \in \mathbb{Z}$. There is a filtration $\emptyset = U_0 \subset U_1 \subset U_2 \subset U_3 \subset U_4 \subset U_5 = \mathbb{R}$, where U_1 is a neighborhood of the interval $[-\frac{1}{4}, \frac{1}{4}]$, $U_2 = U_1 + V_1$, $U_3 = U_1 + V_1 + V_2$, V_1 and V_2 are neighborhoods of the points $-\frac{1}{2}$ and $\frac{1}{4}$, respectively, and $U_4 = (-\frac{3}{4}, \frac{3}{4})$. There are infinite numbers of equilibrium points in U_1, moreover the dynamic in U_1 is similar to the system behavior in U_4. By the same way we can construct a filtration in U_1 which together with $U_2 \subset U_3 \subset U_4 \subset U_5 = \mathbb{R}$, refines the previous one. By repeating this process we get an infinite sequence of refining filtrations which is fine.

A fine sequence of filtrations controls the growth of chain recurrent set Q under perturbation of a dynamical system. More precise, if F_1, F_2, ... is a fine sequence of filtrations then, according to Proposition 81 there is a neighborhood V of f in C^0 topology for a finite sequence of filtrations F_1, F_2, ..., F_l so that the sequence F_1, F_2, ..., F_l is a refined sequence of filtrations for each map $g \in V$. We will prove that for any homeomorphism f there exists a fine sequence of filtrations. Moreover, a fine sequence of filtrations can be constructed by a special sequence of symbolic images generated by a subdivision process.

8.2 Filtration on a Symbolic Image

Let us consider a symbolic image G of a dynamic system.

Definition 87. *A finite sequence $\Phi = \{B_0, B_1, \ldots, B_m\}$ of sets of vertices on the symbolic image G is called filtration if*

$$\emptyset = B_0 \subset B_1 \subset \cdots \subset B_m = Ver(G) \tag{8.1}$$

and for each B_k if the initial vertex i of an edge $i \to j$ lies in B_k then the final vertex j lies in B_k as well.

The second condition means that there is no exit from B_k. Let L_k be a maximal invariant set in B_k, see Fig. 8.5.

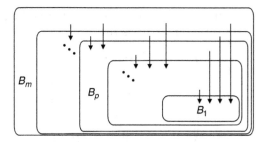

Fig. 8.5. Filtration on symbolic image

Proposition 88. *Each maximal invariant set* $L_k \subset B_k$ *is attractor and*

$$\emptyset = L_0 \subset L_1 \subset \cdots \subset L_m = Ver(G) \tag{8.2}$$

Proof. Let us fix some $k = 1, 2, \ldots, m$ and prove that the exit $Ex(L_k)$ is empty. Conversely, suppose that $i \to j$ be an edge such that $i \in L_k$ and $j \notin L_k$. Consider a path $\omega_1 = \{\ldots, i, j, p, \ldots\}$ through the edge $i \to j$. Since there is no exit from B_k, the positive semi-path $\omega^+ = \{i, j, p, \ldots\}$ lies in B_k. Since $i \in L_k$ and L_k is invariant, there is a path $\omega_2 = \{\ldots, q, i, \ldots\}$ through i which is in L_k. In particular, the negative semi-path $\omega^- = \{\ldots, q, i\}$ lies in B_k. We construct a new path by combining the negative and positive semi-paths ω^- and ω^+. The path $\omega = \omega^- \omega^+ = \{\ldots, q, i, j, p, \ldots\}$ has to be in B_k. The path ω lies in L_k, because L_k is the maximal invariant set in B_k. In particular, $j \in L_k$, a contradiction. Thus $Ex(L_k) = \emptyset$, and each L_k is attractor. We have the inclusions (8.2), because the inclusions (8.1) hold and L_k is the maximal invariant set in B_k.

$$\odot$$

It follows from the proof and Theorem 70 that for each k the set B_k corresponds to a fundamental neighborhood $U = int\{\bigcup M(i), \ i \in B_k\}$ of an attractor of the dynamic system.

Let $\Phi = \{B_0, B_1, \ldots, B_m\}$ be a filtration on a symbolic image. A maximal invariant subset of $B_k \setminus B_{k-1}$ is denoted $J_k(\Phi)$ and we set $J(\Phi) = \bigcup_k J_k(\Phi)$. Let us show that the set of recurrent vertices RV lies in $J(\Phi)$. Fix a recurrent vertex i and denote by $H(i)$ the class of recurrent vertices equivalent to i. It follows from the inclusions (8.1) that there is a set B_l such that $i \in B_l$ and $i \notin B_{l-1}$. The set B_l is a subset of the domain of attraction $D(L_l)$. As B_l has no exit and L_l is the maximal invariant set in B_l, it follows from Proposition 69 that $H(i) \subset L_l$ and $H(i) \cap B_{l-1} = \emptyset$. Hence, $H(i) \subset B_l \setminus B_{l-1}$. The set $H(i)$ is invariant. Since $J_l(\Phi)$ is the maximal invariant set in $B_l \setminus B_{l-1}$, the class $H(i)$ lies in $J_l(\Phi)$. Thus the recurrent vertices RV are in $J(\Phi)$.

A filtration Φ is said to be fine if the set $J(\Phi)$ coincides with the set of recurrent vertices. We prove that there exists a fine filtration on any symbolic image. Denote by H_p, $p = 1, \ldots, s$ the classes of equivalent recurrent vertices. We set $H_p \prec H_q$ if and only if there exists an admissible path from H_p to H_q.

Let the vertices of a symbolic image be renumbered according to Proposition 72. In this case the transition matrix has the form

$$
\Pi = \begin{pmatrix}
(\Pi_1) & \cdots & \cdots & \cdots & \cdots \\
 & \ddots & & & \\
0 & & (\Pi_p) & \cdots & \cdots \\
 & & & \ddots & \ddots \\
0 & & 0 & & (\Pi_s)
\end{pmatrix},
$$

where the elements under the diagonal blocks are zeros, each diagonal block Π_p corresponds to either a class of equivalent recurrent vertices H_p or a non-recurrent vertex. In the last case Π_p coincides with zero. Under this numbering there is no admissible path from a class H_p to a class H_q if $q < p$. As the numbering of rows of the transition matrix coincides with the numbering of the vertices, we have to move in the same class or to go down, as the following scheme shows.

$$
\begin{pmatrix}
H_1 & \downarrow & \downarrow & \downarrow & \downarrow \\
 & \ddots & \downarrow & \downarrow & \downarrow \\
 & & H_p & \downarrow & \downarrow \\
 & & & \ddots & \downarrow \\
 & & & & H_s :
\end{pmatrix}
$$

Hence we cannot get higher then H_k. Thus, each class H_p generates an attractor consisting of the vertices of numbers (indexes) more or equal to the numbers of the ones from H_p. Note that according to chosen numbering of vertices the minimal (by inclusion) attractor H_s has the maximal number (index) s. This numbering is opposite to the one of filtration. To formalize this observation let us introduce the numbers $n(H_p) = \min\{i : i \in H_p\}$, and construct the sets

$$E_p = \{i : i \geq n(H_p)\}, \quad p = 1, \ldots, s, \quad E_{s+1} = \emptyset.$$

Set $B_k = E_p$, where $p = s + 1 - k$, $k = 0, 1, \ldots, s$. We have $B_0 = \emptyset$ and $B_s = Ver(G)$.

Proposition 89. *The finite sequence* $\Phi = \{\emptyset = B_0, B_1, \ldots, B_s = Ver(G)\}$, *defined as above, is fine filtration on the symbolic image G.*

Proof. Fix a number p between 0 and s, define $k = s + 1 - p$ and set $N(H_p) = \max\{i : i \in H_p\}$. Consider a decomposition of the set of vertices $Ver(G)$

$$B_k = \{i; \ i \geq n(H_p)\},$$

$$B_k^* = \{i : i \leq N(H_{p-1})\},$$

$$W_k = \{i : N(H_{p-1}) < i < n(H_p)\}.$$

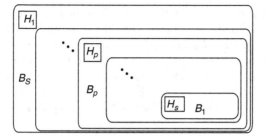

Fig. 8.6. Filtration and transition matrix

According to the construction process, the set B_k contains the classes H_q, $q \geq p$; the set B_k^* contains the classes H_l, $l < p$ and the set W_k contains only the non-recurrent vertices. In this case the transition matrix takes the form

$$
\Pi = \begin{pmatrix} (Y^*) \cdots & \cdots & \cdots & \cdots & \cdots \\ & 0 & & & \\ 0 & & \ddots & \cdots & \cdots \\ & \ddots & & 0 & \\ 0 & & 0 & & (Y) \end{pmatrix},
$$

where the block Y corresponds to B_k, and the block Y^* corresponds to B_k^*. Hence B_k and B_k^* are attractor and repellor, respectively. The set $B_k \setminus B_{k-1}$ coincides with $H_p \cup W_{k-1}$. The maximal invariant set in $B_k \setminus B_{k-1}$ is the class H_p, i.e., $J_k(\Phi) = H_p$. Thus $J(\Phi) = \{\cup H_p : p = 1, \ldots, s\}$ and the sequence $\Phi = \{B_0, B_1, \ldots, B_s\}$ is fine filtration.

⊙

8.3 Fine Sequence of Filtrations

Let us consider a construction of dynamical system filtration. Suppose that a dynamical system is generated by a mapping $f; M \to M$. For a covering $C = \{M(i)\}$ we construct the symbolic image G.

Proposition 90. Let $\Phi = \{B_0, B_1, \ldots, B_s\}$ be a filtration on the symbolic image G. Then the finite sequence $F = \{U_0, U_1, \ldots, U_s\}$, where $U_k = int\{\bigcup M(i) : i \in B_k\}$, is a filtration for the mapping f.

Proof. According to Proposition 88, each maximal invariant set L_k in B_k is an attractor on the symbolic image G. It follows from Theorem 70 that the set $int\{\bigcup M(i) : i \in L_k\}$ is a fundamental neighborhood of an attractor Λ_k. We show that the set $U_k = int\{\bigcup M(i) : i \in B_k\}$ has the property: $f(clU_k) \subset U_k$. Fix some point $x \in \{\bigcup M(i) : i \in B_k\}$. According to Theorem 14, the image $f(x)$ generates an directed edge $i \to j$, $x \in M(i)$, $f(x) \in M(j)$ and $i \in B_k$. As

B_k has no exit, the vertex j is in B_k as well. This means that $f(x) \in \{\bigcup M(i) : i \in B_k\} = clU_k$. Actually, the image $f(x)$ has to be in U_k. Otherwise, if $f(x)$ is on the boundary $clU_k \backslash U_k$ then $f(x) \in M(j)$, with numbers $j \notin B_k$. This implies the existence of the edge $i \to j$, $j \notin B_k$ and we obtain a contradiction. Hence $f(clU_k) \subset U_k$. It follows from the inclusion $B_k \subset B_{k+1}$ that $U_k \subset U_{k+1}$. Since $B_0 = \emptyset$ and $B_s = Ver(G)$, $U_0 = \emptyset$, $U_s = M$. Thus the sequence $F = \{U_0, U_1, ..., U_s\}$ is a filtration for the homeomorphism f.

$\odot$

Now we are ready to apply the subdivision process to obtain a fine sequence of filtration. Consider the following

Algorithm for construction of a fine sequence of filtrations.

1) Let C be an arbitrary finite covering of M by closed cells. Given the covering, the symbolic image G is constructed.
2) The classes H_p of equivalent recurrent vertices are recognized. Denote by

$$d = \max\{diamM(i) : i \text{ is recurrent}\}.$$

3) A fine filtration $\Phi = \{B_0, B_1, \ldots, B_s\}$ on the symbolic image G is obtained by setting $B_k = \{i : i \geq n(H_p), p = s + 1 - k\}$.
4) The filtration $F = \{U_0, U_1, \ldots, U_s\}$ for the dynamical system is defined by setting $U_k = \{\bigcup M(i) : i \in B_k\}$.
5) The cells corresponding to the recurrent vertices are subdivided. The new covering is found.
6) The symbolic image G for the new covering is constructed.
7) Return to the second step.

The described algorithm generates a sequence of symbolic images G_m, fine filtrations Φ_m on each G_m, a sequence of filtrations F_m on M and the sequence of numbers d_m. The following theorem justifies the proposed algorithm.

Theorem 91. *If $d_m \to 0$ as m tends to infinity then the sequence of filtration $\{F_m\}$ is fine.*

Proof. Assume that the subdivision algorithm forms a sequence of filtrations $\Phi_m = \{B_0, B_1, ..., B_s\}$ on the symbolic images G_m. According to Proposition 89 each filtration Φ_m is fine. Hence a maximal invariant set in $B_k \backslash B_{k-1}$ coincides with the class of equivalent recurrent vertices H_p. By Proposition 90 each filtration Φ_m generates a filtration $F_m = \{U_0, U_1, ..., U_s\}$ on the manifold M, where $U_k = \{\bigcup M(i) : i \in B_k\}$. Since the maximal invariant set in $B_k \backslash B_{k-1}$ is H_p, $k = s + 1 - p$, a maximal invariant set $I_k(F_m)$ in $U_k \backslash U_{k-1}$ lies in $\{\bigcup M(i) : i \in H_p\}$. Hence the chain recurrent set Q lies in $P_m = \{M(i) : i \text{ is recurrent}\}$. It follows from Theorem 54 that for each m the inclusion $P_m \supset P_{m+1}$ holds, and if $d_m \to 0$ as $m \to \infty$ then

$$\lim_{m \to \infty} P_m = \bigcap_{m>0} P_m = Q.$$

Thus the sequence of filtrations $\{F_m\}$ is fine.

$\odot$

Corollary 92. *For each homeomorphism f there exists a fine sequence of filtrations.*

9

Structural Graph

One of important practical problems in the theory of dynamical systems is the development of constructive methods for investigation of the global orbit structure. Classical works over a period of time 1960–1980 [28, 48, 93, 110, 115, 118, 137] have shown that the global system dynamics is essentially determined by connections between components of its chain recurrent set. In the present chapter we give the theoretical substantiation of a computer focused method for construction of the structural graph of a dynamical system. A vertex $\{i\}$ of the structural graph corresponds to a component Q_i of the chain recurrent set Q. Each edge $i \to j$ of the graph corresponds to the orbits which have the α-limit set in the component Q_i and the ω-limit set in the component Q_j, i.e. the edge $\{i \to j\}$ corresponds to the connection $\{Q_i \to Q_j\}$. By the definition the structural graph is topological invariant of dynamical system. The structural graph provides a way to obtain information not only about the number of components of Q and connections between them but about the structure of attractors and their areas of attraction as well. The basic assumption in this chapter is that the number of components of the chain recurrent set is finite. The results of this chapter are obtained by D.A. Mizin and G.S. Osipenko.

9.1 Symbolic Image and Structural Graph

Consider a discrete dynamical system generated by a homeomorphism $f : M \to M$ of a compact metric space M, being Q the set of chain recurrent orbits of the system.

Definition 93. *A subset $\Omega \subset Q$ refers to as a component of the chain recurrent set if any two points from Ω can be connected by a periodic ε-orbit for any $\varepsilon > 0$.*

It follows that the chain recurrent set Q can be represented as the union of disjoint closed invariant components Q_i:

$$Q = \bigcup_i Q_i.$$

Let $\{Q_1, Q_2, Q_3, \dots\}$ be components of the chain recurrent set. We say that there is a connection $Q_i \to Q_j$ if there is a point x such that $\alpha(x) \subset Q_i$ and $\omega(x) \subset Q_j$. In other words, there exists an orbit which starts at Q_i and ends at Q_j.

Definition 94. *Let Γ be a graph with vertices $\{i\}$ corresponding to components Q_i and edges $\{i \to j\}$ corresponding to connections $\{Q_i \to Q_j\}$. The graph Γ is called structural graph of the dynamical system f. The transition matrix $A = (a_{ij})$ $(a_{ij} = 1$ if there exists the edge $i \to j$ and $a_{ij} = 0$ otherwise) is called structural matrix of the dynamical system f.*

According to definition, structural graph and its transition matrix are topological invariants of a dynamical system. This means that two systems have identical structural graphs if there exists a continuous and invertible change of coordinates which transforms orbits of one dynamical system to orbits of another one. Let q be the number of components of the chain recurrent set. Then the structural matrix has the size $q \times q$. The following theorem is the main result of this chapter.

Theorem 95. *Let a dynamical system has the finite number of components of the chain recurrent set. Then there is the finite algorithm for the construction of the structural graph.*

Let G be the symbolic image of the map f related to the covering C. Recall that a vertex i is recurrent if there exists an admissible closed path through i. The recurrent vertices i and j are equivalent if there exists an admissible closed path through i and j. Therefore, the set RV of recurrent vertices is decomposed into disjoint classes H_k of equivalent recurrent vertices: $RV = \bigcup_k H_k$. The set RV is similar to the chain recurrent set and classes H_k are similar to its components.

By Proposition 72, the transition matrix of the symbolical image (under the appropriate numbering) takes the form

$$\Pi = \begin{pmatrix} (\Pi_1) \cdots & \cdots & \cdots & \cdots \\ & \ddots & & \\ 0 & (\Pi_k) \cdots & \cdots \\ & \ddots & & \ddots \\ 0 & 0 & (\Pi_s) \end{pmatrix}, \tag{9.1}$$

where each diagonal block Π_k corresponds to either one of classes H_k or some non-recurrent vertex and in this case consists of a zero. Joining the equivalent vertices on G in one vertex, we form the new graph G^*. A vertex k of the graph G^* conforms to a class H_k and the existence of the edge $k \to l$ means that there is an admissible path from H_k to H_l which does not pass through other classes $\{H_m, m \neq k, l\}$.

Definition 96. *The graph G^* is called structural graph of the symbolic image G.*

The transition matrix of G^* has the form

$$\Pi^* = \begin{pmatrix} 1 & \cdots & \cdots & \cdots & \cdots \\ & \ddots & & & \\ 0 & & 1 & \cdots & \cdots \\ & \ddots & & & \ddots \\ 0 & & 0 & & 1 \end{pmatrix}, \tag{9.2}$$

where diagonal entries are units, sub-diagonal entries are zeros, and the matrix size is determined by the number of classes of equivalent recurrent vertices. Clearly, the structural graph of the symbolical image is similar to the structural graph of a dynamical system. However, as the following example shows, these graphs may differ to a large extent.

Example 97. Structural graphs of a linear mapping and its symbolic image.

Let us consider the linear mapping f_λ of the extended real line $(-\infty, +\infty) \cup (\infty = +\infty = -\infty)$ into itself:

$$f_\lambda : x \to \lambda x,$$

$\lambda \in (0, 1)$. Let M be the compactified real line, i.e. there is a point $-\infty = \infty = +\infty$. It is homeomorphic to the unit circle S^1, and ∞ is fixed point of f_λ. Let $\lambda = 1/2$. Assume that the covering C consists of 1-length segments $[n, n+1]$, $n = -3, -2, -1, 0, 1, 2, 3$ and the segment $[3, +\infty) \cup (-\infty, -3] \cup \{\infty\}$. The image of the cell $M(1) = [0, 1]$ intersects $M(1)$ and $M(-1) = [-1, 0]$, i.e. the vertex $i = 1$ is recurrent. The image of the cell $M(2) = [1, 2]$ intersects $M(2)$ and $M(1)$, i.e. the vertex $i = 2$ is also recurrent. Moreover, as $f(M(1)) \cap M(2) = \emptyset$, there is no connection $1 \to 2$ and vertices 1 and 2 are not equivalent. The image of the cell $M(3) = [2, 3]$ intersects $M(1)$ and $M(2)$ and does not intersect $M(3)$. Hence, the vertex $i = 3$ is non-recurrent (transient). Construction of the connections for the cells $M(-1)$, $M(-2) = [-2, -1]$, and $M(-3) = [-3, -2]$ is produced by the same way. Fig. 9.1 depicts a) the symbolic image for the given covering, b) the structural graph of the symbolic image, and c) the structural graph of the dynamical system. We note that classes $H(2)$ and $H(-2)$ of equivalent recurrent vertices are false in the sense that the cells $M(2)$ and $M(-2)$ do not contain chain recurrent points. Moreover, as the mapping f is linear, the process of refining of coverings keep the number of "false" classes. It is not difficult to see that the refinement of the covering changes the symbolic image but keeps the structural graph. This means that by refinement of a covering we cannot destroy "false" classes and construct the structural graph of dynamical system. We shall give a rigorous definition of "false" classes later.

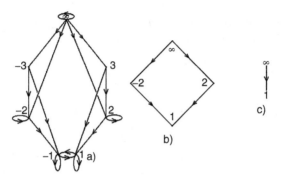

Fig. 9.1. Symbolic image and structural graphs of the mapping $x \to 1/2x$

9.2 Sequence of Symbolic Images

Given a continuous mapping $f : M \to M$, consider a sequence $\{C_t, t \in N\}$ of coverings of the manifold M by cells which are consecutive subdivisions. Denote by $M(z^t)$ cells of the covering C_t, where z^t is the cell index, and d_t the maximal diameter of cells from the covering C_t. Let $\{G_t\}$ be the sequence of symbolic images of f corresponding to the sequence C_t. As was shown in the section 2.6 the process of construction results in two sequences of mappings $\{s_t\}$ and $\{G_t\}$ and the commutative diagram

$$
\begin{array}{ccccccc}
V_1 & \overset{s_1}{\leftarrow} & V_2 & \overset{s_2}{\leftarrow} & V_3 & \overset{s_3}{\leftarrow} \ldots \overset{s^*}{\leftarrow} & M \\
G_1 \downarrow & & G_2 \downarrow & & G_3 \downarrow & & f \downarrow \\
V_1 & \overset{s_1}{\leftarrow} & V_2 & \overset{s_2}{\leftarrow} & V_3 & \overset{s_3}{\leftarrow} \ldots \overset{s^*}{\leftarrow} & M
\end{array}
\tag{9.3}
$$

Each s_t maps an admissible path on an admissible path. Let $\xi^t = \{z^t(k), \ k \in \mathbb{Z}\}$ be an admissible path on the symbolic image G_t. We denote by P_t the space of admissible paths on the symbolic image G_t. The mapping $s_t : G_{t+1} \to G_t$ generates the mapping in the spaces of paths, $s_t(P_{t+1}) \subset P_t$, however, $s_t(P_{t+1}) \neq P_t$, in general. If we fix a path ξ^t on each symbolic image G_t such that $\xi^t = s_t(\xi^{t+1})$ then we obtain the sequence of paths $\{\xi^t \in P_t\}$. Each orbit $T(x_0) = \{x_k = f^k(x_0), \ k \in \mathbb{Z}\}$ generates an admissible path $\xi^t = \{z^t(k), \ x_k \in M(z^t(k))\}$ on G_t and for paths of this kind we have $\xi^t = s_t(\xi^{t+1})$. Such a path is the coding of the orbit $T(x_0)$, which corresponds to the covering C_t. Let Cod_t be the set of the codings of all true orbits of f corresponding to the covering C_t. It is easy to understand that $Cod_t \subset P_t$.

Let us construct the set P_1^k of paths on G_1 which are images of paths admissible on G_k under the mappings s_*, namely,

$$
P_1^2 = s_1(P_2), \ P_1^3 = s_1 s_2(P_3) \ldots, P_1^k = s_1 s_2 \ldots s_{k-1}(P_k) \ldots.
$$

In a similar way we can construct the set P_l^k of paths on G_l which are images of paths admissible on G_k under the mappings s_*, $k > l$. Denote by diam(C) the maximal diameter of cells of the covering C.

Theorem 98. *Let $C_1, C_2, \ldots, C_l, \ldots, C_k, \ldots$ be a sequence of closed coverings of the space M. Assume that each covering of the sequence is a subdivision of the previous one and $diam(C_k) \to 0$ as $k \to +\infty$. Then*

1. $P_l^k \supset P_l^{k+1}$ *for* $k > l$;
2. $Cod_l \subset P_l^k$ *for* $k > l$;
3. $Cod_l = \bigcap_{k>l} P_l^k$.

This theorem follows from Theorem 18 on the strong shadowing property.

9.3 Structural Graph of the Symbolic Image

Let C be a closed covering of the compact M, G be the symbolic image of the covering, $\{H_k\}$ be classes of equivalent recurrent vertices, and G^* be the structural graph of G. We refer to the set $R(x) = \{\bigcup M(i_m) : f^m(x) \in M(i_m)\}$ as the support of the orbit $\{f^m(x)\}$ and to the set $R_k = \{\bigcup M(i), i \in H_k\}$ as the support of the class H_k.

Proposition 99. 1. *If Q_k is a component of the chain recurrent set and $Q_k \cap R_k \neq \emptyset$ then $R_k \supset Q_k$;*

2. *For each component Q_k of the chain recurrent set there exists R_k such that $R_k \supset Q_k$.*

A class H of equivalent recurrent vertices is called *true* if there exists a component Q^* of chain recurrent set that is in the support $R(H)$ of the class H. If the support of the class H does not contain any component Q^*, the class H is called *false*, being vertices on the structural graph G^* are called *false vertices*. The edges $i \to j$ of the structural graph of the symbolic image for which there is no point x with ω-limit set in $Q_j \subset R_j$ and with α-limit set in $Q_i \subset R_i$ are called *false edges*. We show that there is the finite algorithm for construction of a new graph such that its vertices correspond only to those classes of equivalence on the symbolic image whose supports necessarily contain components of the chain recurrent set and all edges are true (that is they correspond to connections between components of the chain recurrent set).

Let us define a new graph G^{**} for which the following requirements are satisfied:

1. Each vertex of the graph G^{**} is a vertex of the structural graph of symbolic image G^* and $k \in Ver(G^{**})$ means that the class H_k of equivalent recurrent vertices is true.
2. For $k, l \in Ver(G^{**})$ let R_k and R_l be supports of classes H_k and H_l, respectively. The graph G^{**} contains the edge $k \to l$ if and only if there exists a point $x \in M$ such that $\omega(x) \subset Q_l$, $\alpha(x) \subset Q_k$, where $Q_l \subset R_l$, $Q_k \subset R_k$ are components of the chain recurrent set and the support $R(x)$ of an orbit passing through x is disjoint from supports of other classes, i.e. $R(x) \cap R_i = \emptyset$ for $i \neq k, l$.

Definition 100. *The graph G^{**} is called true structural graph of the symbolic image G.*

Construction of the graph G^{**}.

Let C_1 be a covering of M by closed cells, C_2 be its subdivision, G_1 and G_2 be the corresponding symbolic images, and G_1^* and G_2^* be their structural graphs. Recall that there is the mapping $s : G_2 \to G_1$, $s(j) = i$ if $m(j) \subset M(i)$, $m(j) \in C_2$, $M(i) \in C_1$. Let $P_1^2 = s(P_2)$ be the set of paths on G_1 which are images of paths on G_2. We construct a new graph G_{12}^* as following. Let $k \in Ver(G_1^*)$. We include the vertex k in $Ver(G_{12}^*)$ only in the case when there is an admissible path $\xi = \{i_n\} \in P_1^2$, where $i_n \in H_k$, for any $n \in \mathbb{Z}$, i.e. the class H_k contains the path which is the image $s(\xi^*)$ of the path ξ^* on G_2. Let us show that in this case the class H_k contains the image $s(H^*)$ of some class H^* of equivalent recurrent vertices on G_2. In fact, by definition we have $\xi = s(\xi^*) \subset H_k$. As the path ξ^* is infinite in both directions and the number of vertices is finite then certainly it has periodic parts at the beginning and at the end. Thus, ξ^* starts in some class A and ends in some class B and the images $s(A)$ and $s(B)$ are necessary contained in H_k. Thus, one can take A or B as a class H^*. We call classes H_k for which $k \in Ver(G_{12}^*)$ marked classes.

Now let us determine the set of edges of the graph G_{12}^*. Let $k, l \in Ver(G_{12}^*)$. We say that the graph G_{12}^* contains the edge $k \to l$ if there exists an admissible path $\xi \in P_1^2$ from the class H_k to the class H_l which does not pass through other marked classes, i.e. there is a path ξ which does not pass through the classes H_i for $i \in Ver(G_{12}^*) \setminus \{k, l\}$.

It follows from the construction of the graph G_{12}^* that $Ver(G_{12}^*) \subset Ver(G_1^*)$. We show that the number of edges of the graph G_{12}^* does not exceed the number of edges of the graph G_1^*. Let $k \to l$ be an edge on the graph G_{12}^*. Then there is an admissible path $\xi \in P_1^2$ on G_1 from the class H_k to the class H_l which does not pass through other marked classes H_i, $i \in Ver(G_{12}^*) \setminus \{k, l\}$. This path either passes through unmarked classes or does not pass. On the structural graph G_1^* there is either a finite length path $k \to \cdots \to l$ or an edge $k \to l$ which corresponds to this path. Thus, one can associate with each edge of the graph G_{12}^* at least one edge on the graph G_1^*. It should be marked that $E(G_{12}^*) \not\subset E(G_1^*)$ in general, where $E(G)$ stands for the set of edges of the graph G. However, the number of edges of the graph G_{12}^* does not exceed the number of edges of the graph G_1^*, i.e. $|E(G_{12}^*)| \leq |E(G_1^*)|$, where $|E|$ stands for the number of elements of a set E.

Similarly, considering the next subdivision C_3 and the set of admissible paths P_1^3 on the graph G_1, we construct the graph G_{13}^*. By Theorem 98, $P_1^3 \subset P_1^2$. Hence, $Ver(G_{13}^*) \subset Ver(G_{12}^*)$ and $|E(G_{13}^*)| \leq |E(G_{12}^*)|$. Consider the sequence of subdivisions $\{C_k\}$ and the sequence of graphs G_{1k}^*, $k \in \mathbb{N}$.

Proposition 101. *Assume that a dynamical system has the finite number of components of the chain recurrent set. Let $\{G_{1k}^*\}$ be the sequence of graphs constructed above. Then there is a positive integer $n_1 \in \mathbb{N}$ such that for every*

$k \geq n_1$ the graph G^*_{1k} coincides with the true structural graph of the symbolic image G_1, i.e. $G^*_{1k} = G^{**}_1$, $k \geq n_1$.

The proposition follows from the fact that numbers of vertices and edges of G^*_{1k} monotonically decrease as k tends to infinity.

9.4 Construction of the Structural Graph

Repeating the same algorithm for the graph G^*_2, we construct the graph G^{**}_2 similarly to the graph G^{**}_1. Let us show that the number of vertices of the graph G^{**}_2 is not less than the number of vertices of the graph G^{**}_1. In fact, if $k \in Ver(G^{**}_2)$ then its support contains necessarily a component Q_k of the chain recurrent set. On the graph G^{**}_1 there is only one vertex l corresponding to this component. The support R_l corresponding to the vertex l may contain several components of the chain recurrent set. On the graph G^{**}_2 several vertices may correspond to these components. Thus, $|Ver(G^{**}_1)| \leq |Ver(G^{**}_2)|$. The same way we can show that the number of edges of the graph G^{**}_2 is not less than the number of edges of G^{**}_1.

Now let us consider the sequence of graphs

$$G^{**}_1, G^{**}_2, G^{**}_3 \ldots.$$

We have the inequalities

$$|Ver(G^{**}_1)| \leq |Ver(G^{**}_2)| \leq \cdots \leq |Ver(\Gamma)|, \tag{9.4}$$

$$|E(G^{**}_1)| \leq |E(G^{**}_2)| \leq \cdots \leq |E(\Gamma)|, \tag{9.5}$$

where Γ is the structural graph of a dynamical system.

Theorem 102. *If a dynamical system has the finite number of component of the chain recurrent set then there is an integer $m > 0$ such that $G^{**}_m = \Gamma$.*

Theorem follows from the fact that numbers of vertices and edges of G^{**}_m monotonically increase and ones are bounded above by $Ver(\Gamma)$.

Thus, under the assumption of finiteness of the number of components of the chain recurrent set we can construct the structural graph of a dynamical system for the finite number of steps.

Calculation scheme.

1. The sequences of coverings C_k and symbolic images G_k are constructed.
2. The mappings $s : G_k \to G_l$, $k > l$ are determined.
3. The graphs G^*_{lk} are constructed. For each l the sequence G^*_{lk}, $k > l$, is stabilized at G^{**}_l; the sequence G^{**}_l is stabilized at Γ.

4. The structural graph Γ is determined by G_{lk}^* for sufficiently large l and k.

Thus, for the mapping f there exists a positive integer k, a finite sequence of subdivisions $C_1, \ldots, C_k$ of the initial covering C, and the corresponding sequence of symbolic images $G_1, \ldots, G_k$ by which the structural graph Γ of a dynamical system is uniquely determined. However we should indicate that there is no an estimate for the number of subdivision k.

Example 103. Structural graph of Ikeda mapping.
Let us consider once again Ikeda mapping:

$$T(x,y) = (d + C_2(x \cos \tau - y \sin \tau), C_2(x \sin \tau + y \cos \tau)), \qquad (9.6)$$

where

$$\tau = C_1 - \frac{C_3}{1 + x^2 + y^2}.$$

Numerical modeling of dynamics of this mapping was carried out at the following values of parameters: $C_1 = 0.4$, $C_2 = 0.9$, $C_3 = 6.0$, $d = 0.8$. If $|C_2| < 1$ then the mapping T is dissipative, i.e. there is a domain D such that any orbit gets into this domain and remains in it. Hence, there exists the global attractor A_g. Adding to $\mathbb{R}^2$ the point at infinity ∞, we obtain compactified plane $\overline{\mathbb{R}^2}$ which is homeomorphic to S^2. The mapping T can be continued on $\overline{\mathbb{R}^2}$ so that the point at infinity ∞ is a source.

For given parameter values all orbits (except for ∞) of the system reach the square $M = [-10, 10] \times [-10, 10]$. Using the sequence of symbolic images we have localized the global attractor as an invariant set maximal in M (see Fig. 9.2). The chain recurrent set of the mapping under study has three components: the strange Ikeda attractor A, the hyperbolic saddle H,

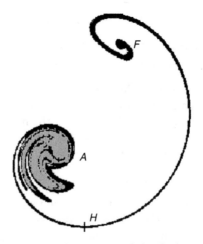

Fig. 9.2. Maximal attractor of the Ikeda mapping

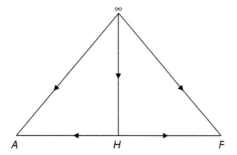

Fig. 9.3. Structural graph of Ikeda mapping

and the stable focus F. The structural graph of the mapping T is shown in Fig. 9.3. The point at infinity is repellor. The structural matrix has the form

$$
\begin{matrix}
\infty \\
H \\
A \\
F
\end{matrix}
\begin{pmatrix}
1\ 1\ 1\ 1 \\
0\ 1\ 1\ 1 \\
0\ 0\ 1\ 0 \\
0\ 0\ 0\ 1
\end{pmatrix} .
$$

In the case examined the initial covering involved one cell M. The subsequent coverings were constructed by dividing each cell of the previous covering into 4 equal parts. The separation of components of the chain recurrent set occurs at the eighth step ($l = 8$) of the process. However, there exist more than twenty false classes (components). At the ninth step the mapping $s : G_9 \rightarrow G_8$ destroys false classes and the stabilization of the structural graph G_k^{**}, $k \geq 8$ happens. Thus, it is enough nine steps to calculate the structural graph.

Let us notice that false components arise as a rule on boundaries of true components and are destroyed by the mapping s at the next subdivision since the new support of a true component together with "glued" false components appears to be inside the previous support. This observation allows to conclude that the stabilization of the structural graph comes immediately after the separation of components of the chain recurrent set.

10

Entropy

The concept "entropy" appeared in works of the originators of statistical mechanics: R. Clausius, J.C. Maxwell, L. Boltzmann in the end of the 19th century. In the middle of the 20th century C. Shennon introduced the entropy in the information theory which deals with the problems of transmission of messages. In [68] A.N. Kolmogorov used this concept in ergodic theory. This work has initiated the entropy theory of dynamical systems which has deep results and extensive applications. By now it has became clear that the entropy is very fine topological invariant of a dynamical system, which may be treated as a measure of chaos or randomness of the system dynamics. In sixties of the last century a wide application of numerical methods for solving ordinary differential equations began. It was a computer simulation that contributed to the fact that the first paper devoted to dynamical chaos has evolved. When investigating a Navie-Stocks system the meteorologist E. Lorenz proved that for some parameters values trajectories of the system in finite time go into a some set (called absorbing) and don't abandon it afterwards. The system has an attractor, which is inside the absorbing set. It turns out that being on the set the system demonstrates rather complex chaotic behavior. There are principal characteristics of quantitative features of specific chaotic systems and entropy is one of them. Entropy measures the complexity of mappings. In this chapter we introduce the notion of topological entropy and discuss the algorithms for its estimation. In more exact terms, the entropy of a dynamical system is evaluated using the entropy of its symbolic image. This method has been implemented to evaluate the entropy of Henon and logistic maps.

10.1 Definitions and Properties

Let M be a compact metric space. Consider a discrete dynamical system generated by a homeomorphism $f : M \to M$. Let $C = \{M(1), .., M(n)\}$ be a finite covering of M. Take into a consideration an N-length orbit of a point x $\{x_k = f^k(x), k = 0, \ldots N - 1\}$ and a coding of the orbit, i.e. a sequence

$\xi(x) = \{i_k, k = 0, \ldots N - 1\}$, where $x_k \in M(i_k)$. The symbols i_k of this sequence correspond to those cells of the covering which the orbit of the point x visits. In this case the sequence $\xi = \{i_k\}$ is said to be allowed coding. It is not difficult to understand that not any index sequence $\{i_k\}$ is allowed coding. To estimate the growth rate of the number of allowed codings depending on N the quantity

$$h = \lim_{N \to +\infty} \frac{\log_b K(N)}{N} \qquad (10.1)$$

is used. Here $K(N)$ is the number of various allowed codings of length N, the basis b is any number greater than 1. It is common practice to use $b = 2$ or $b = e$. It turns out that $h = 0$ for simple systems and $h > 0$ for chaotic ones. In the last case the estimation $K(N) \sim B b^{hN}$ holds, where B is a constant, i.e. the number of various allowed codings increases exponentially with respect to N, being h as the exponent. In this connection the value h is commonly thought as a measure of complexity or randomness of the system under investigation. The existence of the limit (10.1) follows from the next Polia lemma.

Proposition 104. [14] *Let $a_1, a_2, ..$ be a sequence of nonnegative numbers which satisfy the condition*

$$a_{m+n} \leq a_m + a_n \qquad \forall m, n \geq 1.$$

Then $\lim_{n \to \infty} \frac{a_n}{n}$ exists and equals to $\inf_{n \geq 1} \frac{a_n}{n}$.

It should be noted that for the number of allowed codings we have

$$K(m + n) \leq K(m)K(n). \qquad (10.2)$$

Then for $a_N = \log_b K(N)$ it follows from (10.2) that $a_{m+n} \leq a_m + a_n$. Hence, according to the above lemma, there exists the limit in the relation (10.1).

Topological entropy. Let $f : M \to M$ be a homeomorphism and $C = \{M(1), \ldots, M(n)\}$ — a finite open covering of M. Given an element $M(i_0)$, we find the element $M(i_1)$ of the covering C such that $M(i_0) \cap f^{-1}(M(i_1)) \neq \emptyset$. Then we find the element $M(i_2)$ such that $M(i_0) \cap f^{-1}(M(i_1)) \cap f^{-2}(M(i_2)) \neq \emptyset$ etc. To put this another way, consider the elements $M(i_k)$, when $0 \leq i_k \leq n$, such that $M(i_0 i_1 \ldots i_{N-1}) = \bigcap_{k=0}^{N-1} f^{-k}(M(i_k)) \neq \emptyset$. The collection of the constructed sets $M(i_0 i_1 \ldots i_{N-1})$ is denoted by C^N. It is easy to check that from the inclusion $x \in \bigcap_{k=0}^{N-1} f^{-k}(M(i_k))$ follows that $f^k(x) \in M(i_k)$, $k = 0, \ldots, N - 1$. Hence for a point $x \in M(i_0 i_1 \ldots i_{N-1})$ its $N - 1$ iterations belong to the set $\bigcup M(i_k), k = 0, 1, 2, \ldots, N - 1$. The aggregate of all sets from C^N forms a finite open covering of M. As the sets $M(i)$ may intersect to one another, elements of the covering C^N may do it as well. Let $\rho(C^N)$ denotes the cardinality of the minimal subcovering which may be chosen from C^N. In other words, $\rho(C^N)$ is the number of various codings of length N. Let

$$h(C) = \lim_{N \to +\infty} \frac{\log \rho(C^N)}{N},$$

where the limit exists because the sequence $\rho(C^N)$ satisfies conditions of Polia lemma.

Definition 105. *The value*

$$h(f) = \sup_C h(C),$$

where supremum is considered for all open coverings, is called topological entropy of the mapping $f : M \to M$.

In this manner, there is a correspondence between an element $M(i_0, i_1, \ldots, i_{N-1})$ of C^N and the word of length N $[i_0, i_1, \ldots, i_{N-1}]$ which is a coding of some orbit. It should be marked that different orbits may have the same coding. Choosing a minimal subcovering we minimize the number of different allowed codings. With a knowledge of some coding of length N, one can predict the system behavior for a time p using the relation $\frac{|C^{N+p}|}{|C^N|}$, where $|\cdot|$ denotes cardinality of a set. Suppose that $|C^N|$ has the exponential growth rate by N, i.e., $|C^N| \sim Ab^N$, $b > 1$, where A, b are positive numbers. Let q be a positive number. Then for $p \geq \log_b q$ and any N the following relation holds

$$\frac{|C^{N+p}|}{|C^N|} \sim b^p \geq q. \tag{10.3}$$

The formula (10.3) means that there are at least q variants of the future behavior of the system. In this regard the system resembles a random one. Hence the exponential growth of $|C^N|$ is indicative of a complexity of the dynamical system. In this case

$$h(C) = \lim_{N \to +\infty} \frac{\log \rho(C^N)}{N} \sim \lim_{N \to +\infty} \frac{\log |C^N|}{N} \sim \log b > 0.$$

If $|C^N|$ increases slower, for example according to the formula $|C^N| \sim AN^\alpha$, where $\alpha > 0$ then

$$\frac{|C^{N+p}|}{|C^N|} \sim \left(1 + \frac{p}{N}\right)^\alpha \to 1, \quad N \to +\infty. \tag{10.4}$$

In other words, the system history determines a trifle over one variant of its future behavior and in this sense the system is predictable. The relation (10.4) means that the value $|C^N|$ as well as $\rho(C^N)$ does not change practically when N increases, i.e. $h(C) = 0$.

Hence, topological entropy is a measure of randomness for a system: if it equals to zero then the number of trajectories of length N increases moderately when N grows and the future system behavior is predictable. Otherwise the system is considered as non-determinate. Thus, topological entropy is a measure of non-determination of the system.

Example 106. Consider the identity mapping $f = id : M \to M$. Let us prove that the topological entropy of this mapping equals to zero, $h(f) = 0$. Actually, let $C = \{M(1), \ldots, M(n)\}$ be an arbitrary open covering of M. Then for any N the sets $M(11 \ldots 1)$, $M(22 \ldots 2)$, $M(nn \ldots n)$ (the number of indexes in each of them is N) form the minimal collection of sets from C^N, being this collection covers the M. In other words, $\rho(C^N)$ does not depend on N, hence $h(C) = 0$. Being the covering C chosen arbitrary, $h(f) = 0$.

Definition 107. *1. A covering D is said to be refined in a covering C ($D \succ C$), if any element of D is contained in an element of C.*

2. A sequence of open coverings $\{C_n\}$ is said to be exhaustive, if for any covering B there exists n^*, such that $C_n \succ B$ for $n > n^*$.

It should be marked that given above definition of topological entropy is not constructive. Therefore the following proposition may be used to compute this characteristic.

Proposition 108. [4] *If a sequence $\{C_n\}$ of open coverings is exhaustive, then*

$$h(f) = \lim_{n \to \infty} h(C_n).$$

However, in what follows we will use a sequence of special closed coverings C_n of a phase space, because we will estimate the entropy using a symbolic image. As mentioned above, the set $C^N = \{M(i_0 i_1 \ldots i_{N-1})\}$ may be regarded as a set of (allowed) codings of orbits of length N, where $i_0, i_1, \ldots, i_{N-1}$ are codes of cells which the orbits pass through. Thus the entropy is the growth rate of the number of codings of orbits. Hence, it is sufficient to compute the entropy of the space of allowed codings to estimate the entropy of the dynamical system. So, in next section we consider methods of computation of the entropy of a space of sequences, using the terminology from [72].

10.2 Entropy of the Space of Sequences

The space of sequences. Consider a finite set of symbols S, which is called alphabet. Construct bi-infinite sequences using the elements of S:

$$\xi = \{i_n\} = \{\ldots i_{-2} i_{-1} i_0 i_1 i_2 \ldots\},$$

where $i_n \in S, n \in \mathbb{Z}$. Let $S^{\mathbb{Z}}$ be the space of such sequences of symbols from the alphabet S:

$$S^{\mathbb{Z}} = \{\xi = \{i_n\} : i_n \in S, \quad n \in \mathbb{Z}\}.$$

A finite sequence of symbols from S is called the word over S. For example, if $S = \{a, b\}$ then the collection of symbols $u = ababababb$ is the word of length 8.

Let $P \subset S^{\mathbb{Z}}$. A word u is said to be admissible in P, if $u \in P$. Denote the number of admissible words of length N by $K(N)$.

Definition 109. *The value*

$$h(P) = \lim_{N \to +\infty} \frac{\log K(N)}{N}$$

is called the entropy of the space P.

Let G be a graph with vertex set V and adjacency matrix Π. The space of admissible paths coded by symbols from V

$$P_G = \{\xi = \{v_i\} \in V^{\mathbb{Z}} : \pi_{i,i+1} = 1 \quad \forall i \in \mathbb{Z}\}.$$

is called vertex space. If G is a symbolic image, the space P_G can be viewed as the space of allowed words (sequences of symbols) over the alphabet V or the space of admissible paths on the graph G.

Theorem 110. [4] *If Π is the adjacency matrix for G and λ is the maximal eigenvalue of Π then*

$$h(P_G) = \log \lambda.$$

Theorem 110 allows to find the entropy of the space of admissible paths on a symbolic image. However, it is well to bear in mind that not any admissible path on the symbolic image is generated by an orbit of the corresponding dynamical system. In other words, a path on symbolic image may be a false coding. Hence, the number of admissible paths is greater then the number of paths generated by the orbits.

Let $C = \{M(i)\}$ be an arbitrary covering (closed, open) or a partition. A sequence $\{i_n, \ n \in \mathbb{Z}\}$ is said to be allowed coding (or coding), if there exists an orbit $\{x_n = f^n(x_0)\}$ such that $x_n \in N(i_n)$. Let $P(f,C)$ be the space of (allowed) codings induced by C.

Definition 111. *The entropy of the space $P(f,C)$ is called the entropy of the mapping f with respect to the covering C:*

$$h(f,C) = h(P(f,C)).$$

Let C be a closed covering and G be the corresponding symbolic image. As any allowed coding is an admissible path on the symbolic image, $P(f,C) \subset P_G$. Since the space of admissible paths contains the space of codings, we have the estimate

$$h(f,C) \le \log \lambda,$$

where λ is the maximal eigenvalue of the matrix Π.

Proposition 112. [85] *Let $f : M \to M$ be a continuous mapping of the compact M, $\{\widehat{C}_k\}$ be a sequence of partitions with $\mathrm{diam}(\widehat{C}_k) \to 0$ as $k \to \infty$ and $\widehat{C}_{k+1} \succ \widehat{C}_k$. Then*

$$h(f) \le \lim_{k \to \infty} h(f, \widehat{C}_k)$$

We will use the following type of a closed finite covering. The sets $C = \{M(1), \ldots, M(n)\}$ are said to form a closed finite covering for M if the following conditions are fulfilled:

1) $\bigcup_i M(i) = M$;
2) any element is a closing of its interior: $M(i) = \overline{IntM(i)} \neq \emptyset$, $i = 1, \ldots, n$.
3) elements of the covering intersect along their boundary: $M(i) \bigcap M(j) = \partial M(i) \bigcap \partial M(j)$, $i \neq j$, where $\partial M(i)$ denotes the boundary of the set $M(i)$.

The following theorem provides a way to estimate the topological entropy of a mapping f using the sequence of coverings $\{C_k\}$.

Theorem 113. *Let $\{C_k = \{M_k(i)\}\}_{k \geq 1}$ be a sequence of closed finite coverings of a compact M, such that $C_{k+1} \succ C_k$ for any k and $d_k = \text{diam}(C_k) \to 0$ when $k \to +\infty$. Then the following statements are true:*
1) the sequence $h(f, C_k)$ is non-decreasing, i.e. $h(f, C_k) \leq h(f, C_{k+1})$;
2) $h(f) \leq \lim_{k \to +\infty} h(f, C_k)$.

Proof. 1. For a closed covering C_n denote by $\Theta_N(f, C_n)$ the set of allowed codings of length N. Let $\{C_k\}$ be a sequence of closed finite coverings and $C_{k+1} \succ C_k$. Let us show that each coding from $\Theta_N(f, C_k)$ corresponds to coding from $\Theta_N(f, C_{k+1})$. Moreover, the different codings from $\Theta_N(f, C_k)$ correspond to different codings from $\Theta_N(f, C_{k+1})$. Let $[i_0, \ldots, i_{N-1}], [j_0, \ldots, j_{N-1}]$ are in $\Theta_N(f, C_k)$. It means that there exist $x, y \in M$ such that $f^l(x) \in M_k(i_l)$, $f^l(y) \in M_k(j_l)$, where $M_k(i_l), M_k(j_l) \in C_k$, $l = 0, .., N - 1$. Being $C_{k+1} \succ C_k$, there exist $M_{k+1}(i_l) \subset M_k(i_l), M_{k+1}(j_l) \subset M_k(j_l)$ such that $f^l(x) \in M_{k+1}(i_l)$, $f^l(y) \in M_{k+1}(j_l)$, $l = 0, .., N - 1$. In other words, we can match two different words from the set $\Theta_N(f, C_k)$ to two different words from the set $\Theta_N(f, C_{k+1})$. Hence $|\Theta_N(f, C_k)| \leq |\Theta_N(f, C_{k+1})|$ and as a consequence,

$$\frac{\log |\Theta_N(f, C_k)|}{N} \leq \frac{\log |\Theta_N(f, C_{k+1})|}{N}.$$

Passing to the limit we obtain $h(f, C_n) \leq h(f, C_{n+1})$, i.e. the sequence $h(f, C_n)$ is nondecreasing and there is $\lim_{n \to \infty} h(f, C_n)$ (finite or infinite).

2. Construct a partition $\widehat{C}_1$ from the covering C_1 by the following way. If $M(i), M(j) \in C_1$ and $\Gamma := \partial M(i) \cap \partial M(j) \neq \emptyset$, then delete the set Γ from the set $M(i)$ (or $M(j)$). Construct the partition $\widehat{C}_2$ so that it would be refined in the partition $\widehat{C}_1$. To do it we consider an auxiliary partition $C^* := C_2 \cap \widehat{C}_1$ and construct from it the partition $\widehat{C}_2$ by the same way as the partition $\widehat{C}_1$. Thus, we obtain a sequence of refined partitions $\widehat{C}_n$, being their diameters tends to zero (as $diam(C_n) \to 0$) when $n \to \infty$. Let us prove that $h(f, \widehat{C}_n) \leq h(f, C_n)$. Let $\widehat{C}_n = \{\widehat{M}(i_k)\}$ be the partition obtained from the covering C_n. Then the set of elements $\{\widehat{M}(i_k)\}$ is a covering of C_n. Let

$[i_0, \ldots, i_{N-1}] \in \widehat{\Theta}_N(f, \widehat{C}_n)$. Then there exists $x \in M$ such that $f^k(x) \in \widehat{M}(i_k)$, $k = 0, \ldots, N-1$. Hence, $f^k(x) \in \widehat{M}(i_k)$, $k = 0, \ldots, N-1$ and $[i_0, \ldots, i_{N-1}] \in \Theta_N(f, C_n)$. It means that $|\widehat{\Theta}_N(f, \widehat{C}_n)| \leq |\Theta_N(f, C_n)|$ and $h(f, \widehat{C}_n) \leq h(f, C_n)$. Finally, using Proposition 112, we have

$$h(f) \leq \lim_{n \to \infty} h(f, \widehat{C}_n) \leq \lim_{n \to \infty} h(f, C_n).$$

Theorem 113 is proved.

⊙

Hence to obtain an estimate of entropy it is sufficient to consider a closed covering of M and construct its successive subdivisions with diameters tending to zero. Then the limit of the sequence $\{h(f, C_k)\}$ is an upper bound for the topological entropy of the mapping f. To find the numbers $h(f, C_k)$ we use the a concept of symbolic image and methods of symbolic dynamics.

10.3 Entropy and Symbolic Image

Theorem 110 gives a possibility to obtain the entropy of the space of admissible paths of a symbolic image. There are admissible paths which do not correspond to any orbits. Such paths may be called false codings. Our purpose is to reveal false codings.

Earlier the construction of a special sequence of symbolic images was applied, which helps to shadow trajectories of a system at hand. We use the same method to estimate the entropy of the system. Recall the idea of the method briefly.

Let f be a homeomorphism, $C_1 = \{M(i)\}$ — a closed covering of the phase space M, G_1 be a symbolic image relative to the covering C_1. Form a new covering C_2 dividing every cell $M(i)$. Let G_2 be the symbolic image with respect to the covering C_2. Denote by $m(i, k)$ the cells of C_2 such that $\bigcup_k m(i, k) = M(i)$. Mark the vertices of the graph G_2 with (i, k) and construct the mapping $s : G_2 \to G_1$ which maps all vertices (i, k) into the vertex i. It follows from $f(m(i, k)) \cap m(j, l) \neq \emptyset$ that $f(M(i)) \cap M(j) \neq \emptyset$. Hence the mapping s transforms the edge $(i, k) \to (j, l)$ in the edge $i \to j$ and by doing so it maps the graph G_2 in the graph G_1. In particular, periodic paths are transformed into periodic ones. Thus, the mapping s may be conceived as a mapping which associates any path on the graph G_2 with the definite path on G_1. To be exact, if $\gamma = \{(i_n, j_n)\}$ is a path on G_2, then $\xi = s(\gamma) = \{(i_n)\}$ is a path on G_1. It should be noticed that in general the inverse mapping $s^{-1} : G_1 \to G_2$ is not defined.

Let $\Pi = (\pi_{ij})$ be the adjacency matrix for the graph G_1. Construct a space of admissible paths (a space of vertices) P_1. As was mentioned above, any word from P_1 is generated by a path on an appropriate symbolic image. Hence the map $s : G_2 \to G_1$ may be extended to the path map $s : P_2 \to P_1$, where P_2 is the space of admissible paths on the symbolic image G_2. Denote by P_1^2 the image $s(P_2)$.

Proposition 114. *1. Let P_1 be the space of the admissible paths on the symbolic image G_1, $P_1^2 = s(P^2)$. Then*

$$h(P_1^2) \le h(P_1).$$

2. Let P_2 be the space of the admissible paths on the symbolic image G_2, P_1^2 be as above. Then

$$h(P_1^2) \le h(P_2).$$

Proof. 1. Denote by $B_N(P)$ the set of allowed N-length words of P. According to definition of P_1 and P_1^2 we have $B_N(P_1^2) \subset B_N(P_1)$ for $N \in \mathbb{N}$. Hence, $|B_N(P_1^2)| \le |B_N(P_1)|$ and

$$\frac{\log |B_N(P_1^2)|}{N} \le \frac{\log |B_N(P_1)|}{N}.$$

Passage to the limit for $N \to +\infty$ completes the proof.

2. According to definition of P_2 and P_1^2 we have $|B_N(P_1^2)| \le |B_N(P_2)|$. Hence,

$$\frac{\log |B_N(P_1^2)|}{N} \le \frac{\log |B_N(P_2)|}{N}.$$

Passage to the limit for $N \to +\infty$ completes the proof.

$\odot$

Let $C = C_0$ be a closed covering of a domain M and G_0 be the symbolic image with respect to this covering. Consider the sequence of closed coverings $C_0, C_1, C_2, \ldots, C_k, \ldots$ such that the covering C_i is a subdivision of C_{i-1}. Let $G_0, G_1, G_2, \ldots, G_k, \ldots$ be the sequence of the associated symbolic images. As was shown in the previous chapter the following diagram is commutative

$$
\begin{array}{ccccccc}
V_0 & \overset{s_0}{\leftarrow} & V_1 & \overset{s_1}{\leftarrow} & V_2 & \overset{s_2}{\leftarrow} \ldots \overset{s}{\leftarrow} & M \\
G_0 \downarrow & & G_1 \downarrow & & G_2 \downarrow & & f \downarrow \\
V_0 & \overset{s_0}{\leftarrow} & V_1 & \overset{s_1}{\leftarrow} & V_2 & \overset{s_2}{\leftarrow} \ldots \overset{s}{\leftarrow} & M
\end{array}.
$$

Applying that result to the spaces of admissible paths, we have the following chain of mappings

$$P_0 \overset{s_0}{\leftarrow} P_1 \overset{s_1}{\leftarrow} P_2 \overset{s_2}{\leftarrow} \ldots \overset{s}{\leftarrow} \{T_f\},$$

where $\{T_f\}$ is the space of trajectories of the mapping f.

Fix a $l \ge 0$. Construct images of the admissible paths from P_k in P_l, where $k \ge l$:

$$\begin{aligned}
P_l^{l+1} &= s_l(P_{l+1}), \\
P_l^{l+2} &= s_l s_{l+1}(P_{l+2}), \\
&\cdots \\
P_l^{l+r} &= s_l s_{l+1} \cdots s_{l+r-1}(P_{l+r}), \\
&\cdots
\end{aligned}$$

Thus we obtain a double sequence of spaces of paths $P_l^k, l = 0, 1, \ldots, k > l$, which generates a double sequence of the entropies $h_l^k = h(P_l^k)$. For fixed l all the spaces $P_l^k, k > l$ are in the space of admissible paths of the symbolic image G_l. It would appear reasonable that passing to limit for k we could obtain the space of allowed codings.

Summing up, every map s_l transforms an admissible path on an admissible one. Let $\omega_l = \{i_k^l\}, \ k \in \mathbb{Z}\}$ be an admissible path on G_l. Denote by P_l the space of admissible paths on the symbolic image G_l. The map $s_l : G_{l+1} \to G_l$ induces a map in the spaces P_l. It should be marked that generally speaking $s_l(P_{l+1}) \neq P_l$. The next theorem follows from the strong shadowing property.

Theorem 115. *Let $C_1, C_2, \ldots, C_k, \ldots$ be a sequence of closed coverings of a set M such that every next covering is a refinement of the preceding one and $\mathrm{diam}(C_k) \to 0$ as $k \to \infty$. Then for any natural l the following statements are true:*

1) *$P_l^k \supset P_l^{k+1}$ for any $k > l$ and the entropies $h_l^k = h(P_l^k)$ decrease monotonically by k :*

$$h_l^k \geq h_l^{k+1}.$$

2) *the set of the (allowed) codings $Cod_l = P(f, C_l)$ is*

$$\lim_{k \to \infty} P_l^k = \bigcap_{k > l} P_l^k;$$

3) *$h_l = \lim_{k \to +\infty} h_l^k$ increases monotonically by l and*

$$h(Cod_l) \leq h_l = \lim_{k \to +\infty} h_l^k;$$

4) *being h the topological entropy of the mapping f,*

$$h \leq h^* = \lim_{l \to \infty} h_l.$$

10.4 The Entropy of a Label Space

In accordance with Theorem 110 the entropy of the space of admissible paths on a symbolic image can be computed as logarithm of the maximal eigenvalue of adjacency matrix. It turns out that the entropy of the space $P_l^k, \ k > l$

may be obtained using the maximal eigenvalue of some matrix. The spaces P_l^k are represented as spaces of allowed words over the vertices of the graph G_l, allowed words are generated by the paths on the graph G_k. To do so, any edge $i \to j$ of the graph G_k is marked by a vertex α of the graph G_l such that $M(\alpha) \supset m(i)$, where $M(\alpha)$ and $m(i)$ are cells of coverings C_l and C_k respectively. In this case admissible paths of G_k generate allowed words constructed from labels. The obtained space of allowed words may be presented as a graph, being its adjacency matrix makes possible to obtain the required entropy. Let us realize the described plan.

We will use a notion of multigraph. This is a graph in which some edges have common initial and terminal vertices.

Definition 116. *Let MG be a multigraph with vertex set* Ver *and edge set E. Associate a pair of vertices $(i,j) \in$ Ver *with a value π_{ij} which is equal to the number of directed edges from the vertex i to the vertex j. The matrix $\Pi = (\pi_{ij})$ is called the adjacency matrix for MG or the matrix of allowed transitions.*

Notice that for an ordinary graph the adjacency matrix is 0, 1- matrix. For an edge e denote by $i(e) \in$ Ver and $t(e) \in$ Ver its beginning and end respectively.

Let MG be a multigraph with edge set E. The space

$$P_{MG} = \{\xi = \{e_i\} \in E^{\mathbb{Z}} : t(e_i) = i(e_{i+1}) \quad \forall i \in \mathbb{Z}\}.$$

is called edge space. It is easy to understand that P_{MG} can be regarded as a space of allowed words over the alphabet E.

The pair (MG, W), where MG is a multigraph with edge set E and $W : E \to S$ is a function which associates any edge e with a label $W(e)$ from a finite alphabet S is called labeled graph. Different edges may carry identical labels. For example, if $S = E$ and $W(e) = e$, then the function W defines the one-to-one correspondence between labels end edges. In case the alphabet S consist of a single symbol, a function $W : E \to S$ marks all edges with the same label.

Definition 117. *Let $\Gamma = (MG, W)$ be a labeled graph and P_{MG} be the edge space for MG. The set of admissible sequences of labels*

$$P_\Gamma = \{\xi = \{i_n\} \in S^{\mathbb{Z}} : \text{ there exists } \gamma = \{e_n\} \in P_{MG}, \ i_n = W(e_n) \quad \forall n \in \mathbb{Z}\}$$

is called the space of admissible labeled paths. The graph Γ is a presentation of this space.

It should be noticed that the space P_Γ may have several different presentations.

Definition 118. *A graph $\Gamma = (MG, W)$ is called right-resolving, if for any vertex its outgoing edges carry different labels. If a right-resolving graph Γ presents the space of admissible labeled paths P_Γ, then Γ is called right-resolving presentation of P_Γ.*

In what follows a space of admissible labeled paths is termed "the space of labels" for short. To compute the entropy of a space of labels we have to choose a presentation (among all right-resolving ones) with a minimal number of vertices. Such a presentation is called minimal right-resolving presentation.

Computation of the entropy of a space of labels. Let P_Γ be a space of labels and Γ be its presentation. Produce the algorithm of construction of a minimal right-resolving presentation R for P_Γ, which is described in [72]. We have to construct a new labeled graph R whose vertices and edges are called hypervertices and hyperedges respectively. For a minimal right-resolving graph R the following statements hold:

- the space of labels of the graph R coincides with the alphabet of P_Γ;
- for any hypervertex the outgoing edges carry different labels;
- there not exist another presentation for P_Γ with lesser number of vertices.

The algorithm of construction of the minimal right-resolving presentation. Suppose that we have a labeled graph Γ.

1. Consider an empty graph R and add the vertex coinciding with an arbitrary vertex of the graph Γ.
2. Select a hypervertex H of R which does not have outgoing hyperedges. If there is not such a vertex, go to item 6.
3. Let H_1 be the set of all vertices of the graph Γ which are endpoints for the edges such that
   a) the initial points of the edges are vertices forming the hypervertex H;
   b) the edges carry the same label i.
   Add the hypervertex H_1 and the hyperedge $H \to H_1$ labeled by i to the graph R.
4. Repeat item 3 for all labels of the graph Γ.
5. Go to item 2.
6. Delete from R all hypervertices (including all hyperedges) which are not endpoints for any hyperedge.

The following theorem justifies the described algorithm and provides the method of computation of the entropy of the label space.

Theorem 119. [72]

1. *The constructed graph R is the minimal right-resolving presentation of the space P_Γ.*
2. *If P_Γ is the space of labels and $\Gamma = (MG, W)$ is the minimal right-resolving presentation with adjacency matrix Π for MG, then the entropy of the space P_Γ may be computed by the formula*

$$h(P_\Gamma) = \log \lambda_\Pi,$$

where λ is the maximal eigenvalue of the matrix Π.

10.5 Computation of Entropy

Let $C_1, C_2, \ldots, C_k, \ldots$ be a sequence of closed coverings of a set M such that every next covering is a refinement of the preceding one and $\operatorname{diam}(C_k) \to 0$, $k \to \infty$. We obtain a double sequence of spaces of paths $P_l^k, l = 0, 1, \ldots, k > l$, which generates a double sequence of entropies $h_l^k = h(P_l^k)$. Theorems 115 and 119 enable a constructive method to estimate the entropy of a dynamical system, which is in a sequential computation of the values h_l^k. The outline of the computation is the following.

Let C_1 be a closed covering of a phase space M, C_2 be a subdivision of C_1. Let G_1, G_2 be the symbolic images associated with C_1 and C_2 respectively. Denote the vertices and edges for the graphs G_1, G_2 by Ver_1, E_1 and Ver_2, E_2. Starting from the graph G_2 we construct a labeled graph $\Gamma_{12} = (G_2, W_{12})$, where $W_{12} : E_2 \to \mathrm{Ver}_1$ is a function which assigns the label $W_{12}(e)$ from the alphabet Ver_1 to an edge $e \in E_2$. In other words, the vertices of the graph G_1 label the edges of the graph G_2 in the following way: if $(i_1, j_1), (i_2, j_2) \in \mathrm{Ver}_2$ and $(i_1, j_1) \to (i_2, j_2)$ is the edge on the graph G_2 then this edge is labeled by $i_1 \in \mathrm{Ver}_1$. Denote by P^* the space of labels for which the graph Γ_{12} is a presentation.

Proposition 120. *Let P^* be the constructed above space of labels. Then $P^* = P_1^2$.*

Proof. Let $a = \{i_n\} \in P_1^2$. This means that there is $b = \{i_n, j_n\} \in P_2$ such that $a = s(b)$. As the edge $(i_s, j_s) \to (i_t, j_t)$ is labeled with i_s, the equality $a = s(b)$ holds if and only if $a \in P^*$.

$\odot$

In that way, for any $l > 0, k > l$ the space P_l^k coincides with the space of labels for which the labeled graph $\Gamma_{lk} = (G_k, W_{lk})$ is a presentation. Here $W_{lk} : E_k \to \mathrm{Ver}_l$ is a function assigning the label $W_{lk}(e)$ from the alphabet Ver_l to an edge $e \in E_k$. To compute the entropy of the space P_l^k we have to find its minimal right-resolving presentation. According to Theorem 119, there exists an algorithm which constructs the minimal right-resolving presentation R_{lk} for the graph Γ_{lk}. Let Π_{lk} be the adjacency matrix of R_{lk}. Then by Theorem 110 $h(P_l^k) = \log \lambda_{lk}$, where λ_{lk} is the maximal eigenvalue of the matrix Π_{lk}. Finally, from the inequality $h \leq \lim_{k \to +\infty} h(C_k)$ we obtain

$$h(f) \leq \lim_{l \to +\infty} \lim_{k \to +\infty} h(P_l^k) = \lim_{l \to +\infty} \lim_{k \to +\infty} \lambda_{lk}. \tag{10.5}$$

In summary, we obtain

The algorithm of the computation of topological entropy.

1. Choose a closed covering $C = \{C_l\}$ of a compact M and construct the symbolic image G according to C.
2. Construct a subdivision C^* of the covering C ($C^* \succ C$) and the symbolic image G^* associated with C^*.

3. Using the graphs G and G^* form a labeled graph Γ and its minimal right-resolving presentation R.
4. Compute the logarithm of the maximal eigenvalue of the adjacency matrix Π of the graph R and set the entropy value $h = \log \lambda_{max}$.
5. Subdivide the covering C^* and obtain a new covering C^{**}.
6. Construct a symbolic image G^{**} for C^{**}.
7. Go to item 3 using the graphs G^* and G^{**} respectively.

Repeating the process of successive subdivision we obtain a sequence of positive numbers $\{h_{lk} = \log \lambda_{lk}\}$, $k > l \geq 1$. According to the inequality (10.5) we obtain the following results:

- the value $\log \lambda_{lk}$, where the number $k - l$ is sufficiently large, constitutes an estimation of the entropy $h(f, C_l)$ of the mapping f for the covering C_l;
- the value $\log \lambda_{lk}$, where the numbers l and $k - l$ are reasonably large, presents an estimation of the entropy $h(f)$ of the mapping f.

10.5.1 The Entropy of Henon Map

Consider Henon map

$$f(x, y) = (1 - 1.4x^2 + 0.3y, x),$$

where $M = [-1.5, 1.5] \times [-1.5, 1.5]$. Consider the sequence C_k of refinements of the compact M such that

$$C_k = \left\{ \left[-1.5 + \frac{3l}{2^k}, -1.5 + \frac{3(l+1)}{2^k} \right] \times \left[-1.5 + \frac{3p}{2^k}, -1.5 + \frac{3(p+1)}{2^k} \right] \right\},$$

$k \in N$, $l, p = 0, .., 2^k - 1$. This sequence of coverings satisfies the conditions of Theorem 113. Hence, the upper bound for the entropy of Henon map is the following $h(f) \leq \lim_{l \to +\infty} h(C_l) = \lim_{l \to +\infty} \lim_{k \to +\infty} h(P_l^k)$.

The results of computations of values $h(P_l^k)$ are shown in the Table 10.1. For brevity the values $h(P_l^k)$ are denoted by h_{lk}. The dependence h_{lk} on k is shown in Fig. 10.1. The graphics converge to the number $h = 0.45 + \varepsilon$, where $\varepsilon > 0$, i.e. $h(f) \leq h$. The entropy of Henon map is 0.4651. It should be noticed that the same value was obtained in [43] by other way. Fig. 10.2 is a modification of Fig. 10.1: all graphics starts from a common initial point.

10.5.2 The Entropy of Logistic Map

Consider the map

$$f(x) = \lambda x(1 - x),$$

where $M = [0, 1]$. Let $C_k = \{[\frac{l}{2^k}, \frac{l+1}{2^k}]\}$, $k \in N$, $l = 0, .., 2^k - 1$ be the sequence of successive refinements of the compact M.

Table 10.1. The upper bounds for the entropy of Henon map

k	$|\text{Ver}_k|$	$|E_k|$	h_{1k}	h_{2k}	h_{3k}	h_{4k}	h_{5k}
1	4	10	-	-	-	-	-
2	12	54	0.85054	-	-	-	-
3	34	174	0.81455	1.06881	-	-	-
4	78	505	0.69273	0.89793	1.13639	-	-
5	181	1192	0.60009	0.76660	0.96874	1.15955	-
6	453	2885	0.57082	0.65094	0.81607	0.96145	1.15546
7	1108	7216	0.53494	0.57784	0.69467	0.80663	0.97145
8	2588	17836	0.50242	0.52710	0.61530	0.69029	0.81573
9	5915	42069	0.48540	0.49440	0.54327	0.59310	0.69123
10	13338	96921	0.47584	0.47751	0.50856	-	-
11	31534	218644	0.46761	-	-	-	-

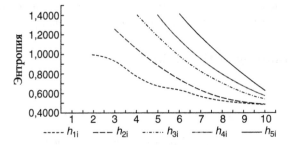

Fig. 10.1. The entropy of Henon map

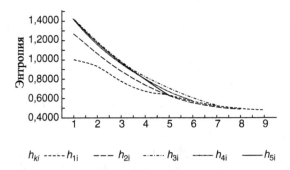

Fig. 10.2. The entropy of Henon map: the graphics are related to a common origin

The behavior of the mapping f is governed to a large extent by the parameter λ. It is known [129] that for $\lambda \leq 3.569$ there exists the finite number
of cycles with periods 2^m and the only attractive cycle which attracts almost
all points from the segment $[0, 1]$. For $\lambda = 3.569$ there exist the cycles with
periods 2^k for any k, being all of them are repelling. In those cases the entropy
of f is equal to zero. When $\lambda > 3.569$ the cycles with periods different from 2^k

Table 10.2. Estimations of the entropy for the map $x \to \lambda x(1-x)$

k	$h_{1k}, \lambda = 3.569$	$h_{1k}, \lambda = 4$	k	$h_{1k}, \lambda = 3.569$	$h_{1k}, \lambda = 4$
2	0.80544	0.85595	9	0.09250	0.69669
3	0.59282	0.80958	10	0.08919	-
4	0.44954	0.75826	11	0.04063	-
5	0.39664	0.73562	12	0.03713	-
6	0.16591	0.71605	13	0.03323	-
7	0.14127	0.70656	14	0.02527	-
8	0.10436	0.69967	15	0.02322	-

appear. For $\lambda = 4$ the whole segment $[0,1]$ is the invariant set. It was shown in [129] that the entropy of an one-dimensional map is not equal to zero if and only if there exist the periodic points which periods are different from numbers of the form 2^k. The results of computation of the sequence $\{h_{1k}\}$ for $\lambda = 3.569, \lambda = 4$ are shown in the Table 10.2. It is easy to see that h_{1k} tends to 0 for $\lambda = 3.569$ whereas h_{1k} tends to 0.69 for $\lambda = 4$.

11

Projective Space and Lyapunov Exponents

In the study of linear systems it is often convenient to introduce the projective space. In particular, for systems on the plane the use of the one-dimensional projective space allows to consider a given mapping as a map of the circle. In this chapter we present material necessary for the use of the projective space P. Namely, we discuss various ways of introducing coordinates and the action of a linear mapping and its base sets on P.

In addition, we introduce such an important characteristic of a dynamical system as Lyapunov exponents. It is known that one of peculiarities of chaotic regimes is the instability of each individual trajectory which belongs to a chaotic attractor. A Lyapunov exponent is a quantitative characteristic of the instability that allows to estimate the fractal dimension of an attractor and the entropy of a dynamical system. Characteristic exponents have been introduced by A.M. Lyapunov to describe properties of linear homogeneous systems of ordinary differential equations and determine stability of the equilibrium, i.e. stability of one trajectory. For a detailed presentation of the theory of Lyapunov exponents see, e.g. [20]. We define Lyapunov exponents for discrete linear systems, namely for trajectories of these systems on the projective space P.

11.1 Definitions and Examples

First of all, let us consider a simple example that clarifies an introduction of the one-dimensional projective space.

Example 121. Let a discrete dynamical system be given by the linear mapping $A : (x, y) \rightarrow (3x, 2y)$. The linear mapping $A : v \rightarrow Av$ generates the mapping $A_s : e \rightarrow Ae/|Ae|$ of the unit circle $S^1 = \{e = (p, q) : \ p^2 + q^2 = 1\}$, where $p = \cos\alpha$, $q = \sin\alpha$, α is the angle between the vector $v = (x, y)$ and the positive semi-axis OX (Fig.11.1,a). In the case considered we have

$$A_s(p, q) = (3p/\sqrt{9p^2 + 4q^2}, 2q/\sqrt{9p^2 + 4q^2}) = (P(p, q), Q(p, q)).$$

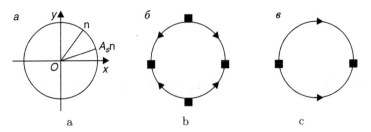

Fig. 11.1. Introduction of the projective space

The mapping A_s generates a dynamical system on the unit circle. Easily seen that A_s has four equilibriums: $(\pm 1, 0)$ and $(0, \pm 1)$. Consider the image of the point (p, q), $p > 0$, $q > 0$ and find the ratio $Q/P = 2/3\, q/p$. From this it follows that $A_s^k e \to (1, 0)$ as $k \to +\infty$ and $A_s^k e \to (0, 1)$ as $k \to -\infty$. Thus, the dynamical system on S^1 has a phase portrait depicted in Fig.11.1,b.

Notice that since $A_s(-e) = -A_s e$, the dynamical system on S^1 is symmetric with respect to the origin. This property allows to identify antipodal points of S^1. Identifying antipodal points of the unit circle S^1 we obtain one-dimensional projective space P^1. The phase portrait of the dynamical system on P^1 is depicted in Fig.11.1,c, the right fixed point corresponds to the points $(\pm 1, 0)$ on S^1, while the left one point corresponds to the points $(0, \pm 1)$. The space P^1 can be viewed as a set of straight lines in $\mathbb{R}^2$ passing through the origin.

Now let us consider the multidimensional case. The projective space P^{d-1} can be defined as a set of straight lines $\{L\}$ in $\mathbb{R}^d$ passing through the origin. Since every straight line intersects twice the unit sphere S^{d-1} at antipodal points then the space P^{d-1} can be obtained by identifying antipodal points of the sphere S^{d-1}.

A general linear mapping $A : v \to Av, v \in \mathbb{R}^d$ can be represented in the form:

$$Av = |Av| \cdot Av/|Av| = r|Ae| \cdot Ae/|Ae| = r|Ae| \cdot A_s(e),$$

where $r = |v|, e = \frac{v}{|v|} \in S^{d-1}$. Thus, the linear mapping $A : v \to Av$ is a product of the mappings

$$e \to A_s(e),$$

$$r \to r\,|Ae|,$$

where the first mappings acts on the unit sphere S^{d-1}, while the second one acts on the positive half-line $\mathbb{R}^+$ and is a multiplication by $|Ae|$. The symmetry of the mapping $A_s(e)$ under the change of the sign: $A_s(\pm e) = \pm A_s(e)$ allows to identify antipodal points and define the dynamical system on the projective space.

11.2 Coordinates in the Projective Space

Consider the plane $\mathbb{R}^2$ with coordinates (x, y). The one-dimensional projective space P^1 is a set of straight lines $\{L\}$ in $\mathbb{R}^2$ passing through the origin. There are three ways to introduce coordinates in this space.

1. Each straight line distinct from the y-axis is uniquely determined by the equation $y = kx$, where $k \in \mathbb{R}$. Thus, the coefficient k can be viewed as the coordinate of the straight line $L(k) = \{(x, y) : x \in \mathbb{R}, y = kx\}$ in the projective space P^1, $k = \tan\alpha$, where α is the angle between the straight line $L(k)$ and the x-axis. For the vertical straight line, $\tan\pi/2 = \pm\infty$. Thus, by adding the infinity point $\pm\infty$ to the real axis $\mathbb{R}$ we get a one-to-one correspondence between points of the projective space P^1 and points of the extended real axis $\mathbb{R} \cup \infty$, the coefficient $k \in \mathbb{R} \cup \infty$ is viewed as a coordinate of the point in P^1. From the topological point of view, the extended real axis coincides with the circle. This way for introducing coordinates in the projective space P^1 is the most simple. The main disadvantage of this method is that it assumes an introduction of "infinity" that produces certain inconvenience in computer implementation.

2. Each straight line $L(k)$ is uniquely determined by the angular coefficient $k = \tan\alpha$, this, in turn, is uniquely determined by the angle $\alpha \in (-\pi/2, +\pi/2)$, $\alpha = \arctan(k)$, (see Fig.11.2,a), for the vertical straight line, $\tan(\pm\pi/2) = \pm\infty$. Thus, by identifying the points $\pm\pi/2$ of the segment $[-\pi/2, +\pi/2]$, we get one-to-one correspondence between points of the circle and P^1. Notice that the method operates with the bounded segment $[-\pi/2, +\pi/2]$. However, this method leads to certain difficulties when one passes to the multidimensional case.

3. Consider two intervals $I_1 = \{(1, y), \ y \in [-1, 1]$ and $I_2 = \{(x, 1), \ x \in [-1, 1]$ with the common point $(1, 1)$ (see. Fig.11.2,b). Each straight line L, except for $y = -x$, intersects $I_1 \cup I_2$ only once. Thus, by identifying the points $(1, -1)$ and $(-1, 1)$ we get one-to-one correspondence between straight lines L and points of the set $I_1 \cup I_2 \setminus \{(1, -1) = (-1, 1)\}$. In order to find coordinates of the straight line $L : y = kx$ in the coordinate system described above, let us consider a vector $v = (x, y)$ on L. If $\max\{|x|, |y|\} = |x|$, then the vector $e_1 = (1, y/x)$ lies on L, $-1 \leq y/x \leq 1$ and $(1, h_1 = y/x) \in I_1$.

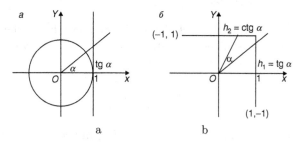

Fig. 11.2. Coordinates in the projective space

If $\max\{|x|, |y|\} = |y|$, then the vector $e_1 = (x/y, 1)$ lies on L, $-1 \leq x/y \leq 1$ and $(h_2 = x/y, 1) \in I_2$. Notice that $h_1 = y/x = \tan\alpha$ and $h_2 = x/y = \cot\alpha$. This method for introducing coordinates is the most tedious. However, it does not use the infinity and can be easily generalized to a multidimensional case.

Multidimensional case. To introduce local coordinates in the projective space P^{d-1} let us consider d copies K_1, ... K_d of the $(d-1)$-dimensional cube $K = \{(e_1, \ ... \ e_{d-1}) : |e_i| \leq 1\}$. Define the mapping of the sets $K_1, ..., K_d$ in $\mathbb{R}^d$ so that the imbedding of the ith copy K_i is of the form

$$(e_1, ..., e_{d-1}) \rightarrow (e_1, ..., 1, ..., e_{d-1}),$$

where 1 takes the ith position. Without loss of generality, we can identify K_i and its image called the disk K_i. The disks $K_1, ..., K_d$ thus constructed intersect each other by points with coordinates that have more than one unity. As an example, the point $(1,1/2,1,-3/4)$ lies in the disks K_1 and K_3 of the space $\mathbb{R}^4$. The disks $K_1, \ ... \ , K_d$ may be thought of as local carts of the projective space P^{d-1}. In fact, consider a straight line $L = \{tn, \ t \in \mathbb{R}, \ n \in S^{d-1}\}$, where S^{d-1} is the unit sphere in $\mathbb{R}^d$. Let $\max\{|n_1|, ..., |n_d|\} = |n_i|$. Then the point

$$(n_1/n_i, ..., n_i/n_i, ..., n_d/n_i) = (e_1, ..., 1, ..., e_{d-1})$$

is an intersection point of the straight line L and the disk K_i. Therefore, to each straight line there corresponds the point on the disks K_i. However, such a correspondence is not one-to-one. For example, to the points $(1,1/2,-1, 3/4)$ and $(-1,-1/2,1,-3/4)$, lying in distinct disks there corresponds the unique straight line as these points are antipodal. Notice that antipodal points lie on the boundaries of the disks K_i. To obtain a one-to-one correspondence we need to identify antipodal points of the disks K_i. In what follows we will construct the symbolic image of mappings of the projective space (projective mappings) and for this we need to have a covering of the projective space. It is reasonable to consider the collection of disks described above as an initial covering $C_0 = \{K_i\}$ of the projective space. More fine coverings will be subdivisions of this covering.

11.3 Linear Mappings

Let us consider a linear mapping $A : \mathbb{R}^2 \rightarrow \mathbb{R}^2$. Suppose that in the coordinates (x, y) the mapping A is of the form $A(x, y) = (ax + by, cx + dy) = (X, Y)$ or, what is the same, is given by the matrix

$$A = \begin{pmatrix} a & b \\ c & d \end{pmatrix}.$$

Find the image of the straight line $L = \{y = kx\}$ under the mapping A. More precise, we will find the angular coefficient $U(k)$ of the image AL. Since the point with coordinates (x, kx) goes into the point $(X, Y) = (ax + bkx, cx + dkx)$, then we have the equality $x = X/(a + kb)$. From this it follows that $Y = (c + dk)/(a + bk) X$. Thus, for the angular coefficient $U(k)$ of the image AL we have

$$U(k) = \frac{c + dk}{a + bk},\tag{11.1}$$

$U(-a/b) = \infty$, and $U(\infty) = d/b$. The formula (11.1) presents the mapping $U : L \rightarrow AL$ that acts in the projective space P^1 in terms of the angular coefficient $k \in R \cup \infty$. The same formula allows to calculate the angle between the image AL and the x-axis. Indeed, since $k = \tan(\alpha)$, then

$$U(\alpha) = \arctan\left(\frac{c + d\tan(\alpha)}{a + b\tan(\alpha)}\right)\tag{11.2}$$

is a presentation of the mapping $U : L \rightarrow AL$ in the angular coordinates, $U(\arctan(-a/b)) = \pi/2$ and $U(\pi/2) = \arctan(d/b)$.

The formula (11.1) allows also to present the mapping $U : L \rightarrow AL$ in the coordinates h_1, h_2 introduced in the previous section. In fact, since $h_1 = \tan(\alpha) = k$ if $|k| \leq 1$ and $h_2 = \cot(\alpha) = 1/k$ if $|k| \geq 1$, then the following formulas

$$U(h_1) = \begin{cases} H_1 = \dfrac{c + dh_1}{a + bh_1} & \text{if } |c + dh_1| \leq |a + bh_1|, \\[2mm] H_2 = \dfrac{a + bh_1}{c + dh_1} & \text{if } |a + bh_1| \leq |c + dh_1|, \end{cases}\tag{11.3}$$

$$U(h_2) = \begin{cases} H_1 = \dfrac{ch_2 + d}{ah_2 + b} & \text{if } |ch_2 + d| \leq |ah_2 + b|, \\[2mm] H_2 = \dfrac{ah_2 + b}{ch_2 + d} & \text{if } |ah_2 + b| \leq |ch_2 + d|, \end{cases}\tag{11.4}$$

present the mapping $U : L \rightarrow AL$ in the coordinates h_1, h_2.

At first sight, formula (11.1) is the most simple. However, it contains the infinity that produces certain difficulties in its computer implementation. The formula (11.2) does not also overcome this difficulty since in calculations the division by zero may occur. Finally, the formulas (11.3) and (11.4) are most complicated. However, they overcome all difficulties described above.

Now let us treat the common case where a linear mapping $A : R^d \rightarrow R^d$ acts in the d-dimensional space. The goal is to present the action of the mapping $PA : P^{d-1} \rightarrow P^{d-1}$ in the local carts $\{K_i\}$. Let a point $e = (e_1, ..., e_d)$ lie in K_i, i.e. $e_i = 1$, $|e_k| \leq 1$, $k \neq i$. Let $Ae = E = (E_1, ..., E_d)$ be its image and $\max\{|E_1|, ..., |E_d|\} = |E_j|$. Then in local coordinates we have

$$PAe = PE = (E_1/E_j, ..., 1, ..., E_d/E_j),\tag{11.5}$$

where "1" takes the jth position, i.e. PAe lies in the disk K_j. It is clear that if $|E_p| = |E_q|$ then either the image PAe enters the common boundary of the disks K_p and K_q, or the antipodal points of these disks correspond to it. Notice that the mapping (11.5) is a direct generalization of (11.3) and (11.4).

11.4 Base Sets on the Projective Space

By base sets of a linear mapping A on the projective space we mean components of the chain-recurrent set of a dynamical system on P^1 generated by A. Without loss of generality, we can assume that the matrix A has the canonical Jordan form, i.e. it can be one of the following matrices:

$$A_1 = \begin{pmatrix} a & 0 \\ 0 & d \end{pmatrix}, \ A_2 = \begin{pmatrix} a & 1 \\ 0 & a \end{pmatrix}, \ A_3 = \begin{pmatrix} a & -b \\ b & a \end{pmatrix}.$$

Action of the matrix A_1. The matrix A_1 has the real eigenvalues a and d and the x- and y-axes as eigenspaces, respectively. Obviously,

$$A_1^n = \begin{pmatrix} a^n & 0 \\ 0 & d^n \end{pmatrix}.$$

From (11.1) we have

$$U^n(k) = (d/a)^n k.$$

Since the projective space P^1 is formed by straight lines which in coordinates in P^1 are given by the angular coefficient k, for brevity, we will refer to straight lines as points $k_1, k_2, \ldots$. The points $k = 0$ and $k = \infty$ of the projective space corresponding to the axes in the Cartesian rectangular coordinate system are fixed for the mapping U.

If $k \neq 0$ and $d/a > 1$, then $U^n(k) \to \infty$ as $n \to +\infty$. Thus, there exist two fixed points of the mapping U: $k = 0$ and $k = \infty$ which form two base sets, other trajectories go from $k = 0$ to $k = \infty$ as $n \to +\infty$.

If $d/a = 1$, then $U(k) = k$, each point is fixed for the mapping U and P^1 forms a single base set.

If $0 < d/a < 1$, then $U^n(k) \to 0$ as $n \to +\infty$, i.e. there are two fixed points $k = 0$ and $k = \infty$ which form two base sets, other trajectories go from $k = \infty$ to $k = 0$ as $n \to +\infty$.

If $-1 < d/a < 0$, then $U^n(k) \to 0$ as $n \to +\infty$. This case is similar to the previous one, except that each iteration yields the change of the sign: $sign\ U(k) = -sign\ k$.

If $d/a = -1$, then $U(k) = -k$, each point, except for $k = 0$ and $k = \infty$ are two-periodic and P^1 forms a single base set.

If $d/a < -1$, then $U^n(k) \to \infty$ as $n \to +\infty$, i.e. there are two fixed points $k = 0$ and $k = \infty$, which are base sets, other trajectories go from $k = 0$ to $k = \infty$ and each iteration changes the sign: $sign\ U(k) = -sign\ k$.

Action of the matrix A_2. The matrix A_2 has the single real eigenvalue a and the x-axis as an eigenspace. We have

$$A_2 = \begin{pmatrix} a & 1 \\ 0 & a \end{pmatrix} = a \begin{pmatrix} 1 & 0 \\ 0 & 1 \end{pmatrix} + \begin{pmatrix} 0 & 1 \\ 0 & 0 \end{pmatrix} = aI + N,$$

where I stands for the identity matrix and $N^2 = 0$, i.e. the matrix N is nilpotent. Easily seen that $U(k) = ak/(a + k)$. Since $N^2 = 0$ then

$$A_2^n = (aI + N)^n = a^n I + na^{n-1}N = a^{n-1}\begin{pmatrix} a & n \\ 0 & a \end{pmatrix}.$$

In particular, $A_2^{-1} = a^{-2}\begin{pmatrix} a & -1 \\ 0 & a \end{pmatrix}$. Obviously, we have $U^n(k) = ak/(a+nk)$. If the sign of a coincides with the sign of nk (this can be achieved by choice of the sign of n) then $U^n(k) \to 0$ as $n \to \pm\infty$. Notice that $U(-a) = \infty$ and $U(\infty) = a$, i.e. "∞" is a transient point. Thus, there is the single fixed point $k = 0$, which forms the single base set, other trajectories begin and end at this point making one turn around the circle.

Action of the matrix A_3. The matrix A_3 has two complex conjugate eigenvalues $z = a \pm bi$. Using the representation $a + bi = \rho(\cos\phi + i\sin\phi)$, where $\rho = \sqrt{a^2 + b^2}$ and $tg\phi = b/a$, we obtain

$$A_3 = \begin{pmatrix} a & -b \\ b & a \end{pmatrix} = \rho\begin{pmatrix} \cos\phi & -\sin\phi \\ \sin\phi & \cos\phi \end{pmatrix},$$

where the last matrix determine a counter-clockwise rotation through the angle ϕ if $\phi > 0$. We have

$$A_3^n = \rho^n\begin{pmatrix} \cos n\phi & -\sin n\phi \\ \sin n\phi & \cos n\phi \end{pmatrix}.$$

If α is an angular coordinate on the projective space then on P^1, A_3 generates the mapping $U^n(\alpha) = \alpha + n\phi$. A periodic point with the period n satisfies the equation

$$U^n(\alpha) - \alpha = k\pi,$$

where k is an integer. From this it follows that for an n-periodic point we have the equality $n\phi = k\pi$, i.e. the angular displacement ϕ is rationally commensurable with π. If so, all points are periodic with the same period. Otherwise, if $\phi = \pi\theta$, where θ is irrational, then the trajectory $T(\alpha) = \{U^n(\alpha), n \in Z\}$ of each point α is dense on P^1. Thus, for the mapping given by the matrix A_3 the space P^1 is a single base set.

11.5 Lyapunov Exponents

The investigation of stability of invariant sets of dynamical systems means the study of the behavior of trajectories near these sets. The simplest invariant sets are fixed (equilibrium) points and periodic solutions. For this kind of invariant sets, the study of system dynamics is based on the linearization of an initial system in a neighborhood of invariant set. For this a method going back to A.M. Lyapunov and A. Poincaré is applied. The method is as follows.

Consider a smooth system of differential equations

$$\dot{x} = f(t, x), \tag{11.6}$$

where $t \in R, x \in M \subset R^n$. Let $x_0(t)$ be a solution of the system (11.6) with the initial data $t = t_0, x = x_0$ whose stability is investigated. The change of coordinates $y = x - x_0(t)$ transforms the system (11.6) to the form

$$\dot{y} = f(t, y + x_0(t)) - f(t, x_0(t)) = \frac{\partial f(t, x_0(t))}{\partial x} y + h(t, y) = A(t)y + h(t, y),$$
$$(11.7)$$

where $\dfrac{\partial f(t, x_0(t))}{\partial x}$ is the Jacobi matrix of the system (11.6) calculated at the solution $x_0(t)$ and the function $h(t, y)$ has the second infinitesimal order at $y = 0$. Thus, a problem of stability of the solution $x_0(t)$ of the system (11.6) reduces to the study of stability of the zero solution $y = 0$ of the system (11.7). The system

$$\dot{y} = A(t)y \tag{11.8}$$

is called a first approximation system for the system (11.7). Under some properties of the system (11.8), the dynamics of (11.8) governs the dynamics of the system (11.6) near the solution $x_0(t)$.

As a problem of stability of the solution $x_0(t)$ involves the study of perturbations of this solution with respect to initial data, one usually considers the variational system

$$\dot{v} = A(t)v, \tag{11.9}$$

where the matrix $A(t)$ is the same as above and $v(t) = \Phi(t)v(0)$, $\Phi(t)$ is a matrix composed by partial derivatives of the solution with respect to initial data, i.e.

$$\Phi(t) = \left(\frac{\partial x_i(t, t_0, x_0)}{\partial x_{0j}} \right), \quad i, j = 1, \ldots, n, \quad \Phi(0) = I.$$

To classify a growth of the norm of the solution $v(t)$ of (11.9), A.M. Lyapunov [75] introduced a collection of the functions $\exp \lambda t$ and the number λ called a Lyapunov exponent of the solution $x_0(t)$ in the direction $v(0)$. Lyapunov named λ *characteristic exponent*, now we say *Lyapunov exponent*. A Lyapunov exponent of the solution $v(t)$ is defined by the equality

$$\lambda(v) = \overline{\lim_{t \to \infty}} \frac{1}{t} \ln |v(t)| = \lim_{t \to \infty} \sup_{\tau \in [t, +\infty)} \frac{1}{\tau} \ln |v(\tau)|. \tag{11.10}$$

The Lyapunov exponent offers the following properties.

Property 1.
$$\lambda(cv) = \lambda(v),$$

where $c \neq 0$ is a constant.

Proof. In fact,

$$\lambda(cv) = \lim_{t \to \infty} \sup_{\tau \in [t, +\infty)} \frac{1}{\tau} \ln |cv(\tau)| = \lim_{t \to \infty} \sup_{\tau \in [t, +\infty)} \frac{1}{\tau} \ln |c||v(\tau)|$$

$$= \lim_{t \to \infty} \sup_{\tau \in [t, +\infty)} \frac{1}{\tau} (\ln |v(\tau)| + \ln |c|) = \lambda(v). \tag{11.11}$$

$$\odot$$

Notice that from the definition of Lyapunov exponents it follows that $\lambda(v)$ is a Lyapunov exponent if and only if for every $\varepsilon > 0$ there exists D such that $|v(t)| \leq D\exp((\lambda(v) + \varepsilon)t)$.

Property 2.

$$\lambda(v_1 + v_2) = \max\{\lambda(v_1), \lambda(v_2)\}.$$

Proof. Set $\lambda_1 = \lambda(v_1)$ and $\lambda_2 = \lambda(v_2)$. As mentioned above, we have the inequalities

$$|v_1| \leq D_1 \exp((\lambda_1 + \varepsilon)t), \quad |v_2| \leq D_2 \exp((\lambda_2 + \varepsilon)t).$$

Clearly, these inequalities hold for $D = \max\{D_1, D_2\}$. Let $\lambda_1 \leq \lambda_2$. Then $|v_1(t) + v_2(t)| \leq |v_1(t)| + |v_2(t)| \leq D\exp((\lambda_2 + \varepsilon)t)$. As ε is arbitrary, the last inequality yields $\lambda(v_1 + v_2) \leq \lambda_2$. To show the opposite inequality $\lambda(v_1 + v_2) \geq \lambda_2$ Assume, on the contrary, that $\lambda(v_1 + v_2) = \lambda < \lambda_2$. Then

$$\lambda_2 = \lambda(v_2) = \lambda(v_1 + v_2 - v_1) \leq \max\{\lambda(v_1 + v_2), \lambda(v_1)\} = \max\{\lambda, \lambda_1\} < \lambda_2.$$

The contradiction completes the proof.
$\odot$

It is known that the maximal number of linearly independent solutions of a linear system of differential equations equals the dimension of the phase space [54]. Since each solution of a linear system can be expressed as a linear combination of linearly independent solutions, by the properties of Lyapunov exponents given above, the number of distinct Lyapunov exponents does not exceed the dimension of the phase space. Since a solution is entirely determined by initial data, characteristic exponents depend on the choice of initial data.

Let $\Phi(t)$ be the principle fundamental matrix of the system (11.9), i.e. $\Phi(0) = I$, where I is the identity matrix. A solution $v(t)$, $v(0) = v_0$, of the system (11.8) can be written in the form [54]

$$v(t) = \Phi(t)v_0.$$

From the last identity it follows that

$$|v(t)| = |\Phi(t)e_0||v_0| = |e(t)||v_0|,$$

where $e_0 = \dfrac{v_0}{|v_0|}$ is a unit vector and $e(t)$ is a solution such that $e(0) = e_0$.
Thus, $\lambda(v) = \lambda(e(t)|v_0|) = \lambda(e)$.

This observation shows that a characteristic exponent is only determined by a one-dimensional subspace spanned by the vector e_0. Such a vector can be viewed as a point of the projective space. In addition, from (11.10) it follows that a characteristic exponent is determined by the ω-limit set of a trajectory. As each solution begins and ends at the chain-recurrent set, all characteristic exponents are also determined by trajectories of the chain-recurrent set. Now let us consider several particular cases.

Equilibriums.

If a vector field is time-independent and the solution $x_0(t)$ is an equilibrium, then we reach to the constant coefficient homogeneous linear system

$$\dot{x} = Ax. \tag{11.12}$$

It is not difficult to verify that in this case Lyapunov exponents coincide with real parts of the eigenvalues of the matrix A. In fact, each solution of the system (11.12) is a linear combination of functions $t^k \exp(\gamma t)$, $t^k \cos bt \exp(at)$, $t^k \sin bt \exp(at)$, where γ and $\mu = a+bi$ are eigenvalues of the matrix A. Direct calculations show that

$$\lambda(t^k \exp(\gamma t)) = \lim_{t\to\infty} \sup_{\tau\in[t,+\infty)} \frac{1}{\tau} \ln |\tau^k \exp(\gamma\tau)| = \gamma,$$

$$\lambda(t^k \cos bt \exp(at)) = \lambda(t^k \sin bt \exp(at)) = a.$$

The following example demonstrates the dependence of Lyapunov exponents on the choice of initial data.

Example 122. Consider the linear system

$$\dot{x} = Ax \equiv \begin{pmatrix} 2 & 1 \\ 3 & 4 \end{pmatrix} x.$$

The matrix A has the eigenvalues $\nu_1 = 1$ and $\nu_2 = 5$ and the eigenvectors $\begin{pmatrix} 1 \\ -1 \end{pmatrix}$ and $\begin{pmatrix} 1 \\ 3 \end{pmatrix}$, respectively. Every solution of the system is given by

$$x(t) = c_1 \exp t \begin{pmatrix} 1 \\ -1 \end{pmatrix} + c_2 \exp 5t \begin{pmatrix} 1 \\ 3 \end{pmatrix}.$$

Let $x(0) = \begin{pmatrix} 1 \\ 3 \end{pmatrix}$. Then

$$x(t) = \exp 5t \begin{pmatrix} 1 \\ 3 \end{pmatrix} \quad and \quad \lambda(x(t)) = \lim_{t\to\infty} \sup_{\tau\in[t,+\infty)} \frac{1}{\tau} \ln \sqrt{10} \exp 5\tau = 5.$$

For $x(0) = \begin{pmatrix} 1 \\ -1 \end{pmatrix}$ the Lyapunov exponent equals 1. If an initial vector $x(0)$ does not coincide with the matrix eigenvectors, then the Lyapunov exponent of the solution equals 5. As an example, for $x(0) = \begin{pmatrix} 1 \\ 1 \end{pmatrix}$ the Lyapunov exponent of the solution

$$x(t) = 1/2 \exp t \begin{pmatrix} 1 \\ -1 \end{pmatrix} + 1/2 \exp 5t \begin{pmatrix} 1 \\ 3 \end{pmatrix}$$

equals 5.

Autonomous systems.

Let us consider a smooth autonomous system of differential equations on a compact K:

$$\dot{x} = f(x). \tag{11.13}$$

Proposition 123. *For each solution of the system (11.13), distinct from an equilibrium, one of its Lyapunov exponents equals 0.*

Proof. Let us consider a solution $\phi(t, x_0)$ of the system (11.13) with the initial data $(0, x_0)$ $(\phi(0, x_0) = x_0)$. By the group property of solutions of autonomous systems we have the identity

$$\phi(t + s, x_0) = \phi(t, \phi(s, x_0)), \tag{11.14}$$

for every $s, t \in \mathbb{R}$, $x_0 \in K$. This property results from the fact that for an autonomous system, the solution shift corresponding to the change of t by $t + s$ is also its solution. The differentiation of (11.14) with respect to s yields

$$\frac{\partial \phi(t + s, x_0)}{\partial(t + s)} \cdot \frac{\partial(t + s)}{\partial s} = \frac{\partial \phi(t, \phi(s, x_0))}{\partial \phi(s, x_0)} \cdot \frac{\partial \phi(s, x_0)}{\partial s}.$$

Taking into account that $\dfrac{\partial(t + s)}{\partial s} = 1$ and setting $s = 0$, we obtain

$$f(\phi(t, x_0)) = \frac{\partial \phi(t, x_0)}{\partial x_0} \cdot f(x_0), \tag{11.15}$$

where $\dfrac{\partial \phi(t, x_0)}{\partial x_0}$ is a Jacobi matrix at (t, x_0). The obtained equality means that the vector field f is invariant for the mapping $\dfrac{\partial \phi(t, x_0)}{\partial x_0}$. Show that $\dfrac{\partial \phi(t, x_0)}{\partial x_0}$ is a fundamental matrix of the variational system corresponding to the system (11.13). In fact,

$$\frac{\partial}{\partial t} \left(\frac{\partial \phi(t, x_0)}{\partial x_0} \right) = \frac{\partial}{\partial x_0} \left(\frac{\partial \phi(t, x_0)}{\partial t} \right) =$$

$$\frac{\partial}{\partial x_0} (f(\phi(t, x_0))) = \frac{\partial}{\partial x} (\phi(x(t, x_0))) \cdot \left(\frac{\partial \phi(t, x_0)}{\partial x_0} \right) = A(t) \left(\frac{\partial \phi(t, x_0)}{\partial x_0} \right).$$

Then, by (11.15), $f(\phi(t, x_0))$ is a solution of the variational system for (11.13). Clearly, this solution is bounded and by supposition is distinct from an equilibrium. Thus, we have

$$\lambda(f(\phi(t, x_0))) = \lim_{t \to \infty} \sup_{\tau \in [t, +\infty)} \frac{1}{\tau} \ln |f(\phi(\tau, x_0))| = 0.$$

This finishes the proof. $\odot$

Let the system (11.13) be two-dimensional and have a periodic (hence, bounded) solution $\phi(t)$ distinct from an equilibrium. The solution $\phi(t)$ has two Lyapunov exponents and one of which equals 0.

Discrete dynamical systems.

Consider a smooth discrete system

$$x_{n+1} = f(x_n). \tag{11.16}$$

For a given x_0, let $\{x_k = f^k(x_0)\}$ be the orbit of x_0. For $n = 0, 1, \ldots$ consider the linear system

$$v_{n+1} = A(x_n)v_n, \tag{11.17}$$

where $A(x_n) = Df(x_n)$. It is not difficult to see that for this system the sequence $\{Df^k(x_0)e\}$, where e is a unit vector, is an orbit. Define a Lyapunov exponent of this orbit as

$$\lambda(x_0, e) = \lim_{k \to \infty} \sup_{n > k} \frac{1}{n} \ln |Df^n(x_0)e|.$$

By reasonings similar to the previous ones, one can prove that the Lyapunov exponent of an orbit is uniquely determined by the initial point x_0 and one-dimensional linear subspace spanned by the vector e, i.e. by a point of the projective space.

As it was shown at the beginning of the chapter, the projective space P^1 can be viewed as the unit sphere $S^2 = \{e : |e| = 1\}$, the antipodal points $+e$ and $-e$ being identified. Summarize our reasonings, we reach to the following conclusion. To find Lyapunov exponents of an orbit of the discrete dynamical system (11.16), along with this system one needs to treat the system (11.17), which, in turn, generates the mapping of the unit sphere:

$$e_{n+1} = U(e_n), \quad \text{where } U(e_n) = \frac{Df(x_n)e_n}{|Df(x_n)e_n|}.$$

The term $|Df^n(x_0)e|$ can be written in the form

$$|Df^n(x_0)e| = |Df(x_{n-1})e_{n-1}||Df^{n-1}(x_0)e|, \tag{11.18}$$

where $e_{n-1} = \dfrac{Df^{n-1}(x_0)e}{|Df^{n-1}(x_0)e|}$. By applying (11.18) recurrently, we obtain

$$|Df^n(x_0)e| = |Df(x_{n-1})e_{n-1}||Df(x_{n-2})e_{n-2}| \ldots |Df(x_1)e_1||Df(x_0)e_0|,$$

where $e_k = \dfrac{Df^k(x_0)e}{|Df^k(x_0)e|}$.

It is easy to verify that the sequence $\{e_k\}$ is an orbit of the point $e_0 = e$ for the dynamical system $e_{n+1} = U(e_n)$ on the unit sphere. A Lyapunov exponent can be written in the form

$$\lambda(x_0, e) = \lim_{k \to \infty} \sup_{n > k} \frac{1}{n} \sum_{k=0}^{n-1} \ln |Df(x_k)e_k|. \tag{11.19}$$

We now show that if the orbit $\{(x_k, e_k)\}$ is p-periodic, then

$$\lambda(x_0, e) = \frac{1}{p} \sum_{k=0}^{p-1} \ln |Df(x_k)e_k|. \tag{11.20}$$

In fact, representing an integer n in the form $n = mp + r$, where $0 \le r < p$, we obtain

$$\frac{1}{n} \sum_{k=0}^{n-1} \ln |Df(x_k)e_k| = \frac{mp}{n} \frac{1}{mp} (m \sum_{k=0}^{mp-1} \ln |Df(x_k)e_k| + R),$$

where

$$R = \sum_{k=mp}^{n-1} \ln |Df(x_k)e_k|$$

and, hence, $|R| < p \max_k |\ln |Df(x_k)e_k||$. Moreover, $m \to \infty$ as $n \to \infty$. From this it follows that

$$\lambda(x_0, e) = \lim_{n \to \infty} \frac{mp}{n} \frac{1}{mp} \left(m \sum_{k=0}^{p-1} \ln |Df(x_k)e_k| + R \right) = \frac{1}{p} \sum_{k=0}^{p-1} \ln |Df(x_k)e_k|,$$

and (11.20) is proved.

It is not difficult to find Lyapunov exponents in the case of a fixed point. Let x_0 be a fixed point of f. A corresponding linear system is of the form $v_{n+1} = Av_n$, where $A = Df(x_0)$. Then $Df^n(x_0) = (Df(x_0))^n$ and, by definition, we obtain $\lambda(x_0, e) = \ln |Df(x_0)e|$. Since the number of Lyapunov exponents does not exceed the dimension of the phase space, Lyapunov exponents of a fixed point equal logarithms of modulus of eigenvalues of the matrix $Df(x_0)$. If for every orbit we know its Lyapunov exponents, these Lyapunov exponents are said to be Lyapunov exponents of a system.

Lyapunov exponents of two-dimensional linear discrete systems. Let us find the Lyapunov exponents for orbits of linear mappings given by the matrices

$$A_1 = \begin{pmatrix} a & 0 \\ 0 & d \end{pmatrix}, \; A_2 = \begin{pmatrix} a & 1 \\ 0 & a \end{pmatrix}, \; \text{and } A_3 = \begin{pmatrix} a & -b \\ b & a \end{pmatrix}.$$

The mapping given by the matrix A_1 on P^1 if $a \ne d$ has two fixed points $(1,0)$ and $(0,1)$, where $(1,0)$ and $(0,1)$ may be viewed as the unit vectors on the coordinate axes. As a fixed point is periodic with the period 1, then $\lambda(1,0) = \ln |a|$ and $\lambda(0,1) = \ln |d|$. Every orbit, distinct from the fixed points, approaches the point $(1,0)$ or the point $(0,1)$. By (11.19) the Lyapunov exponent of such a trajectory equals $\ln |a|$ or $\ln |d|$, respectively.

The mapping given by the matrix A_2 on P^1 has the unique fixed point $(1,0)$ and $\lambda(1,0) = \ln |a|$. It is easy to verify that all other orbits have the same Lyapunov exponent.

Finally, consider the mapping given by the matrix A_3 on P^1. For every vector e, $|e| = 1$, we have $\ln |A_3 e| = \ln \rho$, where $\rho = \sqrt{a^2 + b^2}$, and, hence,

$$\frac{1}{m} \sum_{k=1}^{m} \ln |A_3 e(y_k)| = \ln \rho.$$

Thus, for every orbit $\lambda = \ln \rho$.

Generalization of Lypunov exponents.
To treat global properties of dynamical systems in the next chapter we introduce the following generalization of Lyapunov exponents.

1) On a linear bundle we will consider a system of the form

$$x_{n+1} = f(x_n), \quad v_{n+1} = A(x_n)v_n,$$

where the mapping $A(x)v$ is linear in v and continuous in x.

2) For a p-periodic ε-orbit $\gamma = \{(x_k, e_k)\}$ on the projective bundle the Lyapunov exponent will be defined by the formula

$$\lambda(\gamma) = \frac{1}{p} \sum_{k=0}^{p-1} \ln |A(x_k)e_k| \tag{11.21}$$

3) We will study limits of Lyapunov exponents λ that can be derived from (11.21) as $\varepsilon \to 0$.

Morse Spectrum

Let us consider a mapping $F : E \to E$ on a linear bundle E over a manifold M. Suppose that the mapping F covers a homeomorphism $f : M \to M$ and is linear on fibers. The differential Df is a prototype of the mapping F. Each periodic pseudo-orbit on the base M generates Lyapunov exponents. The Morse spectrum is defined as a limit set of Lyapunov exponents of periodic pseudo-orbits. It turns out that the Morse spectrum is a collection of intervals. Our aim is a method for computation of the Morse spectrum. Therefore we consider the dynamical system induced by F on the projective bundle and construct the symbolic image. Valuable information about the system may come from the analysis of the symbolic image. In particular, a neighborhood of the Morse spectrum can be found. This allows to get exponential estimates for the mapping F. A special sequence of symbolic images is constructed to obtain a sequence of embedded neighborhoods of the Morse spectrum that converges to the spectrum.

12.1 Linear Extension

Let $f : M \to M$ be a homeomorphism of the compact manifold M. Let (E, M, π) be a vector bundle over M, where E is a total space and π is a projector from E onto the base M. Assume that for every $x \in M$ a fiber $E(x) = \pi^{-1}(x)$ is a d-dimensional linear space isomorphic to $\mathbb{R}^d$.

Definition 124. *A homeomorphism F of the total space E is said to be a linear extension of f, if F takes fibers to fibers: $f \circ \pi = \pi \circ F$, i.e., the diagram*

$$
\begin{array}{ccc}
E & \xrightarrow{F} & E \\
\pi \downarrow & & \downarrow \pi \\
M & \xrightarrow{f} & M
\end{array}
$$

commutes, and the restriction $F|_{E(x)} : E(x) \to E(f(x))$ on the fiber $E(x)$ is a linear isomorphism.

The investigation of linear extensions has been motivated by the study of the tangent mapping of a diffeomorphism on the tangent bundle of the manifold [25, 26, 123–125, 127]. As a prototype of linear extension we will keep in mind a diffeomorphism $f : M \rightarrow M$ and its differential $Df = F$ on the tangent bundle $TM = E$. By localization of the Morse spectrum one can constructively test the hyperbolicity or the normal hyperbolicity of a dynamical system.

Now we briefly show as the vector bundle E is associated with a projective bundle P, in this process the linear extension $F : E \rightarrow E$ induces a mapping $PF : P \rightarrow P$. Recall that the $(d - 1)$-dimensional real projective space P^{d-1} is a collection of straight lines in $\mathbb{R}^d$ through the origin 0, and one can be formally defined by the identification of one-dimension subspaces in $\mathbb{R}^d$. For a nonzero vector v we denote by $[v] = y$ the class of equivalence of vectors on the straight line $\{kv : k \in \mathbb{R}\}$. So, $\mathbb{R}^d \setminus \{0\}$ is a bundle over the projective space P^{d-1} with the projection $q(v) = [v]$.

Let (P, M, p) be a bundle over M so that each fiber $p^{-1}(x)$ is a projective space $P^{d-1}(x)$ associated with the fiber $E(x)$ of E. The bundle (P, M, p) is called the projective bundle associated with the linear bundle (E, M, π). The bundles E and P are nontrivial, in general. However there is a procedure of its trivialization. More precise, each vector bundle can be included in a trivial one [9]. Hence, without loss of generality, we can use the following coordinates: (x, v) on E and (x, y) on P, where $x \in M$, $v \in E(x)$, $[v] = y \in P^{d-1}(x)$. In these coordinates the linear extension $F : E \rightarrow E$ of $f : M \rightarrow M$ takes the form

$$F(x, v) = (f(x), A(x)v).$$

Fixing a metric ρ on P^{d-1} we introduce the metric

$$r((x, y), (x', y')) = dist(x, x') + \rho(y, y')$$

on P. The mapping F induces the mapping $PF : P \rightarrow P$ on the projective bundle in the following way. Because for nonzero $v_1, v_2 \in \mathbb{R}^n$ with $[v_1] = [v_2]$ (i.e. for $v_1 = kv_2$) and every x we have $A(x)v_1 = kA(x)v_2$, we can define a mapping $PF(x) : P(x) \rightarrow P(f(x))$ with $PF(x, y) = PF(x, [e(y)]) = [F(x, e(y))] = (f(x), [A(x)e(y)])$, where $e(y)$ is the basis vector in the one-dimensional subspace y of $\mathbb{R}^d$. Obviously PF is a homeomorphism of P. So the diagram

$$
\begin{array}{ccc}
E \setminus \{0\} & \overset{F}{\rightarrow} & E \setminus \{0\} \\
\downarrow q & & \downarrow q \\
P & \overset{PF}{\rightarrow} & P \\
\downarrow p & & \downarrow p \\
M & \overset{f}{\rightarrow} & M
\end{array}
$$

commutes.

Recall that for $\varepsilon > 0$ a finite ε-chain ξ is defined as a finite sequence $x_0, ..., x_m \in M$ of length m with $\rho(f(x_i), x_{i+1}) < \varepsilon$, $i = 0, ...m - 1$. An ε-semi-orbit is defined in the same way.

12.2 Definition of the Morse Spectrum

Let us consider the linear extension $F : E \to E$ and the mapping $PF : P \to P$ of the projective bundle. Let $\xi = \{(x_0, y_0), ..., (x_m, y_m)\}$ be a finite ε-chain for PF. Define the exponential growth rate of ξ by

$$\lambda(\xi) = \frac{1}{m} \sum_{i=0}^{m-1} \ln |F(x_i, e(y_i))|,$$

where $|F(x, v)| = |A(x)v|$, $e(y_i)$ is a basis vector, $|e(y_i)| = 1$. Note that the function $a(x, y) = |A(x)e(y)|$ is well defined, i.e. it does not depend on the choice of $e(y)$, and is a rate of change of the vector length at the point (x, y). Recall that if $\xi = \{(x_0, y_0), (x_1, y_1), ...\}$ is an ε-semi-orbit then

$$\lambda(\xi) = \overline{\lim_{m \to \infty}} \frac{1}{m} \sum_{i=0}^{m-1} \ln a(x_i, y_i)$$

is defined as the characteristic or Lyapunov exponent of the ε-semi-orbit ξ. If $\xi = \{(x_0, y_0), ..., (x_p, y_p) = (x_0, y_0)\}$ is a periodic ε-orbit of a period p, then for the Lyapunov exponent of ξ we have

$$\lambda(\xi) = \overline{\lim_{m \to \infty}} \frac{1}{m} \sum_{i=0}^{m-1} \ln a(x_i, y_i) = \frac{1}{p} \sum_{i=0}^{p-1} \ln a(x_i, y_i).$$

If $\xi = \{(x_0, y_0), (x_1, y_1), ...\}$ is a true orbit, i.e., $(x_i, y_i) = PF^i(x, y)$ then

$$e(x_1, y_1) = \frac{A(x)e(x, y)}{|A(x)e(x, y)|}.$$

Hence, we have

$$|F(x_1, e(x_1, y_1))||F(x, e(x, y))| = |A(f(x))e(x_1, y_1)||A(x)e(x, y)| = |A(f(x))A(x)e(x, y)| = |F^2(x, e(x, y))|.$$

Similarly way we obtain the equality

$$\prod_{k=0}^{p-1} |F(x_k, e(x_k, y_k))| = |F^p(x, e(x, y))|. \tag{12.1}$$

Thus, for the true orbit

$$\lambda(\xi) = \overline{\lim_{m \to \infty}} \frac{1}{m} \sum_{i=0}^{m-1} \ln |A(x_i)e(y_i)| = \overline{\lim_{m \to \infty}} \frac{1}{m} \ln \prod_{k=0}^{m-1} |F(x_k, e(x_k, y_k))|$$

$$= \overline{\lim_{m \to \infty}} \frac{1}{m} \ln |F^m(x, e(x, y))|$$

coincides with the classical Lyapunov exponent.

Denote by CR the chain recurrent set of the associated projective mapping PF.

Definition 125. *The Morse spectrum* $\Sigma(F)$ *of* F *on the chain recurrent set* CR *of the associated projective mapping* PF *is defined as follows*

$$\Sigma(F) = \{\lambda \in R : \text{ there are } \varepsilon_k \to 0 \text{ and finite } \varepsilon_k - \text{chains } \xi_k \text{ of lenths } m_k$$
$$\text{in } CR \text{ with } m_k \to \infty \text{ and } \lambda(\xi_k) \to \lambda \text{ as } k \to \infty\}.$$

F. Colonius and W. Kliemann [25] showed that the Morse spectrum coincides with the periodic Morse spectrum

$$\Sigma_{per}(F) = \{\lambda \in R : \text{ there are } \varepsilon_k \to 0 \text{ and periodic } \varepsilon_k - \text{orbits } \xi_k$$
$$\text{with } \lambda(\xi_k) \to \lambda \text{ as } k \to \infty\}.$$

Thus, the Morse spectrum is a limit set of Lyapunov exponents of periodic ε-orbits on the projective bundle if $\varepsilon \to 0$.

12.3 Labeled Symbolic Image

Now we apply the symbolic image construction to the mapping PF. Let $G(f)$ be a symbolic image of the mapping f with respect to a covering $C(M) = \{m(1), \cdots, m(s)\}$. To construct a symbolic image of the induced mapping $PF : P \to P$ it is convenient to choose a covering $C(P) = \{M(z)\}$ of the projective bundle P which conforms with the covering $C(M)$ so that the projection of each cell is a cell: $p(M(z)) = m(j)$. The covering thus chosen generates a natural mapping h from $G(PF)$ onto $G(f)$ taking the vertex z on the vertex j: $h(z) = j$. Since $PF(M(z_1)) \cap M(z_2) \neq \emptyset$ and $P\pi(M(z_{1,2})) = m(j_{1,2})$ implies $f(m(j_1)) \cap m(j_2) \neq \emptyset$, the directed edge $z_1 \to z_2$ on $G(PF)$ is mapped by h on the directed edge $j_1 \to j_2$ of $G(f)$. Hence, the mapping h takes the directed graph $G(PF)$ on the directed graph $G(f)$, so that the diagram

$$
\begin{array}{ccc}
Ver & \overset{G(PF)}{\to} & Ver \\
\downarrow h & & \downarrow h \\
ver & \overset{G(f)}{\to} & ver
\end{array}
$$

commutes, where Ver and ver stand for the vertices of $G(PF)$ and $G(f)$, respectively.

We construct a labeling of the symbolic image by fixing a value $a[ji]$ for each edge $i \to j$. Let $G(PF)$ be the symbolic image of PF. The existence of an edge $i \to j$ on $G(PF)$ guarantees the existence of a point (x, y) in the cell $M(i)$ so that the image $PF(x, y)$ is in the cell $M(j)$. Obviously, such a point is not unique. By setting $a[ji] = a(x, y)$ we fix a label for the edge $i \to j$. Let a point $(x^*, y^*) \in M(i)$ be such that $PF(x^*, y^*) \in M(j)$. We have

$$|a(x^*, y^*) - a(x, y)| = ||A(x^*)e(y^*)| - |A(x)e(y)|| < \eta(d),$$

where $\eta(d)$ is a modulus of continuity of $|A(x)e(y)|$ and d is the maximal diameter of cells of the covering $C(P)$.

Definition 126. *The pair a symbolic image $G(PF)$ and $\{a[ji]\}$ is called a labeled symbolic image G_{lb}.*

For any p-periodic path $\omega = \{z_1, ..., z_p = z_0\}$ on $G(PF)$

$$a(\omega) = a[z_p z_{p-1}]...a[z_2 z_1]a[z_1 z_0].$$

We call $\sigma(\omega) = (a(\omega))^{\frac{1}{p}}$ a multiplicator and

$$\lambda(\omega) = \frac{1}{p} \sum_{k=1}^{p} \ln a[z_k z_{k-1}] = \ln \sigma(\omega)$$

a characteristic or Lyapunov exponent of the periodic path ω. It is easy to see that if a sequence ω is considered as a periodic path with two different periods p' and p'' then the corresponding Lyapunov exponents coincide.

Definition 127. *The spectrum of the labeled symbolic image G_{lb} is defined as*

$$\Sigma(G_{lb}) = \{\lambda \in R : there\, are\, periodic\, paths\, \omega_k\, such\, that\, \lambda(\omega_k) \to \lambda\, as\, k \to \infty\}.$$

Our first aim is it to suggest a constructive method to compute the spectrum of a labeled symbolic image. The second aim is it to compare this spectrum and the Morse spectrum of a dynamical system.

12.4 Computation of the Spectrum

Let H be a class of equivalent recurrent vertices. A periodic path $\omega = \{z_1, ..., z_p = z_0\}$ is called simple if the vertices $z_1, ..., z_p$ are different, i.e., $z_i \neq z_j$ for $i \neq j$; $i, j = 1, ..., p$. Let $\omega = \{z_1, ..., z_p\}$ be a periodic path. If ω is not simple, then there is a vertex z^* such that $z^* = z_l = z_{l+p_1}$. Consider two finite sequences $\omega^* = \{z_1, ..., z_{l-1}, z_{l+p_1}, ..., z_p\}$ and $\omega^{**} = \{z_{l+1}, ..., z_{l+p_1}\}$. Since there are the edges $z_{l-1} \to z_l = z_{l+p_1}$ and $z_{l+p_1} = z_l \to z_{l+1}$, the sequences ω_1 and ω_2 are periodic admissible paths of periods p_1 and $p_2 = p - p_1$, respectively. Obviously, p_1, $p_2 < p$. In this case we say that the path ω is a sum of the periodic paths ω^* and ω^{**} and write

$$\omega = \omega^* + \omega^{**}.$$

By repeating decomposition of periodic paths we come to decomposition of ω in a sum of periodic paths $\omega_1, ..., \omega_q$ of periods $p_1, ..., p_q$, $p_1 + ... + p_q = p$. Because the periods are positive integers, and the maximal period of components decreases, the decomposition process has to finish. Clearly, the final decomposition $\omega = \phi_1 + \phi_2 + ... + \phi_r$ consists of simple periodic paths. It should be noted, that the simple periodic paths $\phi_1, ..., \phi_r$ may coincide. When a periodic path ω^* is repeated k times in ω, we write

$$\omega = k\omega^* + \omega^{**}.$$

Since a symbolic image has a finite number of vertices, the number of simple period paths is finite. For the class H let $\phi_1, ..., \phi_q$ be all simple periodic paths of periods $p_1, ..., p_q$, respectively. Let

$$\lambda(\phi_j) = \frac{1}{p_j} \sum_{k=1}^{p_j} \ln a[z_k^j z_{k-1}^j]$$

be the characteristic exponent of the periodic path $\phi_j = \{z_1^j, ..., z_{p_j}^j\}$. Suppose that $\omega \subset H$ is a periodic path on $G(PF)$, and $\omega = k_1\phi_1 + ... + k_q\phi_q$ is a decomposition of ω with the period $p = k_1 p_1 + ... + k_q p_q$. Without loss of generality, we can consider that each simple periodic path ϕ_j appears in ω with the coefficient $k_j \geq 0$ (the case $k_j = 0$ means that ω does not actually pass through the simple periodic path ϕ_j). If so, we will say that the simple periodic path ϕ_j is contained in ω with the weight $\mu_j = \frac{k_j \, p_j}{p}$. Obviously, $\sum_{j=1}^{q} \mu_j = 1$.

Proposition 128. *The characteristic exponent of a periodic path $\omega = k_1\phi_1 + ... + k_q\phi_q$ is given by the formula*

$$\lambda(w) = \sum_{j=1}^{q} \mu_j \lambda(\phi_j),$$

where $\mu_j = \frac{k_j \, p_j}{p}$.

Proof. Let ω be a periodic path on the symbolic image $G(PF)$. Suppose that $\omega = k_1\phi_1 + ... + k_q\phi_q$ is a decomposition of the periodic path ω for the period $p = k_1 p_1 + ... + k_q p_q$, where $p_1, ..., p_q$ are periods of simple periodic paths $\phi_1, ..., \phi_q$. For the characteristic exponent of $\omega = \{z_1, ..., z_p\}$ we have

$$\lambda(w) = \frac{1}{p} \sum_{k=1}^{p} \ln a[z_k z_{k-1}] = \frac{1}{p} \sum_{j=1}^{q} k_j \sum_{k=1}^{p_j} \ln a[z_k^j z_{k-1}^j]$$

$$= \frac{1}{p} \sum_{j=1}^{q} k_j p_j \lambda(\phi_j) = \sum_{j=1}^{q} \mu_j \lambda(\phi_j).$$

Thus, the characteristic exponent of ω is the average of the characteristic exponents for simple periodic paths with the weights μ_j, $\sum_{k=1}^{q} \mu_j = 1$. $\odot$

Let

$$\lambda_{\min}(H) = \min\{\lambda(\phi_j), \; j = 1, ..., q\},$$
$$\lambda_{\max}(H) = \max\{\lambda(\phi_j), \; j = 1, ..., q\}$$

be the minimum and the maximum of characteristic exponents for simple periodic paths of the class H. From Proposition 128 it follows

Proposition 129. *For the characteristic exponent $\lambda(\omega)$ of any periodic path ω of the class H we have*

$$\lambda_{\min}(H) \leq \lambda(\omega) \leq \lambda_{\max}(H).$$

Proposition 130. *For every $\lambda \in [\lambda_{\min}(H), \lambda_{\max}(H)]$ there is a sequence of periodic paths $\{\omega_m\}$ in H such that $\lambda(\omega_m) \to \lambda$ as $m \to \infty$.*

Proof. Without loss of generality we can assume that $\lambda_{\min}(H) < \lambda < \lambda_{\max}(H)$. Let $\phi_{\min} = \{z_1^*, ..., z_l^*\}$ and $\phi_{\max} = \{z_1^{**}, ..., z_e^{**}\}$ be simple periodic paths with the characteristic exponents $\lambda_{\min}(H)$ and $\lambda_{\max}(H)$, respectively. Because z_l^* and z_1^{**} are equivalent recurrent vertices, there is a periodic path $\psi = \{z_1 = z_1^*, ..., z_j = z_1^{**}, ..., z_q\}$ through z_1^* and z_1^{**}. Clearly there is a periodic admissible path ω in H of the form

$$\omega = \{\overbrace{\underbrace{z_1^*, ..., z_l^*}, ..., \underbrace{z_1^*, ..., z_l^*}}^{k^*-times}, z_1 = z_1^*, ...$$

$$..., z_j = \overbrace{\underbrace{z_1^{**}, ..., z_e^{**}}, ..., \underbrace{z_1^{**}, ..., z_e^{**}}}^{k^{**}-times}, z_1^{**} = z_j, ..., z_q\}$$

with decomposition

$$\omega = k^* \phi_{\min} + k^{**} \phi_{\max} + \psi.$$

Let p^*, p^{**} and q be periods of $\phi_{\min}$, $\phi_{\max}$ and ψ, respectively. For the weights of $\phi_{\min}$, $\phi_{\max}$ and ψ we have $\mu(\phi_{\min}) = \frac{k^* p^*}{p}$, $\mu(\phi_{\max}) = \frac{k^{**} p^{**}}{p}$ and $\mu(\psi) = \frac{q}{p}$, where $p = k^* p^* + k^{**} p^{**} + q$ is a period of ω. If k^* and $k^{**} \to \infty$ then $\mu(\psi) = \frac{q}{p} \to 0$. Since $\lambda_{\min}(H) < \lambda < \lambda_{\max}(H)$ there is δ, $0 < \delta < 1$, such that $\lambda = \delta \lambda(\phi_{\min}) + (1 - \delta)\lambda(\phi_{\max})$. Let $\{V_m\}$ and $\{W_m\}$ be sequences of integers such that V_m, $W_m \to \infty$ and $\frac{V_m}{W_m} \to \delta$ as $m \to \infty$. Since $0 < \delta < 1$ then $0 < V_m < W_m$. Denote by $\{k_m^*\}$ and $\{k_m^{**}\}$ sequences of integers such that

$$\frac{k_m^* p^*}{k_m^* p^* + k_m^{**} p^{**}} = \frac{V_m}{W_m},$$

i.e.

$$k_m^{**} = p^{**} V_m, \quad k_m^* = p^* (W_m - V_m).$$

Let us consider the described above sequence of periodic paths $\{\omega_m\}$ with the decompositions

$$\omega_m = k_m^* \phi_{\min} + k_m^{**} \phi_{\max} + \psi.$$

We have

$$\mu_m(\phi_{\min}) = \frac{k_m^* p^*}{k_m^* p^* + k_m^{**} p^{**} + q} = \frac{V_m}{W_m} \frac{k_m^* p^* + k_m^{**} p^{**}}{k_m^* p^* + k_m^{**} p^{**} + q} \to \delta \text{ as } m \to \infty,$$

$$\mu_m(\phi_{\max}) = \frac{k_m^{**}p^{**}}{k_m^*p^* + k_m^{**}p^{**} + q} = \left(1 - \frac{V_m}{W_m}\right)\frac{k_m^*p^* + k_m^{**}p^{**}}{k_m^*p^* + k_m^{**}p^{**} + q} \to 1 - \delta \text{ as}$$

$$m \to \infty, \mu_m(\psi) = \frac{q}{k_m^*p^* + k_m^{**}p^{**} + q} \to 0 \text{ as } m \to \infty.$$

Thus, for the characteristic exponent of $\omega_m = k_m^*\phi_{\min} + k_m^{**}\phi_{\max} + \psi$ we have

$$\lambda(\omega_m) = \mu_m(\phi_{\min})\lambda(\phi_{\min}) + \mu_m(\phi_{\max})\lambda(\phi_{\max}) + \mu_m(\psi)\lambda(\psi).$$

Hence,

$$\lambda(\omega_m) \to \delta\lambda(\phi_{\min}) + (1-\delta)\lambda(\phi_{\max}) = \lambda \text{ as } m \to \infty.$$
$$\odot$$

The next theorem follows from Propositions 129 and 130.

Theorem 131. *The spectrum of the labeled symbolic image $\Sigma(G_{lb})$ consists of the intervals $[\lambda_{\min}(H_k), \lambda_{\max}(H_k)]$, where $\{H_k\}$ is the full family of classes of equivalent recurrent vertices of the symbolic image $G(PF)$.*

One has to emphasize that the intervals $[\lambda_{\min}(H_k), \lambda_{\max}(H_k)]$ can intersect each other (see [25, 126]).

12.5 Spectrum of the Symbolic Image

By construction, the label $\{a[ji]\}$ depends on a choice of a point $(x, y) \in M(i)$, $f(x, y) \in M(j)$. Hence, characteristic exponent depends on (x, y) as well. Let us examine the variation of the exponent under admissible variations of labels.

Let $\omega = \{z_1, ..., z_p = z_0\}$ be a periodic path on $G(PF)$. The characteristic exponent of the path ω is defined

$$\lambda(w) = \frac{1}{p}\sum_{k=1}^{p}\ln a[z_k z_{k-1}],$$

where each $a[z_k z_{k-1}] = a[ji]$ is determined by the edge $i \to j$ not by k. More precisely, if the path ω passes through the edge $i \to j$ twice, i.e. $z_{k-1} = i$, $z_k = j$ and $z_{l-1} = i$, $z_l = j$ then $a[z_k z_{k-1}] = a[z_l z_{l-1}] = a[ji]$. Let us consider a more generalized situation where each $\alpha[z_k z_{k-1}]$ depends on k and define the exponent

$$\sigma(w) = \frac{1}{p}\sum_{k=1}^{p}\ln \alpha[z_k z_{k-1}].$$

In other words, if a path ω passes twice through an edge $i \to j$, i.e., $z_{k-1} = i$, $z_k = j$ $z_{l-1} = i$, $z_l = j$, and $k \neq l$ then $\alpha[z_k z_{k-1}]$ and $\alpha[z_l z_{l-1}]$ can differ. Moreover, $\alpha[z_k z_{k-1}]$ is defined as

$$\alpha[z_k z_{k-1}] = a(x, y), \text{ where } (x, y) \in M(z_{k-1}),$$

i.e., $PF(x, y)$ is not required to be in $M(z_k)$.

Definition 132. *For an admissible periodic path* $\omega = \{z_1, ..., z_p = z_0\}$ *be on the symbolic image* $G(PF)$ *we define a nonstationary exponent* $\sigma(w)$ *as follows*

$$\sigma(w) = \frac{1}{p} \sum_{k=1}^{p} \ln \alpha[z_k z_{k-1}],$$

where

$$\alpha[z_k z_{k-1}] = a(x, y), \ (x, y) \in M(z_{k-1}), \ k = 1, ...p.$$

One should note that the nonstationary exponent does not depend on labels of the symbolic image. If the label is fixed then the (stationary) exponent $\lambda(\omega)$ is uniquely determined by ω. Hence, the spectrum $\Sigma(G_{lb})$ depends on labels, and we can vary this spectrum by varying labels. Now we define an other kind of the spectrum that only depends on the symbolic image.

Definition 133. *The set* $\Sigma(G) = \{\sigma \in R : \text{there is a sequence of periodic paths } \omega_k \text{ with nonstationary exponents } \sigma(\omega_k) \text{ such that } \sigma(\omega_k) \to \sigma \text{ as } k \to \infty\}$ *is called a spectrum of the symbolic image* $G(PF)$.

It is evident that the spectrum $\Sigma(G)$ does not depend on labeling. In order to find the spectrum of symbolic image we set

$$\alpha(i) = \min_{(x,y)\in M(i)} a(x, y), \quad \beta(i) = \max_{(x,y)\in M(i)} a(x, y).$$

Since each cell $M(i)$ is compact, there are points (x^*, y^*) and $(x^{**}, y^{**}) \in M(i)$ such that $\alpha(i) = a(x^*, y^*)$ and $\beta(i) = a(x^{**}, y^{**})$. For each periodic path $\omega = \{z_1, ..., z_p = z_0\}$ let

$$\Lambda_{min}(\omega) = \frac{1}{p} \sum_{k=1}^{p} \ln \alpha(z_k), \quad \Lambda_{max}(\omega) = \frac{1}{p} \sum_{k=1}^{p} \ln \beta(z_k)$$

be the maximal and the minimal nonstationary exponents, respectively. If we put $\alpha[z_k z_{k-1}] = \alpha(z_{k-1})$, then for the nonstationary exponent we have

$$\sigma(w) = \frac{1}{p} \sum_{k=1}^{p} \ln \alpha(z_{k-1}) = \Lambda_{min}(\omega).$$

Similarly, if we put $\alpha[z_k z_{k-1}] = \beta(z_{k-1})$ then we obtain $\frac{1}{p} \sum_{k=1}^{p} \ln \beta(z_{k-1}) = \Lambda_{max}(\omega)$. Therefore, we have the estimates

$$\Lambda_{min}(\omega) \leq \sigma(\omega) \leq \Lambda_{max}(\omega), \tag{12.2}$$

$$\Lambda_{min}(\omega) \leq \lambda(\omega) \leq \Lambda_{max}(\omega), \tag{12.3}$$

where $\sigma(\omega)$ is an arbitrary nonstationary exponent, and $\lambda(\omega)$ is an arbitrary characteristic exponent of the labeled symbolic image. Let $\phi_1, ..., \phi_q$ be a full

collection of simple periodic paths of a class H and $p_1, ..., p_q$ be their periods
Set

$$\Lambda_{min}(H) = \min\{\Lambda_{min}(\phi_j),\ j = 1, ..., q\}$$

and

$$\Lambda_{max}(H) = \max\{\Lambda_{max}(\phi_j),\ j = 1, ..., q\}.$$

Theorem 134. *The spectrum $\Sigma(G)$ of the symbolic image $G(PF)$ is a collection of the intervals $\{[\Lambda_{min}(H_k), \Lambda_{max}(H_k)]\}$, where $\{H_k\}$ is the full family of classes of equivalent recurrent vertices on the symbolic image.*

Proof. Let H be a class of equivalent recurrent vertices on the symbolic image. As indicated above, $\Lambda_{min}(\phi_j)$ and $\Lambda_{max}(\phi_j)$ can be realized as non-stationary exponents under a corresponding choice of $\alpha[z_k z_{k-1}]$. Since the number of simple periodic paths is finite, $\Lambda_{min}(H)$ and $\Lambda_{max}(H)$ can be realized in a similar manner. By repeating the proof of Proposition 128, one can show that each nonstationary exponent of a periodic path is an arithmetic mean of nonstationary exponents of simple periodic paths with corresponding weights. Hence, the spectrum $\Sigma(G(PF))$ of the symbolic image $G(PF)$ is in $\bigcup_k [\Lambda_{min}(H_k), \Lambda_{max}(H_k)]$, where $\{H_k\}$ is the full family of classes of equivalent recurrent vertices on the symbolic image $G(PF)$. By repeating the proof of Proposition 130, one can show that each $\lambda \in [\Lambda_{min}(H_k), \Lambda_{max}(H_k)]$ belongs to the spectrum of the symbolic image. Thus,

$$\Sigma(G(PF)) = \bigcup_k [\Lambda_{min}(H_k), \Lambda_{max}(H_k)].$$

$\odot$

Theorem 135. *The spectrum of the symbolic image $\Sigma(G)$ offers the properties:*

1) spectrum of any labeling $\Sigma(G_{lb})$ is in $\Sigma(G)$,
2) Morse spectrum $\Sigma(F)$ is in $\Sigma(G)$.

Proof. 1) From the definition of a nonstationary exponent it follows that each characteristic exponent $\lambda(\omega)$ of the periodic path $\omega = \{z_0, z_1, ... z_p = z_0\}$ is a nonstationary exponent $\sigma(\omega)$ with $\alpha[z_k z_{k-1}] = a[z_k z_{k-1}]$, $k = 1, ...p$. Hence, the characteristic exponent $\lambda(\omega)$ of any labeled symbolic image is in the spectrum of a symbolic image.
2) Let us show that for any periodic ε-orbit

$$\xi = \{(x_0, y_0), (x_1, y_1), ..., (x_p, y_p) = (x_0, y_0)\},\ \varepsilon < r,$$

where r is the lower bound of a symbolic image, there is an admissible periodic path $\omega = \{z_0, z_1, ... z_p = z_0\}$ such that the characteristic exponent $\lambda(\xi) = \frac{1}{p} \sum_{k=0}^{p-1} \ln a(x_k, y_k)$ of ξ is the nonstationary exponent $\sigma(\omega)$. Let the vertices z_k be such that $(x_k, y_k) \in M(z_k)$. Since $\varepsilon < r$, by Theorem 14 the path $\omega = \{z_0, z_1, ..., z_p = z_0\}$ is an admissible periodic path on the symbolic image. By setting $\alpha[z_k z_{k-1}] = |A(x_{k-1})e(y_{k-1})|$, we obtain

$$\sigma(\omega) = \frac{1}{p}\sum_{k=1}^{p} \ln \alpha[z_k z_{k-1}] = \frac{1}{p}\sum_{k=0}^{p-1} \ln |A(x_k)e(y_k)| = \lambda(\xi).$$

Hence, a characteristic exponent of any periodic ε-orbit, $\varepsilon < r$, is in $\Sigma(G)$. Since Morse spectrum is a limit set of periodic ε-orbits as $\varepsilon \to 0$, it is contain in the spectrum $\Sigma(G)$ of the symbolic image:

$$\Sigma(F) \subset \Sigma(G).$$

$\odot$

12.6 Estimates for the Morse Spectrum

In this section we find a collection of intervals containing the Morse spectrum, assuming that the spectrum $\Sigma(G_{lb})$ of a labeled symbolic image is known. Since M is compact, the mapping $A(x)$ has a modulus of continuity $\eta_A(\rho)$ on x. Set

$$\eta(\rho) = \eta_A(\rho) + \max_{x \in M} |A(x)|\rho,$$

$$\theta = \left(\min_{x \in M,\ |e|=1} |A(x)e|\right)^{-1} = \max_{x \in M} |A^{-1}(x)|.$$

Proposition 136. *Let ω be an admissible periodic path on the symbolic image $G(PF)$, $\lambda(\omega)$ be a characteristic exponent for the labeling, and $\sigma(\omega)$ be a nonstationary exponent of a periodic path ω. Then*

$$|\lambda(\omega) - \sigma(\omega)| \le \theta\eta(d),$$

where d is a maximal diameter of cells of the covering $C(P)$.

Proof. Let (x, y) and (x^*, y^*) be two points from a cell $M(i)$. Set $a = |A(x)e(y)|$ and $a^* = |A(x^*)e(y^*)|$. We have

$$|a^* - a| \le ||A(x^*)e(y^*)| - |A(x)e(y)|| \le |A(x^*) - A(x)| + |A(x)||e(y^*) - e(y)|$$
$$\le \eta_A(\rho(x, x^*)) + \max_x |A(x)|\rho_1(y^*, y) \le \eta(d),$$

where $\rho_1(*)$ is a distance on the projective manifold, and d is a maximal diameter of cells of the covering $C(P)$.

Let $\omega = \{z_1, ..., z_p = z_0\}$ be a periodic path on $G(PF)$. Denote by $\lambda = \frac{1}{p}\sum_{k=1}^{p} \ln a[z_k z_{k-1}]$ and $\sigma = \frac{1}{p}\sum_{k=1}^{p} \ln \alpha[z_k z_{k-1}]$ a characteristic exponent and a nonstationary exponent of the path ω, respectively. Since $a[z_k z_{k-1}] = |A(x_{k-1})e(y_{k-1})|$ and $\alpha[z_k z_{k-1}] = |A(x_{k-1}^*)e(y_{k-1}^*)|$, where the points (x_{k-1}, y_{k-1}) and (x_{k-1}^*, y_{k-1}^*) are in the cell $M(k - 1)$, we have the estimate

$$|a[z_k z_{k-1}] - \alpha[z_k z_{k-1}]| \le \eta(d).$$

Hence, the following inequality holds

$$|\lambda - \alpha| \leq \frac{1}{p} \sum_{k=1}^{p} |\ln a[z_k z_{k-1}] - \ln \alpha[z_k z_{k-1}]|$$

$$\leq \left(\max_{P} \frac{1}{|a(x,y)|} \right) \eta(d) = \frac{\eta(d)}{\min_P |a(x,y)|} = \theta \eta(d). \qquad (12.4)$$

According to Theorem 14, each ε-orbit with $\varepsilon < r$ generates an admissible path on the symbolic image $G(PF)$. The next proposition follows from the preceding and Proposition 136.

Proposition 137. *Let $\xi = \{(x_1, y_1), ..., (x_p, y_p) = (x_0, y_0)\}$ be a periodic ε-orbit, $\varepsilon < r$, and the vertices z_k be such that $(x_k, y_k) \in M(z_k)$. Then $\omega = \{z_1, ..., z_p = z_0\}$ is an admissible periodic path and the Lyapunov exponent $\lambda(\xi)$ admits the estimate*

$$|\lambda(\omega) - \lambda(\xi)| \leq \theta \eta(d),$$

where $\lambda(\omega)$ is the characteristic exponent of any labeling of the symbolic image.

For a fixed labeling let $\lambda_{\min}(H_k)$ and $\lambda_{\max}(H_k)$ be the maximum and minimum of characteristic exponents of simple paths from the class H_k. By Theorem 131 for the spectrum of the labeled symbolic image we have

$$\Sigma(G_{lb}) = \bigcup_{k} [\lambda_{\min}(H_k), \lambda_{\max}(H_k)],$$

where $\{H_k\}$ is full family of classes of equivalent recurrent vertices of the symbolic image $G(PF)$.

Theorem 138. *The Morse spectrum $\Sigma(F)$ of the linear extension $F : E \to E$ and the spectrum $\Sigma(G)$ are in*

$$\bigcup_{k} [\lambda_{\min}(H_k) - \theta\eta(d), \lambda_{\max}(H_k) + \theta\eta(d)],$$

where by $\{H_k\}$ is the full family of classes of equivalent recurrent vertices on the symbolic image $G(PF)$, d is a maximal diameter of cells of the covering $C(P)$.

Proof. According to Theorem 135, the Morse spectrum is contained in the spectrum of a symbolic image. Thus, it is sufficient to prove that the last is in the union of intervals described above. Let σ be in the spectrum of a symbolic image. This means that there is a sequence of periodic paths $\omega_m = \{z_1^m, ..., z_{p_m}^m = z_0^m\}$ whose nonstationary exponents

$$\sigma_m = \sigma(\omega_m) = \frac{1}{p_m} \sum_{i=1}^{p_m} \ln \alpha[z_i^m z_{i-1}^m],$$

tends to σ as $m \to \infty$. Let a periodic path ω_m be in a class H. By Theorem 131 the characteristic exponent

$$\lambda(\omega_m) = \frac{1}{p_m} \sum_{i=1}^{p_m} \ln a[z_i^m z_{i-1}^m]$$

of the path ω_m is in the interval $I = [\lambda_{\min}(H), \lambda_{\max}(H)]$ and

$$|\alpha[z_i^m z_{i-1}^m] - a[z_i^m z_{i-1}^m]| \le \eta(d).$$

Since all exponents $\lambda(\omega_m)$ are in the interval $[\lambda_{min}(H), \lambda_{max}(H)]$, by (12.4) each nonstationary exponent $\sigma(\omega_m)$ is in the closed interval $[\lambda_{\min}(H) - \theta\eta(d), \lambda_{\max}(H) + \theta\eta(d)]$. Hence,

$$\lambda = \lim_{m \to \infty} \lambda_m$$

is also in this interval.

$\odot$

Definition 139. *The union of intervals*

$$\widetilde{\Sigma}(G_{lb}) = \cup_k \widetilde{I}_k = \cup_k [\lambda_{\min}(H_k) - \theta\eta(d), \lambda_{\max}(H_k) + \theta\eta(d)],$$

where d is a maximal diameter of cells on the projective bundle and $\{H_k\}$ is the full family of classes of equivalent recurrent vertices, is called the extended spectrum of labeled symbolic image.

Recall that the Hausdorff distance $H(A, B)$ between sets A and B is defined by the formula

$$H(A, B) = \max\{D(A, B), \ D(B, A)\},$$

where

$$D(A, B) = \sup_{u \in B} \rho(u, A) = \sup_{u \in B} \inf_{v \in A} \rho(u, v).$$

Theorem 140. *Let $\{\Sigma(G_{lb}^m)\}$ be a sequence of spectrums of labeled symbolic images $\{G_{lb}^m\}$ with the maximal diameters d^m of cells. If $d^m \to 0$ as $m \to \infty$ then*

$$H(\Sigma(F), \Sigma(G_{lb}^m)) \to 0,$$

$$H(\Sigma(F), \widetilde{\Sigma}(G_{lb}^m)) \to 0,$$

$$H(\Sigma(F), \Sigma(G^m)) \to 0. \tag{12.5}$$

Proof. First we notice that

$$H(\widetilde{\Sigma}(G_{lb}^m), \Sigma(G_{lb}^m)) \leq \theta \eta(d^m),$$

$$H(\Sigma(G_{lb}^m), \Sigma(G^m)) \leq \theta \eta(d^m),$$

$$H(\widetilde{\Sigma^m}, \Sigma(G^m)) \leq \theta \eta(d^m),$$

and $\theta \eta(d^m) \to 0$ as $m \to \infty$. Hence, by the triangular inequality to prove the theorem it is sufficient to show only one of the relation 140 holds. For example let us show that $H(\Sigma(F), \Sigma(G^m)) \to 0$ as $m \to \infty$.

We prove by contradiction that $D(\Sigma(F), \Sigma(G^m)) \to 0$ as $m \to \infty$. On the contrary, assume that there exist $\chi > 0$, a sequence of symbolic images $\{G^m\}$ and $\sigma^m \in \Sigma(G^m)$ such that $\rho(\sigma^m, \Sigma(F)) \geq \chi$. There is a sequence $\{w_j^m\}$ of periodic paths on G^m such that $\sigma(w_j^m) \to \sigma^m$ as $j \to \infty$. By Theorem 14 each periodic path w_j^m generates a periodic ε^m- orbit ξ_j^m on the projective space, $\varepsilon^m \to 0$ as $m \to \infty$. In addition, we can choose ξ_j^m so that its characteristic exponent λ_j^m coincides with $\sigma(w_j^m)$. One can assume that $\rho(\lambda_j^m, \Sigma(F)) \geq \frac{1}{2}\chi > 0$. Because $\{\lambda_j^m\}$ is bounded, there is a convergent subsequence $\lambda_{j(k)}^{m(k)} \to \lambda$ as $k \to \infty$. By definition of the spectrum, the limit λ has to be in $\Sigma(F)$. This contradicts to the inequality $\rho(\lambda_{j(k)}^{m(k)}, \Sigma(F)) \geq \frac{1}{2}\chi$. Thus, $D(\Sigma(F), \Sigma(G^m)) \to 0$ as $m \to \infty$. By Theorem 135, $\Sigma(F) \subset \Sigma(G^m)$. Hence, $D(\Sigma(G^m), \Sigma(F)) = 0$. Thus, $H(\Sigma(F), \Sigma(G^m)) \to 0$ as $m \to \infty$.
$\odot$

Theorem 140 guarantees that an accurate estimate for the Morse spectrum can be obtained if the diameters d of the covering $C(P)$ is sufficiently small. However, we do not have a suitable estimate for the diameter d. Thus, an algorithm constructing a monotone sequence of sets converging to the Morse spectrum is of great practical utility. One of methods is presented below.

12.7 Localization of the Morse Spectrum

Recall that the algorithm for localization of the chain recurrent set includes the following steps.

1. Starting with an initial covering C, the symbolic image G of the map f is found.
2. The recurrent vertices $\{i_k\}$ of the graph G are determined and the closed neighborhood $P = \{\cup M(i_k) : i_k \text{ is recurrent}\}$ of the chain recurrent set is found.
3. Cells corresponding to the recurrent vertices $\{M(i_k) : i_k \text{ is recurrent}\}$ are subdivided and the new covering is obtained.

4. The symbolic image G is constructed for the new covering. Notice that cells corresponding to non recurrent vertices do not participate in the construction of the new covering and the new symbolic image.
5. One goes back to the second step.

Let us modify the algorithm for localization of the chain recurrent set to localize the Morse spectrum of a dynamical system.
The algorithm localizing the Morse spectrum includes of the following steps

1. Starting with an initial covering C, the symbolic image G of the map PDf is found, $\Sigma_0 = R$. Cells of the initial covering may have an arbitrary diameter d_0.
2. Classes of equivalent recurrent vertices $\{H_m\}$ are found.
3. Intervals $I_m = [\lambda_{\min}(H_m) - \theta\eta(d), \lambda_{\max}(H_m) + \theta\eta(d)]$ are determined, $\Sigma_k = \Sigma_{k-1} \cap (\cup_m I_m)$.
4. The cells $M(i)$ corresponding to recurrent vertices are subdivided while other cells are excluded. Thus, a new covering is obtained.
5. The symbolic image G is constructed for the new covering.
6. One goes back to the second step.

Repeating the subdivision process and constructing the intervals I_m, we obtain the sequences $\{d_k\}$ and $\{\Sigma_k\}$.

Theorem 141. *The sequence $\{\Sigma_k\}$ offers the following properties:*
1) each Σ_k contains the Morse spectrum of F,
2) $\Sigma_0 \supset \Sigma_1 \supset ... \supset \Sigma(F)$,
3) if $d_k \to 0$ as $k \to \infty$, then

$$\lim_{k \to \infty} \Sigma_k = \bigcap_k \Sigma_k = \Sigma(F). \tag{12.6}$$

Proof. 1) follows from Theorem 135 and 3) is a corollary of Theorem 140. The statement 2) follows from the equality $\Sigma_k = \Sigma_{k-1} \cap (\cup_m I_m)$.
$\odot$

12.8 Exponential Estimates

Let us apply the results obtained to estimate an action of the mapping F along an ε-orbit $\xi = \{(x_i, y_i)\}$. By Theorem 14, if $\varepsilon < r$ (r is a lower bound of a symbolic image) then there is an admissible path $\omega = \{z_i\}$, with $z_i : (x_i, y_i) \in M(z_i)$ corresponding to the ε-orbit ξ. By Theorem 134, the spectrum of the symbolic image $\Sigma(G)$ consists of the intervals $\{[\Lambda_{min}(H_k), \Lambda_{max}(H_k)]\}$, where $\{H_k\}$ is the full family of classes of equivalent recurrent vertices. The interval $[\Lambda_{min}(H), \Lambda_{max}(H)]$ is naturally called the spectrum of the class H.

Theorem 142. *If the spectrum of a class H is in the interval $[a, b]$ then there exist positive constants K_* and K^* such that for any finite ε-orbit $\xi = \{(x_0, y_0), (x_1, y_1), ..., (x_p, y_p)\}$, $\varepsilon < r$ whose admissible path $\omega = \{z_i\}$, $z_i : (x_i, y_i) \in M(z_i)$ is in H, the following estimate holds*

$$K_* \exp(pa) \leq \prod_{k=0}^{p-1} |A(x_k)e(y_k))| \leq K^* \exp(pb). \tag{12.7}$$

Proof. For a directed edge $i \rightarrow j$ set $\alpha[ji] = |A(x)e(y))|$, where $(x, y) \in M(i)$ and $e(y)$ is a base vector.

First we prove (12.7) for a simple periodic path. Let $\psi = \{z_1, ...z_p = z_0\}$ be a simple periodic path in the class H. Fixing $\alpha[z_k z_{k-1}]$, $k = 1, ..., p$, one can find a nonstationary exponent

$$\sigma(\psi) = \frac{1}{p} \sum_{k=1}^{p} \ln \alpha[z_k z_{k-1}].$$

Because $\sigma(\psi)$ is in the interval $[a, b]$,

$$a \leq \frac{1}{p} \sum_{k=1}^{p} \ln \alpha[z_k z_{k-1}] \leq b.$$

From this it follows that the estimate

$$\exp(pa) \leq \prod_{k=1}^{p} \alpha[z_k z_{k-1}] \leq \exp(pb).$$

Consider now a path $\omega = \{z_0, z_1, ..., z_p\}$ in the class H. As above, let us decompose ω in a sum of simple paths. Let the path ω pass through a vertex z^* twice, i.e., $z^* = z_l = z_{l+p_1}$, $p_1 > 0$. Consider two finite sequences $\omega_1 = \{z_0, z_1, ..., z_{l-1}, z_l = z_{l+p_1}, ..., z_p\}$ and $\omega_2 = \{z_l, z_{l+1}, ..., z_{l+p_1} = z_l\}$. By construction, the path ω_2 is an admissible periodic path of a period p_2. We say that the path ω is a sum of paths ω_1 and ω_2:

$$\omega = \omega_1 + \omega_2,$$

where ω_1 is in general a nonperiodic path. By repeating the decomposition process, we can represent the path ω as a sum of simple periodic paths $\{\psi, \psi_2, ..., \psi_q\} \subset H$ and a path ω_0 with different vertices of the class H:

$$\omega = k_1 \psi_1 + k_2 \psi_2 + ... + k_q \psi_q + \omega_0$$

Clearly, $p = k_1 p_1 + k_2 p_2 + ... k_q p_q + p_0$, $k_i \geq 0$, p_0 is a length of the path ω_0. It is easy to see that p_0 does not exceed than the maximal period of simple periodic paths of the class H, i.e., $p_0 \leq t = \max\{p_j, \ j = 1, ..., q\}$. Let us obtain an estimate of the product

$$\Pi(\omega) = \prod_{k=1}^{p} \alpha[z_k z_{k-1}], \tag{12.8}$$

where $\alpha[z_k z_{k-1}] = |A(x_{k-1})e(y_{k-1})|$, $(x_{k-1}, y_{k-1}) \in M(z_{k-1})$. With this aim we change the position of factors in (12.8) according to the decomposition

$$\omega = k_1 \psi_1 + k_2 \psi_2 + \ldots + k_q \psi_q + \omega_0$$

and get its representation in the form

$$\Pi(\omega) = \prod_{j=1}^{q} \prod_{i=1}^{k_j} \Pi_i(\psi_j) \cdot \Pi(\omega_0).$$

For each factor $\Pi_i(\psi_j)$ we have

$$\exp(p_j a) \leq \Pi_i(\psi_j) \leq \exp(p_j b). \tag{12.9}$$

As the product $\Pi(\omega_0)$ has no more than t factors then

$$(K_{min})^t \leq \Pi(\omega_0) \leq (K_{max})^t, \tag{12.10}$$

where $K_{min} \leq \alpha[z^* z^{**}] \leq K_{max}$, $z^* \to z^{**} \in H$, and $K_{min} \leq 1 \leq K_{max}$. From (12.9) and (12.10) it follows that

$$(K_{min})^t \exp\left(a \sum_{j=1}^{q} k_j p_j\right) \leq \Pi(\omega) \leq (K_{max})^t \exp\left(b \sum_{j=1}^{q} k_j p_j\right).$$

Thus,

$$K_* \exp(pa) \leq \Pi(\omega) \leq K^* \exp(pb), \tag{12.11}$$

where

$$K_* = \begin{cases} (K_{min})^t \exp(-at), & if\ a > 0, \\ (K_{min})^t, & if\ a \leq 0; \end{cases}$$

$$K^* = \begin{cases} (K_{max})^t \exp(-bt), & if\ b < 0, \\ (K_{max})^t, & if\ b \geq 0. \end{cases} \tag{12.12}$$

Hence, (12.7) follows from above.

$$\odot$$

Remark. The constants K_* and K^* in (12.7) are determined by (12.12), where $K_{min} \leq |A(x)e| \leq K_{max}$, $K_{min} \leq 1 \leq K_{max}$, t is the maximal period of simple periodic paths of the class H.

Corollary 143. *Let $\xi = \{(x_k, y_k) = PF^k(x, y), \ k = 0, 1, ..., p\}$ be a finite part of the orbit through a point $(x, y) = (x_0, y_0) \in P$ such that the path $\omega = \{z_k\}: \ (x_k, y_k) \in M(z_k), \ k = 0, 1, ..., p$ corresponding to ξ is in H. Then*

$$K_* \exp(pa)|v| \leq |F^p(x, v)| \leq K^* \exp(pb)|v|,$$

where $v \in L(x, y)$.

Proof. First, note that the estimate (12.7) holds for the orbit ξ. If $(x_1, y_1) = PF(x, y)$ then a basis vector is of the form

$$e(x_1, y_1) = \frac{A(x)e(x, y)}{|A(x)e(x, y)|}.$$

Hence, we have

$$|F(x_1, e(x_1, y_1))||F(x, e(x, y))| = |A(f(x))e(x_1, y_1)||A(x)e(x, y)| =$$

$$|A(f(x))A(x)e(x, y)| = |F^2(x, e(x, y))|.$$

The same way we obtain the equality

$$\prod_{k=0}^{p-1} |F(x_k, e(x_k, y_k))| = |F^p(x, e(x, y))|,$$

where $e = v/|v|$. Then (12.7) takes the form

$$K_* \exp(pa)|v| \le |F^p(x, v)| \le K^* \exp(pb)|v|.$$

$\odot$

12.9 Chain Recurrent Components

Recall that a subset $\Omega \subset CR$ is called a component of the chain recurrent set if each two points from Ω can be connected by a periodic ε-orbit for every $\varepsilon > 0$.

Let $\{U_0 = \emptyset, U_1, ..., U_l\}$ be a filtration on the projective bundle P. By Propositions 78 and 82 the following properties hold:
1) the maximal invariant set in U_k, $k = 0, 1, ..., l$

$$A_k = \{\cap f^n(U_k) : n \in Z^+\}$$

is an attractor and
$$\emptyset = A_0 \subset A_1 \subset \cdots \subset A_l = M,$$

2) the maximal invariant set in $U_k \setminus U_{k-1}$:

$$\Omega_k = \{\cap F^n(U_k \setminus U_{k-1}) : n \in Z\}, \ k = 1, ..., l,$$

is a Morse set and the chain recurrent set CR is in $\cup_{k=1}^l \Omega_k$.
The family $\{\Omega_1, ..., \Omega_l\}$ is called a Morse decomposition.

Theorem 144. [19,25,126] *Let F be a linear extension of a homeomorphism $f : M \to M$ on a vector bundle (E, M, π), and PF be a mapping on the projective bundle (P, M, p) induced by F. If Ω is a component of the chain recurrent set on the base M then*

1) *the chain recurrent set of the restriction* $PF|_{P\pi^{-1}(\Omega)}$ *has* l *components*
$\Omega_1, ..., \Omega_l$, $1 \le l \le \dim E(x)$, $x \in M$, *which form a Morse decomposition,*
2) *each set* Ω_i *determines a (continuous, constant dimensional) subbundle*
E_i *over* Ω

$$E_i = \{v \in \pi^{-1}(\Omega) : v \neq 0 \Rightarrow [v] = y \in \Omega_i\},$$

3) *the following decomposition into a Whitney sum*

$$E|_\Omega = E_1 \oplus ... \oplus E_l$$

holds,
4) *each chain recurrent component* Ω_* *on the projective space* P *is projected onto a chain recurrent component* Ω *of the base* M *which has the form described in 2).*

From 4) it follows that any component Ω_* of the chain recurrent set on the projective bundle P meets each leaf $P\pi^{-1}(x)$ at a projective space that depends continuously on $x \in \Omega = P\pi(\Omega_*)$. J.Selgrade [126] proved that this holds not only for a component of the chain recurrent set but for each Morse set Ω_* on the projective bundle. The decomposition described in 1) – 3) is called the finest Morse decomposition on the projective bundle.

A linear extension $F : E \to E$, $F(x,v) = (f(x), A(x)v)$ is said to be hyperbolic on $\Lambda \subset M$ if there exist invariant subbundles E^s and E^u of $TM|_\Lambda$ and constants $d > 0$ and $\alpha > 0$ such that

$$E|_\Lambda = E^s \oplus E^u,$$
$$|F^n(x)v| \le d|v| \exp(-\alpha n), \quad x \in \Lambda, \; v \in E^s(x), \; n > 0,$$
$$\left|F^{-n}(x)v\right| \le d|v| \exp(-\alpha n), \quad x \in \Lambda, \; v \in E^u(x), \; n > 0.$$

The invariance of subbundles $E^{s,u}$ means $A(x)E^{s,u}(x) = E^{s,u}(f(x))$.

Let $C(M) = \{m(j)\}$ be a covering of the manifold M and $C(P) = \{M(z)\}$ be a conforming covering of the projective space P, i.e., a natural projection of each cell is a cell: $p(M(z)) = m(j)$. We denote by $G(f)$ and $G(PF)$ the symbolic images of f and PF, respectively. As above, for the conforming coverings there is a "natural projection" $h(z) = j$ from $G(PF)$ on $G(f)$, where $P\pi(M(z)) = m(j)$. Moreover, the mapping h maps the directed graph $G(PF)$ on the directed graph $G(f)$.

Theorem 145. *Let* $G(f)$ *and* $G(PF)$ *be symbolic images of the mappings* f *and* PF *for the conforming coverings* $C(M)$ *and* $C(P)$, *respectively. If* H *is a class of equivalent recurrent vertices on the symbolic image* $G(f)$ *and* $\{H_1, ..., H_l\}$ *is a full collection of classes of equivalent recurrent vertices on* $G(PF)$ *which are projected on* H *(i.e.,* $h(H_m) = H$, $m = 1, ..., l$*),* Ω_m *is a maximal invariant set in* $V_m = \{\bigcup M(z), \; z \in H_m\}$ *then*

1) *the family* $\{\Omega_1, ... \Omega_l\}$ *is a Morse decomposition (certain of* Ω_m *may be empty),*

2) number l of classes H_m : $h(H_m) = H$, $\Omega_m \neq \emptyset$ does not exceed $\dim E(x)$,
3) each Ω_m defines a (continuous, constant dimension) subbundle E_m over a component Ω in $V = \{\bigcup m(j), \ j \in H\}$:

$$E_m = \{v \in \pi^{-1}(\Omega) : \ v \neq 0 \Rightarrow [v] = y \in \Omega_m\},$$

4) the following decomposition into a Whitney sum holds

$$E|_\Omega = E_1 \oplus \ldots \oplus E_l,$$

5) if the spectrum of the class H_m is in an interval $[a_m, b_m]$ then for any point $(x, v) \in E_m$ and every integer $p > 0$

$$K_* \exp(pa_m) \, |v| \leq |F^p(x, v)| \leq K^* \exp(pb_m) \, |v|,$$

where the constants K_* and K^* are given by (12.12),
6) if 0 is not in the spectrum of classes $\{H_m\}$ then the linear extension F is hyperbolic over Ω,
7) for each component Ω of the chain recurrent set on the base M there exists $d_0 > 0$ so that for every covering $C(P)$ with the maximal diameter of cells $d < d_0$ the full family $\{H_m\}$ of equivalent recurrent vertices on $G(PF)$ induces (see 1)) the finest Morse decomposition over Ω:

$$\{\Omega_1, \ldots, \Omega_l\}, \quad \Omega_m \neq \emptyset$$

Proof. The statement 1) follows from 2)-4). The statement 2) follows from Theorem 144. The statements 3) and 4) are proved in [19], p. 117. Corollary 143 implies of 5). The statement 6) follows from 5). The statement 7) follows from Theorem 144 and the statement 2) of Theorem 53.

⊙

12.10 Linear Programming

Theorem 145 reduces the estimation of the Morse spectrum to finding simple closed paths with minimal and maximal characteristic exponents for each class H_k. In our situation, it is not reasonable to look over all such paths and try to find the optimal one because the number of such paths can sharply increase when we decrease diameters of cells. Instead, we will use another approach that reduces the problem under consideration to a problem of linear programming.

We begin with the related problem of linear programming and necessary definitions. Suppose that a finite directed graph $\langle M, N \rangle$ is given, where M is the set of vertices and N is the set of arcs (edges). To each arc $j \in N$ we associate a number c_j the characteristic of the arc. In our case $c_j = \ln a_j$. So we have the labeled symbolic image. In graph theory, a simple periodic path is named a contour. To each contour of a graph we associate the characteristic of the contour which is equal to the mean value of characteristics of arcs of the contour. We need to find a contour with minimal characteristic. The case of maximal characteristic is easily obtained by replacing the sign of character- istics of arcs. In [121], this problem was reduced to a special problem of linear programming and a method (in principle, a version of the simplex method) for solving this problem was suggested there.

Let us consider the problem of finding nonnegative x_j, $j \in N$, that satisfy the conditions

$$\sum_{j \in N} x_j = 1, \qquad \sum_{j:e(j)=i} x_j - \sum_{j:b(j)=i} x_j = 0, \quad i \in M,$$

where $b(j)$ and $e(j)$ stand for starting point and the ending points of the arc j, respectively and minimize the linear function

$$\sum_{j \in N} c_j x_j.$$

We can treat x_j as a flow on the arc j. The first assumption means that the sum of flows over all arcs is equal to 1. The second one means that the flow is closed, i.e. for each vertex i the entering flow is equal to the leaving flow. In this problem, each extreme point of the convex polyhedron of feasible solutions is an elementary closed flow, i.e., a set of variables x_j such that all nonzero values coincide, the sum of them is equal to 1, and the corresponding arcs j are arcs of some contour. Thus, if m is the number of arcs of a contour, then $x_j = 1/m$ and we have

$$\sum_{j \in N} c_j x_j = 1/m \sum_{j \in N} c_j = 1/m \sum_{j=1}^{m} \ln a_j.$$

The value of the objective function on the contour coincides with the charac- teristic of the corresponding contour. In other words, the contours with the maximal and minimal mean values provide us with the maximal and minimal values of the characteristic Lyapunov exponents of periodic paths on the la- beled symbolic image. Note that the method of [121] was modified because we have to solve problems of a huge size.

1. To simplify the logical scheme of the method, the graph $\langle M, N \rangle$ is pre- processed as follows: we extract strongly connected components, i.e., classes of equivalent recurrent vertices. This operation is linear with re- spect to the number of arcs [114]. Then the problem is solved for each component separately.

2. According to the method, a feasible basic solution is not computed ex- plicitly for every strongly connected component, but a considerable part of the solution is reconstructed by a method which is similar to B. Levit's method [71] for solving the shortest path problem.

3. A very effective method is used to construct the initial feasible basic solution.

To proceed further, we assume that the preliminary preparation mentioned in item 1 has been made and the graph $\langle M, N \rangle$ is strongly connected, i.e., for any two vertices of the graph there is a connecting path.

Methods similar to the simplex method serve for finding extreme points of the polyhedron of feasible solutions. Therefore, in our case, such a method provides us with a contour with minimal characteristic. However, we do not deal with extreme points of a polyhedron, but with more complicated constructions of feasible basic solutions to the problem of linear programming. Such a solution is defined by a basis set $N' \subset N$. It is easy to show that N' is a basis set if and only if the graph $\langle M, N' \rangle$ is a tree with a single additional arc and this arc generates a (single) contour in the graph. This contour is the supporting set of the every extreme point. The remaining arcs of the basic solution are used to find the values of dual variables.

Recall that, along with the main (direct) problem of linear programming, the following dual problem is stated and solved: find a variable z and variables v_i, $i \in M$, such that

$$z + v_{e(j)} - v_{b(j)} \leq c_j, \quad j \in N,$$

and maximize z. The maximum of z coincides with the minimum of the objective function. This means that z must be a characteristic of the desired contour.

In the simplex method, to the basic collection of indices N' one associates the following system of equations:

$$v_{e(j)} - v_{b(j)} = c_j - z, \quad j \in N'.$$

This system has a unique solution up to an additive constant for the variables v (which allows us to call them *potentials*). We have to solve this system at each iteration. To solve the system, it is convenient to start with determining z as the characteristic of the basic contour and after that, ignoring one of the arcs of the contour and defining v for one of the vertices, subsequently find (in the same way as in the potential method for the transportation problem of linear programming) the remaining potentials on the rest of the basis tree. If z and v are such that

$$v_{e(j)} - v_{b(j)} \leq c_j - z, \quad j \in N,$$

then the solution of the direct problem is optimal and a contour with the minimal characteristic is found.

However, the above computation scheme can be significantly simplified and made more effective. We say that a tree is a *directed tree with root i_0* if it contains all paths from i_0 to any other vertex. Using this definition and recalling that we deal with a strongly connected graph, we modify the algorithm as follows. First, for the basic graph $\langle M, N' \rangle$ we consider a directed tree with an additional arc that goes to the root vertex i_0. Second, we construct a directed tree at each iteration trying to preserve the required inequalities. To construct the tree, we use the technique developed for solving the shortest path problem. The known Dijkstra method [39] is not suitable in this case because the values $c_j - z$ can be of different signs. A method due to Levit [71] is more suitable in our situation. This method uses a close technique but is more flexible.

To construct the tree, we divide the set M of vertices into three subsets: $M = M_0 \cup M_1 \cup M_2$, where M_0 is the set of vertices that have been processed by the algorithm, M_1 is the set of vertices waiting for processing, and M_2 is the set of vertices that are not yet reached by the algorithm. The vertices M_1 are ordered. At any time, we have a directed tree with the set of vertices $M_0 \cup M_1$ and with the root at some vertex i_0; moreover, for every vertex in $M_0 \cup M_1$ the potential v is computed (possibly, this is not a final result). We say that a vertex i_1 *precedes* a vertex i_2 if either i_1 lies on the contour or there exists a path joining i_1 and i_2.

Algorithm for constructing the tree and the contour.

1. For a given contour $\langle M_c, N_c \rangle$ compute its characteristic z. Choosing $i_0 \in M_c$ and setting $v[i_0] = 0$, find the remaining $v[i]$ for $i \in M_c$ by the formula $v_{e(j)} = v_{b(j)} + c_j - z$. Then form the tree from the contour by eliminating a (single) arc incoming to i_0.
2. Include all vertices of M_c into M_1 by forming their list.
3. Act as follows (vertex processing) whenever M_1 is nonempty.

   3.1. Eliminate the first vertex i_1 from the list and transform it to M_0.

   3.2. For every arc j such that $b(j) = i_1$ put

   $$w = v[i_1] + c_j - z$$

and set $i_2 = e(j)$.

If $i_2 \in M_2$, then move the vertex i_2 from M_2 to M_1 by placing it at the end of the list. Then set $v[i_2] = w$ and join the arc j to the tree.

If $i_2 \in M_0 \cup M_1$, compare w with $v[i_2]$.

In the case $w \geq v[i_2]$ complete the arc processing and go to the step 3.2.

Otherwise $w < v[i_2]$ verify whether the vertex i_2 precedes the vertex i_1 in the tree.

If i_2 precedes i_1 then the arc j closes the contour with a characteristic less than z. The iteration of our "simplex method" is complete. The new contour is taken as the basic one and the algorithm is repeated, i.e. go to the step 1.

If i_2 does not precede i_1, set $v[i_2] = w$, include the arc j into the tree by replacing some other arc with the endpoint i_2. If i_2 is already located in M_0, transform it to M_1 for repeated processing by placing it at the beginning of the list (to recalculate the potentials on the tree starting from i_2 and down). Go to the step 3.1.

4. Finaly, we arrive to the stage where $M_1 = \emptyset$ and $M_2 = \emptyset$, i.e. the system of potentials satisfying the required inequalities is constructed and the basic contour has the minimal characteristic.

To choose the initial basic contour, the following procedure is rather effective.

1. For every vertex $i \in M$ choose an arc $j(i)$ starting at i with the minimal c_j. Let N_s be the set of arcs thus chosen. On the graph $\langle M, N_s \rangle$ we construct strong components. Notice that the strong components are contours in the case considered. For every vertex i we calculate its *degree* $s[i]$, i.e. the number of the above-selected arcs incoming to this vertex.
2. Form the list of vertices of degree zero.
3. While the list is not empty, act as follows.

   3.1. Eliminate the first vertex i_1 from the list.

   3.2. Eliminate $j(i_1)$ from N_s.

   3.3. Decrease the degree of the vertex $i_2 = e(j(i_1))$ by 1. If the degree obtained is zero, then include i_2 into the list.

4. When the list becomes empty, the arcs in N_s form isolated contours. Among these contours choose one with a minimal characteristic.

An application of this algorithm is available in the next chapter.

13

Hyperbolicity and Structural Stability

The aim of this chapter is substantiation of a constructive method for verification of hyperbolicity and structural stability of discrete dynamical systems. The structural stability was originated by A.A. Andronov and L.S. Pontryagin 1937 [5]. The hyperbolicity of the chain recurrent set is a necessary condition for the structural stability. We will show that hyperbolicity is tested by calculation of the Morse spectrum which can be found for a given accuracy by the symbolic image. If the Morse spectrum does not contain 0, then the chain recurrent set is hyperbolic and the system is Ω-stable. To verify the structural stability it is convenient to use a so-called complementary differential $\widehat{D}f(x) = ((Df(x))^*)^{-1}$, where A^* stands for the adjoint operator. A diffeomorphism f is shown to be structurally stable if and only if the Morse spectrum of $\widehat{D}f$ does not contain 0 and for the complementary differential there is no connection $CR^+ \to CR^-$ on the projective bundle, where CR^+ and CR^- denote the chain recurrent components for the positive and negative parts of the Morse spectrum. These conditions are verified by an algorithm based on the symbolic image of the complementary differential.

13.1 Hyperbolicity

Let $f : M \to M$ be a diffeomorphism of a compact manifold $M \subset \mathbb{R}^n$ and let ρ be a distance on M. The space of diffeomorphisms is endowed with a topology generated by the C^1-distance ρ_1 defined as follows. The C^0-distance ρ_0 between f and g is defined by $\rho_0(f, g) = \max_{x \in M} \rho(f(x), g(x))$. The norm of the differential $Df : TM \to TM$ (TM is the tangent bundle of M) is defined by

$$\|Df\| = \max_{x \in M} |Df(x)| = \max_{x \in M} \max_{|v|=1} |Df(x)v|.$$

The C^1-distance between smooth mappings f and g ρ_1 is defined by $\rho_1(f, g) = \rho_0(f, g) + \|Df - Dg\|$.

We recall that $p \in M$ is a wandering point if there exists a neighborhood V of p and an integer k_0 such that $f^k(V) \cap V = \emptyset$ for $|k| > k_0$. Otherwise we say that p is nonwandering. We write Ω or $\Omega(f)$ for the set of nonwandering points of f. It is well known [110] that $\Omega(f)$ is compact, invariant, and contains $\alpha-$ and $\omega-$limit points. Hence, each trajectory starts and ends at Ω, i.e., the dynamics of a system is ultimately concentrated on Ω.

Definition 146. *A diffeomorphism f is said to be $\Omega-$stable if there exists $\delta > 0$ such that the inequality $\rho_1(f, g) < \delta$ implies the existence of a homeomorphism $h : \Omega(f) \to \Omega(g)$ such that $hf = gh$.*

The chain recurrent set Q is known [19] to be invariant, closed and contains periodic, homoclinic, and nonwandering trajectories, i.e., $\Omega \subset Q$. However, in general, $\Omega \neq Q$. If a chain recurrent point is not periodic and $\dim M > 1$, then there exists a perturbation of f in the C^0-topology as small as one likes for which this point is periodic [113]. One may say that a chain recurrent point generates periodic trajectories by C^0-perturbations [129, 131]. V. Pliss [115] defined $\pi-$set of points generating periodic trajectories of C^1-perturbations. From the Pugh Closing lemma [110] it follows that the $\pi-$set contains Ω and is contained in Q. It should be noted that there is a constructive method of localization of the chain recurrent set Q, while a method for construction of the $\pi-$set or Ω is unknown.

Definition 147. *A diffeomorphism f is said to be $Q-$stable if there exists $\delta > 0$ such that the inequality $\rho_1(f, g) < \delta$ implies the existence of a homeomorphism $h : Q(f) \to Q(g)$ such that $hf = gh$.*

The main condition for Ω and Q-stability is the hyperbolicity.

Definition 148. *Let Λ be an invariant set of f. A diffeomorphism f is said to be hyperbolic on $\Lambda \subset M$ if there exist invariant subbundles E^s and E^u of $TM|_\Lambda$ and constants $d > 0$ and $\alpha > 0$ such that*

$$TM|_\Lambda = E^s \oplus E^u,$$
$$|Df^n(x)v| \leq d|v| \exp(-\alpha n), \quad x \in \Lambda, \ v \in E^s(x), \ n > 0,$$
$$\left|Df^{-n}(x)v\right| \leq d|v| \exp(-\alpha n), \quad x \in \Lambda, \ v \in E^u(x), \ n > 0.$$

The first condition $TM = E^s \oplus E^u$ means that any vector v can be written in form $v = v^s + v^u$, $v^s \in E^s$, $v^u \in E^u$, and $E^s \cap E^u = 0$. The invariance of subbundles $E^{s,u}$ means that $Df(x)E^{s,u}(x) = E^{s,u}(f(x))$. The main condition for Ω-stability is the following axiom A.

Axiom A. *We say that f satisfies Axiom A if its nonwandering set Ω is hyperbolic and periodic points are dense in Ω.*

S. Smale showed [137] that if f satisfies Axiom A, then Ω decomposes in a finite disjoint union of closed subsets (base sets) which are invariant and transitive: $\Omega = \Omega_1 \bigcup \Omega_2 \bigcup ... \bigcup \Omega_k$. The transitivity of Ω_i means that there exists a dense trajectory in Ω_i. We say that there exists a connection $\Omega_i \to \Omega_j$

if there is a trajectory outside Ω with α-limit set in Ω_i and ω-limit set in Ω_j. A cycle in Ω is a sequence of connections $\Omega_{i_1} \to \Omega_{i_2} \to ... \to \Omega_{i_s} = \Omega_{i_1}$, $s \geq 1$. Summarizing the results of S. Smale [138], M. Shub [131], J. Palis [110], J. Franke and J. Selgrade [42], we obtain the following theorem.

Theorem 149. *The following statements are equivalent:*

1. *f is $\Omega-$stable.*
2. *f satisfies axiom A and has no cycles.*
3. *f is $Q-$stable.*
4. *f is hyperbolic on the chain recurrent set.*

Thus, to guarantee Ω- or Q-stability we have to check hyperbolicity on the chain recurrent set. We notice that if f is Ω-stable, then the chain recurrent set, the nonwandering set, and the $\pi-$set coincide with the closure of the set of periodic points.

Theorem 150. *The following statements are equivalent:*

1) *a diffeomorphism f is hyperbolic on the chain recurrent set Q,*
2) *the Morse spectrum of the differential Df does not contain 0.*

Proof. The implication 2) $\Rightarrow$ 1) follows from Theorem 145. Now we prove that if f is hyperbolic on the chain recurrent set $Q \subset M$, then the Morse spectrum $\Sigma(Df)$ does not intersect any interval $[-\beta, \beta]$, where $0 < \beta < \alpha$, α is the exponent of hyperbolicity.

By Definition 125, we have to consider an ε-chain $\xi = \{(x_k, y_k)\}$ in the chain recurrent set CR on the projective bundle P. Since f is hyperbolic on Q, the projection of unstable subbundles $[E^u|_Q] = PE^u$ is an attractor, and the projection of stable subbundles $[E^s|_Q] = PE^s$ is its complementary repellor [19, 42]. Thus, $CR \subset PE^u \bigcup PE^s$. Denote by $d = \min\{\rho(u, v), \ u \in PE^u, \ v \in PE^s\}$ the distance between PE^u and PE^s, $d > 0$. Let us show that if $\varepsilon < d$, then ξ is in PE^u or in PE^s. Suppose, on the contrary, that there exist (x_k, y_k) in PE^u and (x_{k+1}, y_{k+1}) in PE^s. The invariance of PE^u implies $F(x_k, y_k) \in PE^u$. Since $\rho(F(x_k, y_k), (x_{k+1}, y_{k+1})) < \varepsilon$ and $(x_{k+1}, y_{k+1}) \in PE^s$ then $d \leq \rho(F(x_k, y_k), (x_{k+1}, y_{k+1})) < \varepsilon$. This contradicts $d > \varepsilon$.

Thus, the chain ξ is in PE^s or in PE^u. Let the ε-chain $\xi = \{(x_1, y_1), (x_2, y_2)...(x_N, y_N)\}$ of length N be in PE^s and $e_i = e(y_i)$ be a basis vector. Let us fix an integer $q > 0$ and split the chain ξ into m pieces of length q and a residual chain of length r, $0 \leq r < q$. The number N takes the form $N = mq + r$. Since the projective bundle P is compact, we have

$$|\ln \prod_{i=1}^{q} |Df(x_i)e_i| - \ln |Df^q(x_1)e_1| \, | \leq \delta(q, \varepsilon),$$

where $\delta(q, \varepsilon)$ tends to zero uniformly on (x_i, e_i) as the length q is fixed and $\varepsilon \to 0$. By the definition of hyperbolicity, we have

$$|Df^q(x)e| \le d\exp(-\alpha q),$$

where $e \in E^s$. The exponential growth rate of the chain $\xi(e) = \{(x_k, e_k)\}$, $e_k = e(y_k)$, yields the estimate

$$\lambda(\xi) = \frac{1}{N} \ln \left(\prod_{k=1}^{N} |Df(x_k)e_k| \right)$$

$$= \frac{1}{N} \ln \left(\left(\prod_{p=0}^{m-1} \left(\prod_{k=pq+1}^{(p+1)q} |Df(x_k)e_k| \right) \right) \prod_{k=mp+1}^{N} |Df(x_k)e_k| \right)$$

$$= \frac{1}{N} \left(\sum_{p=0}^{m-1} \ln \prod_{k=pq+1}^{(p+1)q} |Df(x_k)e_k| + \sum_{k=mq+1}^{N} \ln |Df(x_k)e_k| \right)$$

$$+ \frac{1}{N} \sum_{p=0}^{m-1} \ln |Df^q(x_{pq+1})e_{pq+1}| - \frac{1}{N} \sum_{p=0}^{m-1} \ln |Df^q(x_{pq+1})e_{pq+1}|$$

$$= \frac{1}{N} \sum_{p=0}^{m-1} \left(\ln \prod_{i=1}^{q} |Df(x_{(p-1)q+i})e_{(p-1)q+i}| - \ln |Df^q(x_{pq+1})e_{pq+1}| \right)$$

$$+ \frac{1}{N} \sum_{k=mq+1}^{N} \ln |Df(x_k)e_k| + \frac{1}{N} \sum_{p=0}^{m-1} \ln |Df^q(x_{pq+1})e_{pq+1}|$$

$$\le \frac{m}{N} \delta(q, \varepsilon) + \frac{rC}{N} + \frac{m}{N}(\ln d - \alpha q),$$

where $|Df| \le C$. We have

$$\lambda(\xi) \le -\alpha + \frac{N - mq}{N} \alpha + \frac{\ln d}{q} + \delta(q, \varepsilon) + \frac{rC}{N} = -\alpha + \frac{\ln d}{q} + \delta(q, \varepsilon) + \frac{r(C + \alpha)}{N} \le$$

$$-\alpha + \frac{\ln d}{q} + \delta(q, \varepsilon) + \frac{C + \alpha}{m}.$$

Let us choose q such that $\frac{\ln d}{q} < \frac{\alpha - \beta}{6}$. For a chosen q let $\varepsilon > 0$ be such that $\delta(q, \varepsilon) < \frac{\alpha - \beta}{6}$. Finally, let m be such that $\frac{C + \alpha}{m} < \frac{\alpha - \beta}{6}$. Thus, we get

$$\lambda(\xi) \le -\alpha + \frac{\alpha - \beta}{2} < -\beta.$$

Similarly, one can prove that if $\xi \in PE^u$ then $\lambda(\xi) \ge \alpha - \frac{\alpha - \beta}{2} > \beta$.

$\odot$

The next theorem follows from Theorems 145 and 150.

Theorem 151. *A diffeomorphism f is hyperbolic on the chain recurrent set Q if and only if there exists $d_0 > 0$ such that the extended spectrum of the symbolic image for a covering $C(P)$ with maximal diameter $d < d_0$ does not contain 0.*

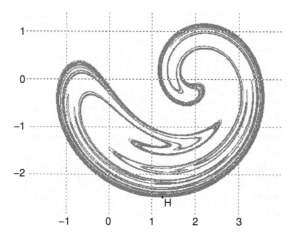

Fig. 13.1. Global attractor A_g and unstable manifolds $W^u(H)$ of H

The Morse spectrum $\Sigma(F)$ of a linear extension F is contained in

$$\Sigma = \bigcup_k [\lambda_{\min}(H_k) - \theta\eta(d), \lambda_{\max}(H_k) + \theta\eta(d)],$$

where $\{H_k\}$ is the full collection of classes of equivalent recurrent vertices on the symbolic image $G(PF)$ and d is the maximal diameter of cells of the covering $C(P)$. Theorem 145 allows to estimate the Morse spectrum. The set Σ is called the extended spectrum of a symbolic image.

Theorem 151 provides a way for the verification of hyperbolicity for a sufficiently fine covering C. However, we have no estimate of the maximal diameter d_0. So, the verification of structural stability should be achieved by the construction of a sequence of symbolic images applying a subdivision of coverings.

Example 152. Verification of hyperbolicity of the modified Ikeda type mapping.
Consider a plane mapping J of the form

$$J : (x, y) \rightarrow (r + a(\cos\tau - y\sin\tau), b(x\sin\tau + y\cos\tau)), (x, y) \in \mathbb{R}^2,$$

where $\tau = C_1 - C_3/(1+x^2+y^2)$. For the standard Ikeda mapping, $a = b = C_2$ and $0 < C_2 < 1$, i.e. J preserves orientation and decreases the area. We consider the mapping J with $r = 2$, $C_1 = 0.4$, $C_3 = 6$, $a = -0.9$, $b = 0.9$. It is not difficult to see that J reverses orientation, decreases area. This implies that in a bounded part of the plane J has the global attractor A_g of zero Lebesgue measure. Numerical simulations show that there exist the unique saddle fixed point H at approximately $(1.3815, -2.4746)$ and the unique saddle 2-periodic orbit P at approximately $(0.2338. - 0.7031)$, $(1.9995, 0.6681)$ whose stable $W^s(P)$ and unstable $W^u(P)$ manifolds intersect, respectively,

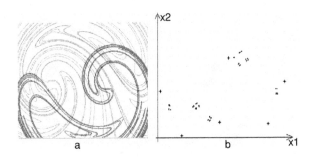

Fig. 13.2. The chain recurrent set $Q(R) = Q_1 + Q_2$ in the rectangle $R = [-1.1, 3.4] \times [-1.5, 1.8]$. The 6-periodic orbit Q_1 marked by '+' and the nontrivial invariant set Q_2 with fractal structure

Table 13.1. Estimates of the Morse spectrum over Ω_1

Component 1	Component 2	Angular size of the cells
$[0.359, 0.587]$	$[-0.829, -0.535]$	0.31
$[0.488, 0.565]$	$[-0.698, -0.659]$	0.15
$[0.508, 0.544]$	$[-0.762, -0.687]$	0.078
$[0.520, 0.528]$	$[-0.741, -0.723]$	0.0196
$[0.522, 0.523]$	$[-0.735, -0.733]$	0.00122
$[0.522, 0.523]$	$[-0.734, -0.733]$	0.00030

unstable $W^u(H)$ and stable $W^s(H)$ manifolds of H forming a heteroclinic cycle. The global attractor A_g (see Fig. 13.1) is a closure of $W^u(H)$ and seems to be non-hyperbolic because of the tangencies between stable and unstable manifolds of H and P (see Fig. 13.2,a).

We will treat hyperbolicity of the maximal chain recurrent set $Q(R)$ contained in the rectangle $R = [-1.1, 3.4] \times [-1.5, 1.8]$ (see Fig. 13.2,a). More precisely, let I be the maximal invariant set in the rectangle R and $Q(R)$ be the chain recurrent set of the restriction $J|_I$. Clearly, R is an isolating set for $Q(R)$. Under the initial partition 10×10 of the base R, the initial size of cells is 0.46×0.33. After several subdivisions (for size of the cell 0.0071×0.0051), $Q(R)$ is decomposed into two components Ω_1 and Ω_2 (see. Fig. 13.2,b) to the strongly connected components of the symbolic image. The set Ω_1 (marked by "+") is a periodic orbit of period 6, while Ω_2 has nontrivial fractal structure.

For a better localization of Ω_1, subdivisions were made up to the cells size became 0.00045×0.00032. In the projective bundle over Ω_1 we obtain the following: 1) two components of chain recurrent set and 2) spectrum of symbolic image consists of intervals $[0.522, 0.523]$ and $[-0.734, -0.733]$ (see Table 13.1). The angular size of cells in the projective space is finished to 0.0003, so the distance between the Morse spectrum and the spectrum of symbolic image does not exceed 0.01. It follows from this that the Morse spectrum over Q_1 lies in the segments $[0.51, 0.53]$ and $[-0.74, -0.72]$. To support the validity

Table 13.2. Estimates of the Morse spectrum over Ω_2

Component 1	Component 2	Angular size of the cells
$[-0.327, -0.177]$	$[0.454, 0.755]$	0.78
$[-0.908, -0.558]$	$[0.574, 0.756]$	0.39
$[-1.011, -0.653]$	$[0.652, 0.796]$	0.19
$[-1.071, -0.779]$	$[0.621, 0.812]$	0.098
$[-1.054, -0.820]$	$[0.624, 0.787]$	0.049
$[-1.034, -0.838]$	$[0.627, 0.792]$	0.0254
$[-1.019, -0.842]$	$[0.630, 0.794]$	0.0122
$[-1.007, -0.840]$	$[0.631, 0.794]$	0.00613
$[-1.006, -0.842]$	$[0.632, 0.793]$	0.00306
$[-1.005, -0.842]$	$[0.632, 0.793]$	0.00153
$[-1.004, -0.843]$	$[0.632, 0.793]$	0.00076
$[-1.004, -0.843]$	$[0.632, 0.793]$	0.00038

of the estimates obtained we use the 6-period orbit P_6 (1.0847, -1.0732), (2.7889, -1.1242), (-0.2626, -1.4846), (3.3560, 0.0508), (-1.0124, -0.2235), (1.3964, 0.7116) to find the eigenvalues of the differential DJ^6 along the orbit. We obtain $\gamma_1 = -23.098$ and $\gamma_2 = -0.012$. The Lyapunov exponents calculated by $\lambda = \frac{1}{6} ln|\gamma|$ are $\lambda_1 = 0.523$ and $\lambda_2 = -0.734$. So we have λ_1 and λ_2 are in the segments $[0.51, 0.53]$ and $[-0.74, -0.72]$. This implies the correctness of the Morse spectrum over Q_1.

For the set Ω_2 the subdivisions in the base were continued up to the cells size 0.00089×0.00058. Then the iterations of the labeled symbolic image were constructed up to the angular cells size 0.00038. As in the case of the first component the Morse spectrum of the system differs from the spectrum of the symbolic image 0.01. Thus, we obtain two intervals $[0.632, 0.793]$ and $[-1.004, -0.843]$ containing the Morse spectrum. Table 13.2 presents the sequence of approximations. To support the validity of the obtained estimates we use few periodic orbits in Q_2. In particular, for the 2-periodic orbit P_2 (0.2385, -0.7024), (1.9989, 0.6691) the differential DJ^2 over P_2 has the eigenvalues $\gamma_1 = -0.134$, and $\gamma_2 = -4.888$ and the Lyapunov exponents $\lambda_1 = -1.004$ and $\lambda_2 = 0.793$. For the 4-periodic orbit P_4 (-0.6836, -0.6319), (0.7312, -0.9389), (1.6003, 0.72792), (3.0613, -0.1713) the Lyapunov exponents are $\lambda_1 = -0.843$ and $\lambda_2 = 0.633$. The invariant set Q_2 contains a 6-periodic orbit P_6^*: (-0.7116, -0.7048), (1.8306, 0.8938), (0.2549, -0.5614), (1.7655, 0.550), (0.7136, -1.0552), (3.1004, -0.3206). Its Lyapunov exponents are $\lambda_2 = 0.677$ and $\lambda_1 = -0.887$. Thus, the Lyapunov exponents of the orbits P_2, P_4, and P_6^* are in the segments $[0.62, 0.80]$ and $[-1.01, -0.83]$.

From the above-obtained results it follows that the Morse spectrum of J over $Q(R)$ does not contain 0, and thus the chain recurrent set $Q(R)$ is hyperbolic. Numerical simulations of the Ikeda type mapping with nontrivial dynamics testifies the effectiveness of the proposed algorithm.

13.2 Structural Stability

Now we describe an algorithm for verification of structural stability by applied symbolic dynamics methods. This verification includes two main steps: estimate for the Morse spectrum of the complementary differential $\widehat{D}f$ and verification of the lack of connection $CR^+ \rightarrow CR^-$ between positive and negative components of the chain recurrent set.

Definition 153. *A diffeomorphism f is said to be structurally stable if for any $\varepsilon > 0$ there exists $\delta > 0$ such that the inequality $\rho_1(f, g) < \delta$ implies the existence of a homeomorphism $h : M \rightarrow M$ such that $fh = hg$ and $\rho_0(h, id) < \varepsilon$, where id is the identity mapping of M.*

For $x \in M$ let $TM(x)$ be the tangent space of M at x. The stable $S(x)$ and unstable $U(x)$ subspaces of f at x are defined as

$$S(x) = \{v \in TM(x) : |Df^n(x)v| \rightarrow 0 \text{ as } n \rightarrow +\infty\},$$
$$U(x) = \{v \in TM(x) : |Df^n(x)v| \rightarrow 0 \text{ as } n \rightarrow -\infty\}.$$

Definition 154. *We say that the transversality condition is fulfilled on M if*

$$TM(x) = S(x) + U(x)$$

for any $x \in M$.

Theorem 155. [19, 79] *If the transversality condition is fulfilled, then*

1) $\Lambda = \{x \in M : TM(x) = S(x) \oplus U(x)\}$ is closed and invariant,
2) f is hyperbolic on Λ, and $S(x) = E^s(x)$, $U(x) = E^u(x)$, $x \in \Lambda$,
3) the chain recurrent set Q is in Λ.

The following theorem is a consequence of results obtained by J. Robbin [118], C. Robinson [119], and R. Mané [79, 80].

Theorem 156. *A diffeomorphism f is structurally stable if and only if the transversality condition is fulfilled on M.*

Thus, Theorem 156 solves the problem of structural stability. However, application of this result is limited by difficulties of verification of the transversality condition [96]. Even application of the classical result on the structural stability of plane vector fields by A.A. Andronov and L.S. Pontryagin [5] requires to recognize the behavior of stable and unstable separatrices of saddles. Here we will give necessary and sufficient conditions of structural stability different from those of Theorem 156.

13.3 Complementary Differential

First, let us recall some notions of the theory of linear operators. As usual, the dual or complementary space of a vector space E is defined as the vector space E^* of linear mappings $E \to \mathbb{R}$ [54]. The dual vector space E^* is isomorphic to E via an inner product $\gamma : E \to E^*$, $\gamma_v(u) = \langle v, u \rangle$. This allows us to identify E^* and E. If $T : E \to E$ is a linear operator, then its adjoint or dual is a linear map $T^* : E \to E$ defined by $\langle Tv, u \rangle = \langle v, T^*u \rangle$.

As mentioned above, the differential $Df : TM \to TM$ is an example of a linear extension of f. The differential Df takes the tangent space $TM(x)$ at x onto the tangent space $TM(f(x))$ at $f(x)$. An other linear extension named complementary differential is defined as follows:

$$\widehat{D}f(x) = ((Df(x))^*)^{-1} : TM(x) \to TM(f(x)).$$

Actually, the complementary differential acts in the dual space TM^* which is identified with TM. By definition, the complementary differential covers f. The main property of the complementary differential is that the inner product persists under the action of the differential and the complementary differential [19]:

$$\langle Dfv, \widehat{D}fu \rangle = \langle v, u \rangle.$$

Theorem 157. *The following statements are equivalent.*

1) A diffeomorphism f is hyperbolic on Q.
2) $\widehat{D}f$ is hyperbolic on Q.
3) The Morse spectrum of $\widehat{D}f$ does not contain 0.

Proof. The equivalence of 1) and 2) is proved in [19] and [123]. The equivalence of 2) and 3) can be proved in the same reasoning as in Theorem 150.
$\odot$

Theorem 157 provides a way to verify hyperbolicity (transversality) on the chain recurrent set. We suppose that the extended spectrum of a symbolic image does not contain 0. Then a dynamical system is hyperbolic on the chain recurrent set, and each class H_k of equivalent recurrent vertices generates the spectral interval

$$[a_k, b_k] = [\lambda_{\min}(H_k) - \theta\eta(d), \lambda_{\max}(H_k) + \theta\eta(d)],$$

which is either positive (if $a_k > 0$) or negative (if $b_k < 0$) The sets

$$H^+ = \left\{ \bigcup H_k, \ a_k > 0 \right\}, \ H^- = \left\{ \bigcup H_k, \ b_k < 0 \right\}$$

are called positive and negative sets of recurrent vertices. From Theorem 53 it follows that

$$P = \left\{ \bigcup M(i) : i \text{ is recurrent} \right\} = \left\{ \bigcup M(i) : i \in H^+ \cup H^- \right\}$$

is a closed neighborhood of the chain recurrent set on the projective bundle. In particular,

$$P^+ = \left\{ \bigcup M(i) : i \in H^+ \right\} \text{ and } P^- = \left\{ \bigcup M(j) : j \in H^- \right\}$$

are closed neighborhoods of $CR^+ = CR \cap P^+$ and $CR^- = CR \cap P^-$, respectively.

Non-recurrent trajectories

Definition 158. [19,127] *The complementary differential is said to have only trivial bounded trajectories if any bounded trajectory*

$$\{(x_{n+1}, v_{n+1}) = (f(x_n), \widehat{D}f(x_n)v_n), \ n \in Z\}$$

is a zero trajectory, i.e. $v_n = 0$.

Theorem 159. [18,80] *The transversality condition is fulfilled on M if and only if the complementary differential has only trivial bounded trajectories.*

It should be emphasized that hyperbolicity on the chain recurrent set may be verified by the Morse spectrum of the differential or the complementary differential (see Theorem 157). However, to verify the transversality condition on a non-recurrent set, we have to apply the complementary differential (see Theorem 159). To verify the transversality condition on the non-recurrent set, we consider the symbolic image G of the complementary differential on the projective bundle $P\widehat{D}f : P \to P$. Suppose that the extended spectrum Σ of the complementary differential does not contain 0. Let H^+ and H^- be the positive and the negative sets of recurrent vertices on G. The sets $P^+ = \{\bigcap M(i), i \in H^+\}$ and $P^- = \{\bigcap M(i), i \in H^-\}$ are closed neighborhoods of CR^+ and CR^-, respectively.

For a subset Ω of the projective bundle P let

$$E(\Omega) = \{v \in TM : v \neq 0 \Rightarrow [v] = y \in \Omega\}$$

be the subspace spanned over Ω. Since the Morse spectrum Σ of the complementary differential does not contain 0, Σ splits into two disjoint parts: the positive one Σ^+ and negative one Σ^-. Moreover, there exists an interval $[-\beta, \beta]$, $\beta > 0$, such that $\Sigma \cap [-\beta, \beta] = \emptyset$. The chain recurrent set $CR \subset P$ of the projection of the complementary differential $P\widehat{D}f$ breaks up into two disjoint parts CR^+ and CR^- such that $CR^+ + CR^- = CR$ and the Morse spectrum of $\widehat{D}f|_{E(CR^+)}$ is positive $\Sigma(\widehat{D}f|_{E(CR^+)}) = \Sigma^+$ and the Morse spectrum of $\widehat{D}f|_{E(CR^-)}$ is negative $\Sigma\widehat{D}f|_{E(CR^-)}) = \Sigma^-$. The sets CR^+ and CR^- are the positive and negative chain recurrent sets of $P\widehat{D}f$ on P.

Let $\xi = \{(x_k, v_k)\}$ be a nonzero trajectory of $\widehat{D}f$. Denote by $[\xi] = \{(x_k, y_k)\}$, $y_k = [v_k]$, the projection of ξ on P. Suppose that $[\xi]$ is non-recurrent. The ω-limit set $\omega([\xi])$ and the α-limit set $\alpha([\xi])$ of $[\xi]$ are in the chain recurrent set CR. We will say that $[\xi]$ starts at CR^+ if $\alpha([\xi]) \subset CR^+$ and $[\xi]$ starts at CR^- if $\alpha([\xi]) \subset CR^-$. Analogously, $[\xi]$ ends at CR^+ (CR^-) if $\omega([\xi]) \subset CR^+$ (CR^-). Let $[\xi]$ end at CR^+. Similarly to the proof of Theorem 150, we obtain that $\lambda(\xi) \geq \beta > 0$. Then the positive semi-trajectory ξ^+ is unbounded. Analogously, if $[\xi]$ ends at CR^-, then $\lambda(\xi) \leq -\beta < 0$, and the positive semi-trajectory ξ^+ is bounded. In the same way, we establish that if $[\xi]$ starts at CR^+, the negative semi-trajectory ξ^- is bounded, and if $[\xi]$ starts at CR^-, the negative semi-trajectory ξ^- is unbounded. Thus, we reach to the following conclusion.

Proposition 160. *A nonzero non-recurrent trajectory ξ is bounded if and only if $[\xi]$ starts at CR^+ and ends at CR^-.*

The stable and unstable subspaces of the complementary differential at $x \in M$ are defined as

$$\widehat{S}(x) = \{v \in TM(x) : \ \left|\widehat{D}f^n(x)v\right| \to 0 \text{ as } n \to +\infty\},$$

$$\widehat{U}(x) = \{v \in TM(x) : \ \left|\widehat{D}f^n(x)v\right| \to 0 \text{ as } n \to -\infty\},$$

respectively. We set $\widehat{S} = \{\widehat{S}(x), \ x \in M\}$ and $\widehat{U} = \{\widehat{U}(x), \ x \in M\}$. The next proposition follows from the aforesaid.

Proposition 161. *Let the Morse spectrum of $\widehat{D}f$ do not contain 0. Then*

1) the projection of the unstable subspaces is the unstable manifold of CR^+:

$$[\widehat{U}] = W^u(CR^+) = \{(x, y) : \ \alpha(x, y) \subset CR^+\},$$

2) the projection of the stable subspaces is the stable manifold of CR^-:

$$[\widehat{S}] = W^s(CR^-) = \{(x, y) : \ \omega(x, y) \subset CR^-\}.$$

13.4 Structural Stability Conditions

Theorem 162. [19, 125] *The complementary differential has only trivial bounded trajectories if and only if $[\widehat{U}] = W^u(CR^+)$ is an attractor and $[\widehat{S}] = W^s(CR^-)$ is its complementary repellor.*

We say that there exists a connection $CR^+ \to CR^-$ if there is a trajectory ξ of $P\widehat{D}f$ such that its α-limit set $\alpha(\xi)$ is in CR^+ and its ω-limit set $\omega(\xi)$ is in CR^-. As shown above, if the Morse spectrum does not contain 0, the connection $CR^+ \to CR^-$ exists if and only if there is a bounded trajectory of the complementary differential.

Theorem 163. *A diffeomorphism f is structurally stable if and only if*

*1) the Morse spectrum of the complementary differential does not contain
0 and*

*2) the unstable manifold $W^u(CR^+)$ is an attractor and the stable manifold
$W^s(CR^-)$ is its complementary repellor
or the condition 2) can be replaced by 2*) there exists no connection
$CR^+ \to CR^-$.*

Proof. Sufficiency. Suppose that the Morse spectrum of the complementary
differential does not contain 0, the unstable manifold $W^u(CR^+)$ is an attractor, and the stable manifold $W^s(CR^-)$ is its complementary repellor. By
Theorems 156 and 159, it suffices to prove that the complementary differential
has only trivial bounded trajectories. Since the Morse spectrum is closed and
does not contain 0, there exists a segment $[-\beta, \beta]$, $\beta > 0$, having no common
points with the Morse spectrum.

Let $\xi = \{(x_k, v_k)\}$ be a nonzero trajectory of the complementary differential $\widehat{D}f$ and let $[\xi] = \{(x_k, y_k)\}$, $y_k = [v_k]$, be its projection on the bundle P.
If $[\xi]$ is in the chain recurrent set $CR = CR^+ + CR^-$, then ξ is hyperbolic
and ξ is unbounded. Suppose that $[\xi]$ is non-recurrent. If the trajectory ξ is
bounded, $[\xi]$ has to start at CR^+ and to end at CR^-. We have the connection
$CR^+ \to CR^-$, i.e., $W^u(CR^+) \cap W^s(CR^-) \neq \emptyset$. The last inequality contradicts 2). Thus, the trajectory ξ is unbounded. From Theorems 156 and 159 it
follows that the transversality condition holds, and f is structurally stable.

Necessity. Suppose that f is structurally stable. By Theorems 155 and 156,
f is hyperbolic on the chain recurrent set Q. By Theorem 157, the Morse spectrum of the complementary differential does not contain 0. Since the Morse
spectrum is closed, there is a segment $[-\beta, \beta]$, $\beta > 0$, which does not intersect
the Morse spectrum. So the Morse spectrum consists of two disjoint parts Σ^+
and Σ^-. The chain recurrent set $CR \subset P$ splits into two parts CR^+ and CR^-.
By Theorem 159, the complementary differential has only trivial bounded trajectories. In the same way, we can show that $W^u(CR^+) \cap W^s(CR^-) = \emptyset$ and
any trajectory outside $W^u(CR^+) \cup W^s(CR^-)$ has to start at CR^- and to end
at CR^+. Then $W^u(CR^+)$ is an attractor and $W^s(CR^-)$ is its complementary
repellor, and there is no connection $CR^+ \to CR^-$.

$\odot$

13.5 Verification Algorithm

We suppose that the extended spectrum of a symbolic image does not contain
0. Then a dynamical system is hyperbolic on the chain recurrent set, and each
class H_k of equivalent recurrent vertices generates the spectral interval

$$[a_k, b_k] = [\lambda_{\min}(H_k) - \theta\eta(d), \lambda_{\max}(H_k) + \theta\eta(d)],$$

which is either positive (if $a_k > 0$) or negative (if $b_k < 0$) The sets

$$H^+ = \{\bigcup H_k, \ a_k > 0\}, \ H^- = \{\bigcup H_k, \ b_k < 0\}$$

are positive and negative sets of recurrent vertices.

Proposition 164. *If there is no path from H^+ to H^- on the symbolic image then there is no connection $CR^+ \to CR^-$.*

Proof. On the contrary, suppose that there exists an orbit $\{(x_n, y_n) = P\widehat{D}f^n(x)e(y)\}$ such that $(x_n, y_n) \to CR^+$ as $n \to -\infty$ and $(x_n, y_n) \to CR^-$ as $n \to +\infty$. Since P^+ is a neighborhood of CR^+, there exists k such that $(x_n, y_n) \in P^+$, $n \le k$. In particular, $(x_k, y_k) \in M(i_k)$, $i_k \in H^+$. Similarly, there exists $m > k$ such that $(x_m, y_m) \in M(i_m), i_m \in H^-$. The finite orbit $\{(x_k, y_k), ..., (x_n, y_n), ..., (x_m, y_m)\}$ generates an admissible path of the form $\omega = i_k \to ... \to i_n \to ... \to i_m$ on the symbolic image, where i_n is defined by the inclusion $(x_n, y_n) \in M(i_n)$. This means that ω is a path from H^+ to H^-. We get a contradiction.

⊙

Definition 165. *We say that the connection $H^+ \to H^-$ holds if there is an admissible path from H^+ to H^-.*

Theorem 166. *If an extended spectrum of the symbolic image G of $P\widehat{D}f$ does not contain 0 and there is no connection $H^+ \to H^-$, then f is structurally stable.*

Proof. Let the conditions of the theorem hold for f. The Morse spectrum of the complementary differential $\widehat{D}f$ does not contain 0. By Theorem 157, the diffeomorphism f is hyperbolic on the chain recurrent set $Q \subset M$. The sets P^+ and P^- are closed neighborhoods of the positive and negative chain recurrent sets CR^+, CR^- of $P\widehat{D}f$ on the projective bundle P. It follows from Proposition 164 that there is no connection $CR \to CR^-$ for $P\widehat{D}f$. Then the conditions 1) and 2*) of Theorem 163 hold and, hence, f is structurally stable.

⊙

Thus, the structural stability can be verified by the following algorithm:

1) The complementary differential $\widehat{D}f$ is determined.
2) A covering C of the projective bundle P is chosen.
3) The labeled symbolic image G of $P\widehat{D}f : P \to P$ is constructed.
4) The extended spectrum of G is found.
5) If the spectrum does not contain 0, the positive and negative sets of recurrent vertices H^+ and H^- are determined.
6) The lack of connection $H^+ \to H^-$ is verified.

Theorem 167. *If a diffeomorphism f is structurally stable, then there exists $d_0 > 0$ such that for any symbolic image G with cells of maximal diameter $d < d_0$ the extended spectrum does not contain 0 and there is no connection $H^+ \to H^-$.*

Proof. Let f be structurally stable. From Theorem 157 it follows that the Morse spectrum $\Sigma(\widehat{D}f)$ does not contain 0. Moreover, there exists $\beta > 0$ such that $\Sigma(\widehat{D}f) \cap [-\beta, \beta] = \emptyset$. From Theorem 145 it follows that there is $d_1 > 0$ such that for any symbolic image G with cells of the maximal diameter $d < d_1$, the extended spectrum does not contain 0. If the Morse spectrum does not contain 0, the lack of connection $H^+ \to H^-$ is equivalent to the existence of the attractor-repellor pair $W^u(CR^+)$ and $W^s(CR^-)$ by Theorem 163. From Theorem 71 it follows that for any attractor-repellor pair there exists $d_2 > 0$ such that this pair can be constructed by symbolic image with cells of the maximal diameter $d < d_2$. In particular, such symbolic image has no connection $H^+ \to H^-$. It remains to chose d_0 such that $0 < d_0 \le d_2 \le d_1$.

$\odot$

Theorem 167 guarantees verification of the structural stability for a sufficiently fine covering C. However, we have no estimation of the maximal diameter d_0. This forces us to verify the structural stability by construction of a sequence of symbolic images applying a subdivision of coverings.

14

Controllability

The methods of applied symbolic dynamics were effectively used for a qualitative research of a global behavior of dynamical systems without control. The study of the controllability is realized by a comparison of a dynamical control system and its symbolic image. The symbolic image has many features of the initial dynamical control system. The method does not require any preliminary information of the system and uses only information of general character. In particular, the local controllability along trajectories is required. As shown in [65], this property is typical and can be easily checked. All necessary estimations can be made by methods of numerical and symbolic calculations. The main result states that a dynamical control system is controlled if and only if its symbolic image is controlled. The controllability of the symbolic image can be tested by a computer.

14.1 Global and Local Control

We consider a family of vector fields $f_u, u \in U$ with a phase space M and a controlling space U. Usually it is possible to describe (at least locally) a dynamical control system by the smooth system of differential equations

$$\dot{x} = f(x, u). \tag{14.1}$$

Let T be a time interval on which the control system operates. The set of the function $\mathcal{U} = \{u = u(t) : T \to U\}$ is called the set of admissible controls. Let $D((0, \tau], u(t), x_0)$ be a shift operator along trajectories of the system (14.1) from the initial point x_0 with the control $u(t)$ at the time τ. An admissible constant control is denoted by $u(t) = \bar{u} = const$. Let the trajectory $\{x(t), u(t)\}$ satisfy the equation (14.1).

Definition 168. *A control system is controlled from the point x_0 to the point x_1 at the time τ if there exists the control $u = u(t)$ on the interval $(0, \tau]$ such that $D((0, \tau], u, x_0) = x_1$. In this case we write*

$$x_0 \xrightarrow{u,\tau} x_1.$$

Definition 169. *A control system is said to be globally controlled in M if it is controlled from any point x_0 to any point x_1 at some time τ, which depends on x_0.*

Definition 170. *A control system (14.1) is said to be locally controlled along the trajectory $\{\overline{x}(t), \overline{u}\}, t \in (0, t_1]$ if for any $t \in (0, t_1]$ we have*

$$\overline{x}(t) \in \text{Int}\{x | x = D((0, t_1], u, x_0), \quad u \in \mathcal{U}\},$$

where by $\text{Int} A$ is the interior of A.

The last means that the system is controlled from a point of the trajectory $\{\overline{x}(t), \ t \in (t_0, t_1]\}$ to each point of a neighborhood of the trajectory. The property of local controllability along the trajectory $\overline{x}(t)$ is valid if the linearized system

$$\dot{\Delta x} = \frac{\partial f}{\partial x}\Delta x + \frac{\partial f}{\partial u}\Delta u, \quad \Delta x = x - \overline{x}, \ \Delta u = u - \overline{u}$$

is controlled. The property of local controllability along a trajectory can be checked by a calculation of the rank of the matrix constructed on the given family of vector fields (see [140]).

Now we define more precisely the set of admissible controls. Let $U_0 = \{u_1, \ldots, u_N\}, U_0 \subset U$ be a finite set of admissible constant controls.

First, we suppose that the set $\mathcal{U}$ of admissible controls contains such set

$$\mathcal{U}_0 = \{\overline{u} | \overline{u} : T \to U_0\} \tag{14.2}$$

of piecewise constant controls that for any interval $(\underline{t}, \overline{t}) \subset T$ there exists a partition $\underline{t} = t_0 < t_1 < \cdots < t_M = \overline{t}$ so that $\overline{u}(t) = u_i, \ u_i \in U_0$, for $t \in (t_i, t_{i+1}], \ i = 1, \ldots, M - 1$. Thus, it is possible to consider a polysystem

$$\dot{x} = f(x, u_i), \quad x \in M, \quad i = 1, \ldots, N. \tag{14.3}$$

for the set $\mathcal{U}_0$ of admissible controls.

Secondly, we suppose that the set $\mathcal{U}$ of admissible controls contains the set $\Delta\mathcal{U}$ of local controls:

$$\Delta\mathcal{U} = \{\Delta u \in C | \exists \underline{t}, \overline{t} \quad \Delta u : [\underline{t}, \overline{t}] \to \Delta U, \quad \Delta U \subset U, \quad \Delta u(\underline{t}) = 0\}, \tag{14.4}$$

$\Delta U = \cup_{k=1}^{N}\Delta U_k$, where $\Delta U_k = \{u \mid |u - u_k| < \delta(u_k)\}, \ u_k \in U_0$ and $\delta(u_k)$ are constants. The local controls $\Delta u_k = u - u_k$ are given [65] by the formula

$$\Delta u_k(t) = v^*(t)\eta,$$

where η is an n-dimensional constant vector, $v(t) = \theta^{-1}(t)b(t)$, $\theta(t)$ is the fundamental matrix of the system $\dot{\theta}(t) = A(t)\theta(t), \theta(0) = I$,

$$A(t) = \frac{\partial f}{\partial x}(\overline{x}_k(t), u_k)), \quad b(t) = \frac{\partial f}{\partial u}(\overline{x}_k(t), u_k),$$

where $\overline{x}_k(t) = \overline{x}(x_0, u_k)$ is a trajectory starting from the point x_0 with the constant control u_k. We note that the numbers $\delta(u_k)$ are defined by the constant vectors η. Now, we make the basic assumption about a control system.

Basic Assumptions.

The control system (14.1) is locally controlled along any trajectory of a form $\overline{x}(x_0, \overline{u}, t)$ starting from the point x_0, where $x_0 \in M$, $\overline{u} \in U_0$. The local controllability is ensured by the controls belonging to $\Delta\mathcal{U}$. In particular, the local controllability means that each trajectory $\overline{x}(x_0, u_k, t)$, $u_k \in U_0$ beginning at the point x_0 lies in some open controlled cone with the vertex at x_0. This cone is filled by the trajectories of the control system. We denote by $\Delta x(x_0, u_k, t)$ the greatest possible radius of the open sphere centered at the point $\overline{x}(x_0, u_k, t)$ and laying in the cone. This means that for any x so that

$$|x - \overline{x}(x_0, u_k, t)| < \Delta x(x_0, u_k, t) \tag{14.5}$$

there exists the local control $\Delta u_k(s)$, $\Delta u_k(0) = 0$, $0 \le s \le t$, so that the control

$$\hat{u}_k(s) = u_k + \Delta u_k(s) \tag{14.6}$$

moves x_0 to x at the time t. Thus $\Delta x(x_0, u_k, t)$ is a radius of the admissible local control from x_0 at the time t with the constant control u_k.

Definition 171. *The value*

$$\rho(\Delta t) = \min_{y \in M, k} \Delta x(y, u_k, \Delta t)$$

is called the radius of the local control at the time Δt.

Because the function $\Delta x(y, u_k, t)$ is continious and M is a compact, the radius of the local control is positive. We will use controls of the form $\hat{u} = \overline{u} + \Delta u$, where $\overline{u} \in \mathcal{U}_0$ and $\Delta u \in \Delta\mathcal{U}$. Thus the control is decomposed in two parts; constant $-\overline{u}$ and local $-\Delta u$. We consider $\overline{u}$ as a basic control and Δu as a correcting control.

14.2 Symbolic Image of a Control System

Now we construct the symbolic image of a control system. We cover the set M by cells with the maximal diameter d. Since M is compact, it is possible to take the finite covering:

$$C = \{M(1), \ldots, M(n)\}. \tag{14.7}$$

Let us consider the family of the shift operators generated by the polysystem (14.3):

$$F_k : M \to M, \quad F_k = D((0, \Delta t], u_k), \quad u_k \in U_0, \tag{14.8}$$

where Δt is fixed. The set F_k of the operators realizes a shift on trajectories of vector fields of equations (14.3) at the time Δt. We say that there is a transition from a cell $M(i)$ to a cell $M(j)$, if there is some k so that $M(j) \cap F_k(M(j)) \neq \emptyset$. Let us denote this as

$$i \xrightarrow{u_k} j \text{ or } M(i) \xrightarrow{u_k} M(j). \tag{14.9}$$

Definition 172. *The symbolic image of a polysystem (14.3) is a directed graph G for which a set of vertices $\{i\}$ corresponds to the set of cells $\{M(i)\}$ and the set of edges $i \to j$ corresponds to the set of the transitions (14.9). Thus, each edge $i \to j$ gets the control transition u_k.*

One should note that the symbolic image G is framed by the constant controls u_k.

Definition 173. *The symbolic image G is said to be controlled from the vertex i to the vertex j if there exists an admissible chain of the transitions so that*

$$i = i_0 \xrightarrow{u_{k_1}} k_1 \xrightarrow{u_{k_2}} \cdots \xrightarrow{u_{k_s}} i_s = j.$$

The symbolic image G is said to be controlled if one is controlled from i to j for every pair (i, j) of the vertices.

In the first case the framed graph G has an admissible path from i to j and piecewise constant controls that garantee the transitions

$$M(i) = M(i_0) \xrightarrow{u_{k_1}} M(i_1) \xrightarrow{u_{k_2}} \cdots \xrightarrow{u_{k_s}} M(i_s) = M(j).$$

It the second case the graph G contains an admissible path from i to j for every pair (i, j), i.e., G is a strong component.

14.3 Test for Controllability

Let us fix two arbitrary points $x_*, x^* \in M$. It follows from Theorem 14 that if the polysystem (14.7) is controlled from the point $x_* \in M(i_*)$ to the point $x^* \in M(i^*)$ then the symbolic image is also controlled from i_* to i^*. The conditions of the converse statement is given by the following theorem.

Theorem 174. *Suppose that the system (14.1) is locally controlled along trajectories with the constant controls $u_k \in U_0$, i.e., Basic Assumption is fulfilled, and the radius of the local control is so that*

$$d + q \leq \rho,$$

where by d is the diameter of the covering and q is the largest diameter of the images $F_k(M(i)), \quad i = 1, ..., n.$

1. *If there is a path from the vertex i to the vertex j on the symbolic image G and the points $x_i \in M(i)$, $x_j \in M(j)$ then the system (14.1) is controlled from the point x_i to point x_j. Moreover, if there is the chain of transitions*

$$i = l_0 \xrightarrow{u_1} l_1 \xrightarrow{u_2} \cdots \xrightarrow{u_s} l_s = j$$

on the symbolic image then the system is controlled by the transitions

$$x_i = x_0 \xrightarrow{\hat{u}_1} x_1 \xrightarrow{\hat{u}_2} \cdots \xrightarrow{\hat{u}_s} x_s = x_j$$

in the phase space M, where $x_k \in M(l_k)$, $k = 0, \ldots s$.
2. *The system is controlled if and only if the symbolic image is controled.*

Proof. The second statement follows from the first one. Let us prove the first statement. Suppose that the symbolic image G has a path between the vertices i and j, and $x_i \in M(i)$ and $x_j \in M(j)$. It follos that there exists an admissible chain of transitions

$$i = l_0 \xrightarrow{u_1} l_1 \xrightarrow{u_2} \cdots \xrightarrow{u_s} l_s = j$$

on the symbolic image so that $x_i \in M(i)$ and $x_j \in M(j)$. Let a sequence $x_i = x_0, x_1, \ldots x_s = x_j$ be so that $x_k \in M(l_k)$. It follows from Theorem 14 that the constructed sequence is a finite (q+d)-trajectory, i.e., the distance between $F_k(x_k)$ and x_{k+1} is less then $q + d$. By the definition of the radius of the local control there exists a local control Δu_k so that the control

$$\hat{u}_k = u_k + \Delta u_k$$

moves the point x_k to x_{k+1}. Thus the system is controlled by the transitions

$$x_i = x_0 \xrightarrow{\hat{u}_1} x_1 \xrightarrow{\hat{u}_2} \cdots \xrightarrow{\hat{u}_s} x_s = x_j$$

$$\odot$$

15

Invariant Manifolds

15.1 Stable and Unstable Manifolds

Discrete systems. Let p be a fixed point for the diffeomorphism $f : \mathbb{R}^n \to \mathbb{R}^n$. To study the behavior of the orbits near p we use the linearized map $x \to Ax$, where A is the Jacobian matrix $Df(p) = [\partial f_i / \partial x_j](p)$ at the point p. The point is hyperbolic if the modulus of the eigenvalues of the matrix A are different from 1. In this case the eigenvalues λ are splitted into two parts: The stable ones for which $|\lambda| < 1$ and the unstable ones for which $|\lambda| > 1$. Let $\{v_1, ..., v_k\}$ be the eigenvectors corresponding to the stable part and $\{v_{k+1}, ..., v_n\}$ the ones corresponding to the unstable part. The eigenspace E^s spanned by the vectors $\{v_1, ..., v_k\}$ is called stable and the eigenspace E^u spanned by $\{v_{k+1}, ..., v_n\}$ is called unstable. According to construction $E^s + E^u = \mathbb{R}^n$, $E^s \cap E^u = \emptyset$.

The following theorem describes the structure of the invariant manifolds in a neighborhood of the hyperbolic point of the diffeomorphism. Different variants of this theorem was proved Lyapunov, Hadamar and Perron.

Theorem 175. [6] *Let p be a hyperbolic fixed point of the diffeomorphism f. Then there exist local stable and unstable manifolds*
$$W_{loc}^s(p) = \{x \in U : f^k(x) \to p \text{ for } k \to +\infty\},$$
$$W_{loc}^u(p) = \{x \in U : f^k(x) \to p \text{ for } k \to -\infty\}$$
in some neighborhood U of the point p. The subspaces E^s and E^u are tangent spaces to the manifolds $W_{loc}^s(p)$ and $W_{loc}^u(p)$ at the point p.

The local manifolds $W_{loc}^s(p), W_{loc}^u(p)$ are uniquely defined in the sense that for different neighborhoods V_1, V_2 they coincide in the intersection $V_1 \cap V_2$. The neighborhood U can be considered as the interior of a sphere of radius small enough and the local manifolds form discs of dimensions k and $n - k$.

The global stable and unstable manifolds of the fixed point p are defined in the following way:
$$W^s(p) = \{x \in R^n : f^k(x) \to p \text{ for } k \to +\infty\} = \bigcup_{k \in Z_-} f^k(W_{loc}^s(p)),$$

$W^u(p) = \{x \in R^n : f^k(x) \to p \text{ for } k \to +\infty\} = \bigcup_{k \in Z_+} f^k(W^u_{loc}(p))$,

where $Z_- = \{0, -1, -2, \ldots\}, Z_+ = \{0, 1, 2, \ldots\}$.

In other words, the global stable manifold is the image of the local stable manifold for backward iterates and the global unstable manifold is the image of the local unstable manifold for forward iterates. The manifolds constructed in that way are one-to-one images of the spaces $\mathbb{R}^k, \mathbb{R}^{n-k}$ where this embedding can have quite a complicated form.

Continuous systems.

The stable and unstable manifolds are defined analogously for continuous systems. Let p be an equilibrium point of the system $\dot{x} = f(x)$ of differential equations, where $x \in \mathbb{R}^n$. To study the behavior of the orbits close to p we use the linearized system $\dot{x} = Ax$, where A is the Jacobian matrix $Df(p) = [\partial f_i / \partial x_j](p)$ at the point p. The point p is hyperbolic if the real parts of the eigenvalues of the matrix A are different from zero. In this case the eigenvalues λ are splitted into two parts: The stable ones for which $Re\lambda < 0$ and the unstable ones for which $Re\lambda > 0$. Let $\{v_1, \ldots, v_k\}$ be the eigenvectors corresponding to the stable part and $\{v_{k+1}, \ldots, v_n\}$ the ones corresponding to the unstable part. The eigenspace E^s spanned by the vectors $\{v_1, \ldots, v_k\}$ is called stable and the eigenspace E^u spanned by $\{v_{k+1}, \ldots, v_n\}$ is called unstable.

Theorem 176. [103] *Let p be a hyperbolic equilibrium point for the system $\dot{x} = f(x)$. Then there exist local stable and unstable manifolds*

$W^s_{loc}(p) = \{x \in U : \phi(t, x) \to p \text{ for } t \to +\infty\}$,

$W^u_{loc}(p) = \{x \in U : \phi(t, x) \to p \text{ for } t \to -\infty\}$, *in a neighborhood U of the point p, where $\phi(t, x)$ is the solution to the system with $\phi(t, 0) = x$.*

The subspaces E^s and E^u are tangent spaces to the manifolds $W^s_{loc}(p)$ and $W^u_{loc}(p)$ at the point p.

The global stable and unstable manifolds at the equlibrium point p are defined analogously:

$W^s(p) = \{x \in \mathbb{R}^n : \phi(t, x) \to p \text{ for } t \to +\infty\} = \bigcup_{k \in R_-} \phi(t, W^s_{loc}(p))$,

$W^u(p) = \{x \in \mathbb{R}^n : \phi(t, x) \to p \text{ for } t \to -\infty\} = \bigcup_{k \in R_+} \phi(t, W^u_{loc}(p))$,

where $R_- = \{t \leq 0\}, R_+ = \{0 \geq t\}$.

Homoclinic trajectories.

Let the point p be a hyperbolic fixed point of a diffeomorphism. The invariant manifolds $W^s(p)$ and $W^u(p)$ not only determine the local dynamics but also essentially effect the global dynamics. In particular, this effect can be noticed when stable and unstable intersect. We say that two manifolds W_1 and W_2 intersect transversally at the point $q \in W_1 \cap W_2$ if the sum of tangent spaces $TW_1(q) + TW_2(q)$ at the point q form the whole space $\mathbb{R}^n$. To understand this we consider the following example.

Example 177. Heteroclinic points.

We consider the diffeomorphism $f : (x, y) \to (x + y, y + \cos(x + y))$. This map has the fixed points $A(-3\pi/2, 0)$ and $A(\pi/2, 0)$. Numerical experiments show that the stable manifold $W^s(A)$ intersects the unstable manifold $W^u(B)$ at a point q.

The points of the intersection $W^s(A) \cap W^u(B)$ are called heteroclinic. The existence of such a point of intersection forces the manifolds $W^s(A)$ and $W^u(B)$ to intersect at points on the backward orbit $\{f^k(q),\ k < 0\}$, with the limit point B. The stable manifolds $W^s(A)$ and $W^s(B)$ of different points cannot intersect and because of that the stable manifold $W^s(A)$ must be stretched (and oscillating) along $W^s(B)$ as shown in the Fig. 15.1. The unstable manifolds $W^u(B)$ and $W^u(A)$ have similar behavior close to the point B.

When H. Poincarè studied the motion of planets he showed that cases where the points A and B coincide are possible, that is there is a hyperbolic point A with intersecting stable and unstable manifolds. H. Poincarè called such trajectories double asymptotic. Now such intersection points $W^s(A) \cap W^u(A)$ are called homoclinic.

The dynamics around a homoclinic point is very complicated, hard to predict and has stochastic character. Such behavior is called *chaos*.

Example 178. Homoclinic orbits.

As an example we consider the map of the form

$$(x, y) \to (x + y + ax(1 - x), y + ax(1 - x)),$$

where $a = 1.35$. The fixed point $A(0, 0)$ is hyperbolic with stable and unstable manifolds intersecting transversally (more exact the right hand separatrices) at a point H with coordinates $(1.3837, 0.000)$ at an angle 0.099 radians, see Fig. 15.2.

S Smale [136] showed that near to a transversal homoclinic trajectory there exists an invariant set Ω and the trajectories of the points of this set can be coded by all possible sequences of two symbols. In particular, periodic sequences correspond to periodic trajectories of the same period. This implies

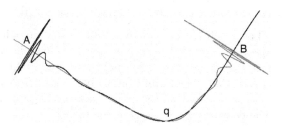

Fig. 15.1. Heteroclinic orbits

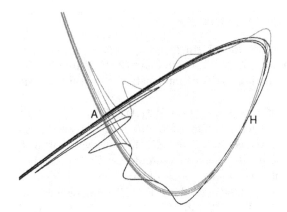

Fig. 15.2. Homoclinic orbits

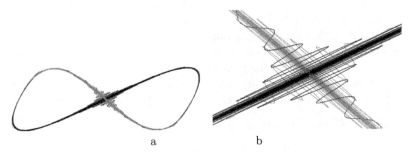

a   b

Fig. 15.3. Double homoclinic hyperbolic orbit

that there exist periodic trajectories of any large period and the set of periodic points are dense in Ω. The invariant set Ω is homeomorphic to a product of two cantor sets. Thus the existence of a transversal homoclinic point implies chaotic dynamics.

Example 179. Double homoclinic hyperbolic orbit.

We consider one example more, where the left and right separatrices of the stable and unstable manifolds form two homoclinic cycles. The map f has the form

$$(x, y) \to (x + y + ax(1 - x^2), y + ax(1 - x^2)),$$

where $a = 0.4$. The origin $(0,0)$ is a hyperbolic point and its left (right) separatrices intersect, see Fig. 15.3,a. Such a pair of homoclinic cycles generate very complicated dynamics near the fixed point, see Fig. 15.3,b.

One should notice that such complicated dynamics arises in systems of dimension 2 at the same time as the maps have good properties and are simple enough. The coordinate functions of the map

$$(x, y) \to (x + y + ax(1 - x), y + ax(1 - x)),$$

are, for example, polynomials of second order. The map preserves areas because its Jacobian $\det(Df)$ is equal to 1.

Our aims are to give methods for constructing the global manifolds we described, to calculate the coordinates of homoclinic points and also to estimate the angle between the stable and unstable manifolds where they intersect.

15.2 Local Invariant Manifolds

We notice that the unstable manifold of some point for the inverse map f^{-1} is a stable manifold for the point for the map f itself. The construction of the stable manifold is analogous to the construction of unstable manifolds. The difference is that in the construction of the unstable manifold we use the map f itself but in the construction of the stable manifold we use the inverse map f^{-1}. Thus we consider only the construction of unstable manifolds. We consider a hyperbolic fixed point of the diffeomorphism $f : \mathbb{R}^n \to \mathbb{R}^n$. We put the origin at the fixed point O and we choose the stable and unstable subspaces $E^s(0)$ and $E^u(O)$ as the coordinate planes. Let $x \in E^u(O)$ and $y \in E^s(0)$.

Now we can formulate a constructive theorem about the unstable manifold. Constructivity means that the proof allows us to construct a sequence of manifolds converging to the manifold we seek for.

Theorem 180. [102] *Let f be a diffeomorphism and 0 a hyperbolic fixed point. There exists a neighborhood $U = \{|x| < \delta\} \times \{|y| < \delta\}$ of the point 0 such that the local unstable manifold $W^u_{loc}(0)$ has the form*

$$W^u_{loc}(0) = \{(x,y) : |x| < \delta, \ y = h(x)\}$$

in U. Moreover, the iterates of any manifold L given in the form

$$L = \{(x,y) : |x| < \delta, \ y = p(x)\}$$

converge to the unstable manifold in C^1-topology, that is
$$W_k = f^k(L) \bigcap U \to W^u_{loc}(0) \text{ for } k \to \infty.$$

The convergence in the C^1-topology means that the manifolds $W_k = f^k(L) \bigcap U$ have the form

$$W_k = \{(x,y) : |x| < \delta, \ y = p_k(x)\}$$

and the C^1-norm of the difference of the functions

$$||p_k - h|| = \sup_{|x| < \delta} \{|p_k(x) - h(x)|, \ |Dp_k(x) - Dh(x)|\}$$

converges to zero for $k \to \infty$.

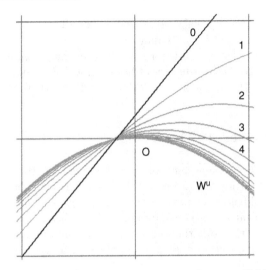

Fig. 15.4. The convergence to invariant manifold

Thus we get an algorithm for computing approximations for the unstable manifold and the approximations will be as good as the computer and the time of computation allow. In practical computations, the manifold L is chosen as a hyperplane. For example, if the unstable manifold is a one dimensional curve then L can be taken as a segment intersecting the stable manifold transversally.

Example 181. The convergence to invariant manifold.

We consider the diffeomorphism of the plane

$$f(x,y) = (1.1x - 0.1y\sin(x),\ 0.7y - 0.5x^2).$$

The origin is a hyperbolic fixed point. We take the neighborhood $U = [-0.5, 0.5] \times [-0.5, 0.5]$. The unstable manifold of the fixed point coincides with the x-axis and the stable one with the y-axis. The intial segment L does no pass through the origin, it is denoted by the number 0 in the Fig. 15.4, the following iterates are denoted by numbers 1, 2, 3, 4 and the local unstable manifold is denoted by W^u. The figure shows that the convergence described in the theorem is good enough and already the tenth iterate is not a bad approximation for the unstable manifold.

Anyhow for our purposes the construction of local manifolds is not sufficient and because of that we start to construct global manifolds.

15.3 Global Invariant Manifolds

In order to define the distance between manifolds we introduce coordinates in a neighborhood of the global unstable manifold and define the distance between this manifold and the approximations to it. Let M be a smooth submanifold of $\mathbb{R}^n$ of dimension m. At each point $x \in M$ we consider a hyperplane $N(x)$ of complementary dimension $n - m$ such that $N(x)$ is transversal to M at the point x. Without loss of generality we can assume that $N(x)$ is smoothly dependent of x. It is clear that the choice of $N(x)$ is not unique. From the construction follows that the set $E = \{N(x),\ x \in M\}$ has the structure of a vector fiber bundle on the manifold M and it is natural to identify the manifold M with the null section $\{(x, 0)\}$. We construct a special neighborhood of the manifold M in the following way. We define $(x \oplus y) = x + y$, where $x \in M, y \in N(x)$. We fix an arbitrary $\varepsilon > 0$. The set $U_\varepsilon = \{(x \oplus y) : x \in M, y \in N(x), |y| < \varepsilon\}$ is called a tubular neighborhood of the manifold M.

Proposition 182. [143] *For any compact manifold M there exists an $\varepsilon^* > 0$ such that the map $(x, y) \to (x \oplus y)$, $|y| < \varepsilon$ induces smooth coordinates in a tubular neighbourhood of M for any $\varepsilon : 0 < \varepsilon < \varepsilon^*$.*

Now we can define what is meant by saying that some manifolds are C^1-near to M and we can also define the distance between such manifolds. Let N_ε be a tubular neighborhood of M. Let the manifold $M_1 \subset N_\varepsilon$ be given in the form

$$M_1 = \{x \oplus h(x) :\ h(x) \in N(x)\},$$

where $x \oplus h(x) : M \to \mathbb{R}^n$ is a smooth map. Such M_1 are said to be in a C^1-neighborhood of M. The C^1-distance is defined as the C^1-norm of the map

$$\rho(M, M_1) = ||h||_1 = ||h|| = ||Dh|| = \sup_M |h(x)| + \sup_M \max_{|v|=1} |Dh(x)v|.$$

The global manifold W^u is in general not compact. Anyhow, without loss of generality, we can assume that the local unstable manifold W^u_{loc} is a closed disc with diameter small enough. The manifold M is said to be a compact part of the global unstable manifold W^u if M is the image of the local manifold for the kth iterate, that is $M = f^k(W^u_{loc})$. One can say that the compact part of the unstable manifold is a disc of greater, but finite diameter. The following theorem is fundamental for the algorithm for construction of the global unstable manifold.

Theorem 183. *Let 0 be the hyperbolic fixed point of the diffeomorphism $f : \mathbb{R}^n \longrightarrow \mathbb{R}^n$ and let the coordinates described above be introduced in a neighborhood of the point 0. Then the iterates of any manifold L of the form*

$$L = \{(x, y) :\ |x| < \delta,\ y = p(x)\}$$

converge to the compact part M of the unstable manifold W^u in the C^1-topology, that is $W_m = f^k(L) \bigcap N_\varepsilon \to M$ for $m \to \infty$. The iterates converge pointwise to the global unstable manifold W^u.

The theorem is a corollary of Theorem 180 about the local manifold and the uniform convergence of the k-th iterate f^k on the compact set. Analogous theorems are valid for the stable manifold W^s if the inverse map f^{-1} is used instead of f.

The numeric realization of the algorithms have the following limiting factors.

1) The manifold M is numerically given by the coordinates of the points $A_p \in M$ and the number can be great enough but finite.
2) The number of numerical iterates is also finite.
3) The iterates $f^m(A)$ of a given point $A \in L$ leave a neighborhood of the equilibrium point and the tubular neighborhood of any compact part M of the unstable manifold.

15.4 Separatrices for a Hyperbolic Point

We consider the methods and problems of the numerical realizations of the given algorithms in an example where we construct the separatrices of a hyperbolic fixed point of a diffeomorphism of the plane. In the computer realization it is natural to present a curve in the form of a broken line with corners on the curve. The method above for constructing the iterates of a curve reduced to the method of iterating the corners of the broken line. Anyhow, in general, the length of each segment of the broken line grows when iterated. Consequently the constructed broken line cannot be a good approximation to the separatrice for high iterates. Another difficulty is that iterates high enough of the point leave the tubular neighborhood of the compact part of the unstable manifold. To avoid these difficulties we can add more corners to the broken line and look at the iterates not leaving the tubular neighborhood. To add more corners we take a number $h > 0$ small enough and will construct the broken line so that the length of any segment will not exceed this number. The choice of the parameter h is done by the user to get required accuracy of the approximation. If for some iterate the length of a segment exceeds h, we will add more points so that the lengths will not exceed h. Let us look at this procedure in more details. Assume that the approximating broken line has a segment $[AB]$ such that the distance between $f(A)$ and $f(B)$ are greater than h (see Fig. 15.5). Then a new point C on the segment $[AB]$ is added and we will assume that the broken line has two segments $[AC]$ and $[CB]$ instead of the segment $[AB]$. Usually the point C is the midpoint of the segment. If the distances between the points $f(A)$ and $f(C)$ and between the points $f(B)$ and $f(C)$ are less than h then the dividing procedure is stopped, otherwise it is necessary to divide all remaining parts of length greater than h. Thus a new broken line where all segments have length less than h is constructed.

The parameter h can be chosen not only from exactness point of view. For best graphical realization one can use that curves and broken lines are

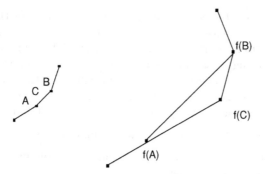

Fig. 15.5. The dividing procedure

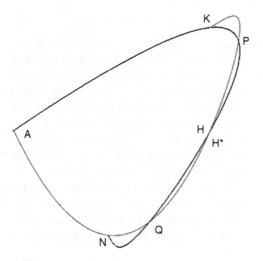

Fig. 15.6. Calculating homoclinic points

indistinguishable on the screen (that is, on the screen we get an entire curve) if the distance between the corners of the broken line is less than one pixel. In other words, on the screen the broken line will look continuous. To avoid unnecessary calculations we can exclude the part of the broken line with non-interesting images outside the chosen neighborhood. The curves in our figures are constructed in this way.

Calculating homoclinic points. To find the coordinates of a homoclinic point we proceed in the following way. By the method above we construct the approximations for $L^s(A)$ and $L^u(A)$ for the stable and unstable separatrices for the hyperbolic point A we have chosen. We find the points of intersection different from A. Such points approximate the homoclinic points. The user chooses one of these points, say H. Thus we get two parts $L^s[AH]$ and $L^u[AH]$ on the approximating stable and unstable separatrices. All other parts of the constructed separatrices can be excluded. The length of the unstable curve

$f(L^u[AH])$ increases for positive iterates and the length of the stable curve $f^{-1}(L^s[AH])$ also increases for negative iterates. The two constructed curves have five intersection points P, Q, K, N and H^*. The middle point H^* near to H is the following approximation to the homoclinic point. Repeating this procedure we get a sequence of points approximating the homoclinic point. For each step we can get an estimate for the angle at the intersection between the stable and unstable curves $f^{-1}(L^s(A))$ and $f(L^u(A))$ at the point H^*.

Example 184. Approximations for the homoclinic points.

We use this method to construct the separatrices of the hyperbolic point for the system generated by the map

$$(x, y) \rightarrow (x + y + ax(1 - x), y + ax(1 - x)),$$

where $a = 1.35$.

The fixed point $A(0,0)$ is hyperbolic. By the method we build the approximations $L^s(A)$ and $L^u(A)$ for the separatrices $W^s(A)$ and $W^u(A)$. The right parts of these separatrices intersect at 8 points. The coordinates of the points and the intersection angles are given in Table *. We chose the point H with the coordinates $(1.38, 0.0)$. The choice can be realized in two ways, in an analytic way by giving the coordinates and graphically. To do this it is enough to move the mouse pointer near to H. Then let the computer choose the nearest intersection point and cut away unnecessary parts from the constructed separatrices and obtain new approximations L^s and L^u.

The image $f(L^u)$ and preimage $f^{-1}(L^s)$ are constructed at the next stage and again the points and intersection angles are found. In particular, the intersection angle at the point $H(1.3837, 0.000)$ is estimated as 0.0992 radians. The given values for the coordinates and angles are stabilized after the forth iterate.

The method allows to find the parameters where there is a tangency between the stable and unstable separatrices. In this case one speaks about the existence of a homoclinic tangency point. It is known [91] that a homoclinic tangency point generates even more complicated dynamics than for transversal intersections.

Table *

Approximations for the homoclinic points		
x	y	Angle
0.666939952599	0.416901386849	0.587379298048
0.959852011034	0.571550174086	0.401180709137
1.383708506121	0.716775451922	0.442693793725
1.583426354446	0.623572155556	0.459078110142
1.383712323195	0.000001269755	0.099161518436
0.950843544939	−0.623581041515	0.223296974712
0.666938368677	−0.716778322800	0.873901092401
0.388301312760	−0.571550964455	0.695533208837

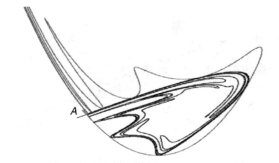

Fig. 15.7. A homoclinic tangency point

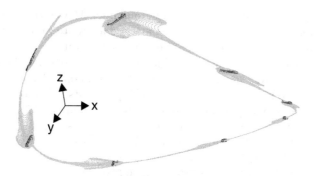

Fig. 15.8. The Möbius band

Example 185. A tangency between the stable and unstable manifolds.

We found that the stable and unstable separatrices has a tangency for the map $f(x, y) = (bx + y + ax(1 - x), y + ax(1 - x))$ for $a = 1.511717565$ and $b = 0.93$. (See Fig. 15.7)

The method we described for construction of iterates of curves can be successfully applied to the construction of manifolds in the space. As an example we consider a discrete system modeling the dynamics of a three dimensional food chain.

Example 186. The dynamics of a food chain.

The dynamical system is generated by the map

$$(x, y, z) \to (axe^{-y}/(1 + x\max\{e^{-y}, g(z)d(y)\}),\ bxyg(y)e^{-z}g(dyz),\ cyz),$$

where $g(x) = (1 - e^{-x})/x$, x is the size of the vegetation, y the size of the herbivore population and z the size of the carnivore population. This example is chosen from [73] where $c = d$. We choose the parameters $a = 4$, $b = 1$, $c = 3$, $d = 4$, that is, we study a modification of the system from the original paper. For these parameters the system has a two dimensional invariant manifold M which is Lyapunov stable (see Fig. 15.8). Below we show that M

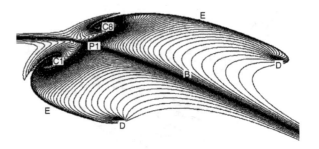

Fig. 15.9. A piece of the Möbius band

is homemorphic to a Möbius band. The manifold M was constructed using iterates of a specific chosen curve. This method allows to find the form of the manifold and get information about the dynamics on it.

To show that M is a non-orientable manifold we consider the restriction of the dynamical system on this manifold. The manifold consists of eight similar parts (see Fig. 15.9) Each part is mapped to its neighbor so that after eight iterates it is mapped into itself. Thus the manifold $M = \bigcup M_i$, $i = 1, 2, ..., 8$ consists of eight homeomorphic submanifolds. On M we found the eight periodic hyperbolic orbit P of period 8:

P1(0.963,0.434,0.522), P2(1.532,0.131,0.683),
P3(2.293,0.081,0.272), P4(2.716, 0.130, 0.070),
P5(2.751, 0.302, 0.024), P6(2.430,0.690,0.021),
P7(1.776, 1.150, 0.048), P8(1.108, 1.035, 0.1670).

Examining the map f^8 we found that the exact coordinates of P1 are

x=0.9610743649183222, y=0.4379625398371493, z=0.5200789910152929

and the eigenvalues of the differential $Df^8(P2)$ are:

$$\lambda_1 = -1.12863, \quad \lambda_2 = 0.893163, \quad \lambda_3 = -0.0808189.$$

Moreover the eigenvectors where found:

$$e_1 = (0.502661, -0.403353. - 0.764616),$$
$$e_2 = (0.205838, -0.834893, -0.510475),$$
$$e_3 = (-0.854039, 0.0768188, 0.514506)$$

The invariant subspace spanned by e_1 and e_2 is a tangent space $TM(P1)$ to M at the point $P1$. The differential $Df^8(P1)$ restricted to $TM(P1)$ changes orientation because

$$\det(Df^8(P1)|_{TM(P1)}) = \lambda_1 \lambda_2 < 0.$$

At the same time, from continuity follows that the map $f : M_i \rightarrow M_{i+1}$ saves orientation. Such a contradiction is possible only if the manifold M is non-orientable. There exists only one non-orientable two-dimensional manifold which is the Möbius band.

The unstable manifold of an 8-periodic hyperbolic orbit P end at 16-periodic source C. Thus $W^u(P1)$ ends at C1(0.9343, 0.4711, 0.4689) and C8(0.9790, 0.4340, 0.5483). The map f^8 changes orientation and maps C1 to C8. The eigenvalues of the differential $Df^{16}(C1)$ where found:

$$\lambda_{1,2} = 0.923074 \pm i0.207652, \quad \lambda_3 = 0.00846026.$$

Thus C is a stable focus on the manifold M as seen in the figure. Moreover, the orientations of the rotations at the points $C1$ and $C8$ are the opposite. Using this fact it is also possible to prove that M is non-orientable because along the band M we preserve the rotation orientation on the curve C, and moving across M we can change the rotation orientation.

There exist an 8-periodic hyperbolic orbit B on M through the point $B1 = 2.47929, 0.644334, 0.0226021$ where the differential $Df^8(B1)$ has the eigenvalues:

$$\lambda_1 = 1.11521, \quad \lambda_2 = -0.774957, \quad \lambda_3 = -0.113216$$

The corresponding eigenvectors are:

$$e_1 = (-0.190361, 0.70823, 0.679833),$$
$$e_2 = (0.0971897, 0.0770054, 0.992282),$$
$$e_3 = (0.514434, 0.505962, 0.692358).$$

Moreover there exists a 16-periodic hyperbolic orbit E through the point E1(2.75395, 0.29734, 0.0286872) with eigenvalues

$$\lambda_1 = 0.923074 + 0.207652i, \quad \lambda_2 = 0.9233074 - 0.0207652i, \quad \lambda_3 = 0.00846026.$$

The unstable manifold of the orbit B intersects with the stable manifold of the orbit P so that $W^u(B)$ oscillates around $W^s(P)$. Hereby, the unstable manifold $W^u(B)$ approaches the unstable manifold of A without hitting it, and eventually also leads to C, see Fig. 15.10. The system has two 16-periodic stable orbits C and D which are minimal attractors.

15.5 Two-dimensional Invariant Manifolds

The two dimensional manifold M is well approximated by the polytope P such that any side is a triangle. Thereby it is possible to assume that the corners of such a polytope are on the given manifold. Here is a direct analogy to the approximation of M with a broken line L which corners are on the given curve. We will not consider the technical part of such approximation because in our case the polytope P is constructed together with the iterates of the chosen surface. Therefore we now consider the construction technology of the iterates of the surface and its approximations. To control the accuracy of the approximation we fix a parameter $h > 0$, which is an estimate for the

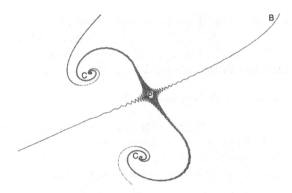

Fig. 15.10. Oscilation of the unstable manifold $W^u(B)$

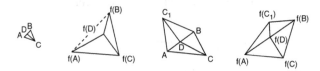

Fig. 15.11. Approximation of surface by polytope

length of the edges of the triangle T on the polytope P. The parameter h is used to get the necessary accuracy of approximations. Let the initial polytope P_0 be T_0. We will assume that the image of the triangle ΔABC for f is a new triangle $\Delta f(A)f(B)f(C)$. It is clear that this triangle does not coincide with the real image $f(\Delta ABC)$. However, if the corners of the triangle ΔABC are on the surface M then the corners of the triangle $\Delta f(A)f(B)f(C)$ are on the image $f(M)$ of the manifold M. If after an iterate the length of one of the sides exceeds h then we divide the triangle so that the lengths do not exceed h. We consider this procedure in detail.

Let there be an edge $[AB]$ in the triangle ΔABC such that the distance $\rho(f(A), f(B))$ is greater than h (see Fig. 15.11). In this case we add a point D on the edge $[AB]$ and assume that instead of the triangle ΔABC we have two triangles ΔADC and ΔDBC. If the edges of the triangles $\Delta f(A)f(D)f(C)$ and $\Delta f(D)f(B)f(C)$ are less than h the dividing procedure is stopped, otherwise we have to again divide an edge with length greater than h.

To optimize the calculations the procedure of adding more points and triangles ought to be simplified. We notice that if the edge $[AB]$ is not on the boundary of the polytope then it also belongs to another triangle ΔABC_1. Then adding a new point to $[AB]$ divides two triangles ΔABC and ΔABC_1. Because of this it is more comfortable to work with the edges instead of triangles. This reduces the number of operations and need for memory. It is convenient to save the data in the form of a graph with the coordinates of the corners. The parameter h can be chosen not only from the exactness point of view. If only the corners of the polytope are plotted on the screen then

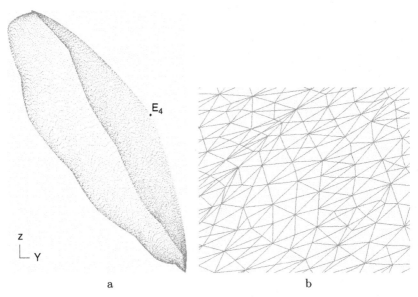

Fig. 15.12. Construction of the invariant manifold $W^u(E_4)$

the surface we get will more transparency for greater h than for smaller. The polytope can also be plotted with its edges which might be useful. In order not to do unnecessary calculations we can exclude the part of the surface which is not of interest, for example, if it leaves the region we study. As an example we apply the described technique to construct an unstable manifold of the food chain.

Example 187. Construction of the two-dimansional invariant manifold.

Let us consider the food chain dynamics

$$(x,y,z) \rightarrow (axe^{-y}/(1 + x\max\{e^{-y}, g(z)d(y)\}),\ bxyg(y)e^{-z}g(dyz),\ cyz),$$

where $g(x) = (1 - e^{-x})/x$. We choose the parameters $a = 2.97, b = 1, c = d = 4$. For these parameters the system has single fixed point E_4 in the domaine $\{x > 0,\ y > 0,\ z > 0\}$. The point E_4 is hyperbolic with 2-dimensional unstable invariant manifold. This manifold is shown in the Fig. 15.12,a where we can see the vertices of an approximating polytope with the parameter $h = 0.01$. The Fig. 15.12,b shows a local structure of the approximating polytope.

16

Ikeda Mapping Dynamics

The Ikeda map I we study is given by

$$I : z \mapsto R + C_2 z \exp(i(C_1 - C_3/(1 + |z|^2))), \quad z \in C, \qquad (16.1)$$

where C is the complex plane of the variable $z = x + iy$ and R, $C1$, $C2$, and $C3$ are real constants (mapping parameters). The Ikeda map occurs in the modeling of optical recording media (crystals) [60]. The numerical results obtained to date (see [94], [120], [48], [37], [144]) show that under certain parameter values the Ikeda map exhibits highly complicated dynamical behavior. In particular, the Ikeda map can have infinitely many hyperbolic periodic orbits, which are located in a bounded part of C, and a strange attractor (the Ikeda attractor). We also consider the modifications of the Ikeda map like mappings reversing orientation and hyperbolic. The aim of the chapter is to give an analysis of the topological structure of orbits by symbolic dynamics methods (the package ASIDS) and by iterations of curves (the package Line). We also present an analysis of orbit behavior near fixed and periodic points and of bifurcations that lead to chaotic attractors as parameters vary.

16.1 Analytical Results

In this section we give some simple analytical results on the Ikeda map we need in the sequel. In the real notation the Ikeda map takes the form

$$I : (x, y) \mapsto (R + C_2(x \cos \tau - y \sin \tau), C_2(x \sin \tau + y \cos \tau)), \quad (16.2)$$

where $\tau = C_1 - C_3/(1 + x^2 + y^2)$. Some obvious properties of the Ikeda map are listed below.

1. The map I can be viewed as a composition of the three diffeomorphisms T_1, T_2, and T_3 of the plane onto itself:

$$I = T_3 \circ T_2 \circ T_1,$$

where $T_1(x,y) = (x \cos\tau - y \sin\tau, x \sin\tau + y \cos\tau)$ is a rotation through the angle $\tau = \tau(r), r^2 = x^2 + y^2$, $T_2(u,v) = (C_2 u, C_2 v)$ is a linear homothetic, and $T_3(s,t) = (R + s, t)$ is a translation along the real axis.

2. If $C_2 > 0$ then I is an orientation preserving diffeomorphism of the plane onto itself.

3. If $|C2| < 1$ then the map I is dissipative, i.e. there exists an $h > 0$ such that
$$\lim_{n\to\infty} \sup ||I^n(x,y)|| < h$$
for each point (x,y).

4. If $|C_2| < 1$ then every disk $K_r = \{(x,y) : x^2 + y^2 < r^2\}$ with the radius $r > |R|/(1 - |C2|)$ is mapped into itself, i.e. $I(K_r) \subset int\, K_r$.

5. For every point (x,y) the Jacobian of I is of the form $\det DI(x,y) = C_2{}^2$. Thus, if $|C_2| < 1$ then I contracts the area, i.e. for the Lebesque measure of every bounded measurable set U we have
$$mes\, I(U) \le mes\, U.$$

Let $|C_2| < 1$. The properties listed above imply the following:

1. Every bounded invariant set $U(I(U) = U)$ is contained in the disk $K(r^*)$ with the radius $r^* = |R|/(1 - |C_2|)$. Let A_g be the maximal bounded invariant set of I contained in $K(r^*)$:
$$A_g = \bigcap_{n=0}^{\infty} I^n(K(r^*)).$$

It is well known that the set A_g is closed connected and asymptotically stable in the large, i.e. A_g is a global attractor. By 5, A_g has measure zero: $mes\, A_g = 0$.

2. The behavior of orbits of I is entirely determined by the behavior of orbits from A_g. In particular, periodic, nonwandering, and chainrecurrent orbits of I are contained in A_g. Results of numerical explorations mentioned above indicate that under certain parameter values the diffeomorphism I can have infinitely many hyperbolic periodic orbits with periods tending to infinity. This leads to the existence of homoclinic orbits and indecomposable continua in A_g. The last means that A_g has a very intricate topological structure.

16.2 Numerical Results

Numerical simulations of the dynamical behavior of the map I have been carried out with $C_1 = 0.4$, $C_2 = 0.9$, $C_3 = 6.0$. The parameter R takes the values within the segment $[0; 1.1]$ increasing by $R = 0.01$. For each value of R, phase portraits are indexed by small letters a, b, anew. The obtained values will be given in approximations. Results of the numerical study are the following.

As R increases from 0 to approximately 0.367, the global attractor A_g is a single asymptotically stable fixed point, i.e. I offers the convergence property.

16.2.1 $R = 0.3$

The Ikeda map has the fixed point $A_0(0.1766, 0.2298)$. This fixed point attracts all other orbits.

16.2.2 $R = 0.4$

The Ikeda map has three fixed points: the fixed point $A_0(0.2280, 0.2568)$, the hyperbolic saddle point $H_0(3.0508, -1.6442)$, and the stable focus $S_0(3.7763, 0.8930)$, see Fig. 16.1,a where the global attractor of the map is shown. The unstable manifold $W^u(H_0)$ of H_0 consists of two separatrices; the limit set of the left separatrix is the sink $A_0{}^* = A_0$ and the limit set of the right one is S_0. However, while S_0 is a regular focus, the sink $A_0{}^*$ has a sufficiently complicated topological structure (see Fig. 16.1,b). The stable manifold $W^s(H_0)$ of the saddle H_0 separates the basins $W^s(A_0{}^*)$ and $W^s(S_0)$ of the attractors $A_0{}^*$ and S_0.

16.2.3 $R = 0.5$

While R increases from $R = 0.4$ to $R = 0.5$ the sink A_0^* bifurcates to the attractor A which when $R = 0.5$ contains the sink $A_0(0.2784, 0.2734)$, the period

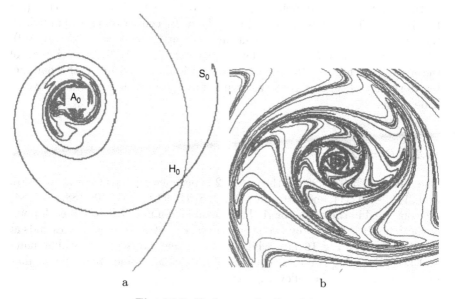

Fig. 16.1. Ikeda map for $R = 0.4$

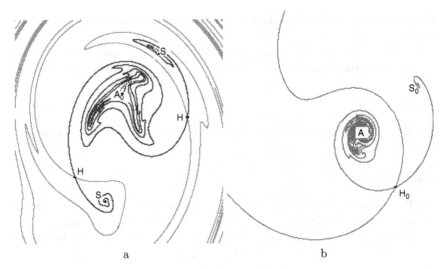

Fig. 16.2. Ikeda map for $R = 0.5$

2 sink $S(0.0897, -0.7195), (0.6758, 0.6141)$, and the period 2 hyperbolic saddle $H(1.0017, 0.0376), (-0.2517, -0.4987)$ (see Fig. 16.2,a). The unstable separatrices $W^u(H)$ of H ends at A_0 and S. The closure of the unstable manifold $W^u(H)$ (colored dark) coincides with the attractor $A = W^u(H) + A_0 + S$. The stable manifold $W^s(H)$ (colored light) separates the basins of attraction of A_0 and S. The basin boundary of A is formed by the stable manifold $W^s(H_0)$ of the hyperbolic fixed point H_0 at approximately $(2.2330, -2.3346)$ (see Fig. 16.2,b). The unstable manifold $W^u(H_0)$ of H_0 consists of two separatrices, the left one ends at A and the right one ends at the sink $S_0(3.5231, 2.1942)$. The closure of $W^u(H_0)$ is the global attractor $A_g = W^u(H_0) + A + S_0$ of the map. This form of the global attractor is preserved up to the parameter value $R = 1$, except that the structure of the attractor A varies over a wide range.

16.2.4 $R = 0.6$

The sink $A_0(0.3397, 0.2809)$, the period 2 hyperbolic orbit $H(1.0094, -0.1100)$, $(-0.2110, -0.4211)$, and the period 2 sink $S(0.5997, 0.6757), (0.2188, -0.7184)$ are contained in the attractor A. The unstable manifold $W^u(H)$ of each point of the orbit H is formed by two separatrices, one of these separatrices ends at the sink A_0, (see Fig. 16.3,a), while the other one intersects the stable manifold $W^s(H)$, giving rise to a sequence of homoclinic points. Some homoclinic points are listed in the following list:

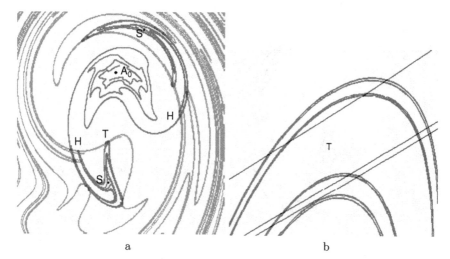

Fig. 16.3. Ikeda map for $R = 0.6$

$(x = 0.192905, y = -0.358028); (x = -0.208911, y = -0.421623);$

$(x = 0.193456, y = -0.357745); (x = -0.210681, y = -0.421222);$

$(x = -0.196432, y = -0.424266); (x = -0.210709, y = -0.421216);$

$(x = -0.197692, y = -0.420132); (x = -0.210990, y = -0.421152);$

$(x = -0.211047, y = -0.421139); (x = -0.210997, y = -0.421150);$

$(x = -0.208703, y = -0.421670); (x = -0.211034, y = -0.421142).$

The Figure 16.3,b where the stable $W^s(H)$ and unstable $W^u(H)$ manifolds are depicted, indicates the transverse character of intersections of these manifolds near T. Since at $R = 0.5$ the manifolds $W^u(H)$ and $W^s(H)$ are disjoint then there exists a parameter value $R^*, 0.5 < R^* < 0.6$, such that the manifold $W^u(H)$ is tangent to the manifold $W^s(H)$. The stable manifold $W^s(H)$ of the orbit H forms the boundary of the basins of attraction of the sink A_0 and the period 2 attractor A_2, which contains the period 2 sink S. In Fig. 16.4,c is shown the basin of attraction of A_2 (colored white grey). Its component containing the point $(0.2188, -0.7184)$ of the sink S is shown in Fig. 16.4,d.

The attractor A_2 is a closure of the unstable manifold $W^u(P)$ of the period 6 hyperbolic orbit $P(0.1869, -0.5785)$, $(0.3556, 0.7053)$, $(0.2818, -0.7800)$, $(0.6249, 0.6969)$, $(0.1343, -0.7635)$, $(0.8751, 0.4730)$. Each connected component of $W^u(P)$ consists of two separatrices, the one ends at the sink S, while the other one ends at the chaotic attractor A_3 (see Fig. 16.4,e). A_3 contains the attractor A_4, induced by the unstable manifold $W^u(Q)$ of the period 6 orbit Q. In Fig. 16.4,f are shown the point $(0.2056, -0.4874)$ of the orbit Q (depicted as a black dot) and its stable and unstable manifolds. The attractor A_4 is a

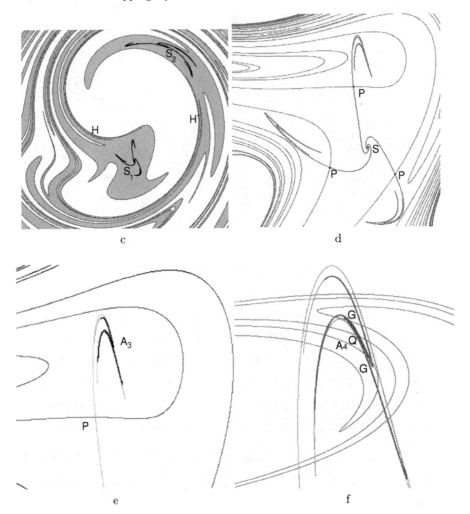

Fig. 16.4. Ikeda map for $R = 0.6$

closure of the unstable manifold $W^u(Q)$, which ends at the period 12 sink G. Fig. 16.4,f presents also two points $(0.2022, -0.4816)$ and $(0.2095, 0.4953)$ of the orbit G.

It is interesting to note that the stable and unstable manifolds are tangent at Q forming a sink. The global attractor A is a closure of the unstable manifold of the orbit $H : A = W^u(H) + A_2 + A_0$. The stable manifold $W^s(H_0)$ of the hyperbolic point $H_0(1.7660, -2.4891)$ is the common boundary of basins of attraction of A and the sink $S_0(3.3064, 2.8382)$. The displacement of A, H_0 and S_0 is similar to that in the cases $R = 0.5$ and $R = 0.7$.

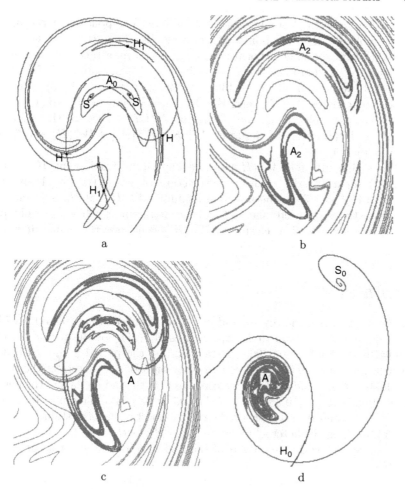

Fig. 16.5. Ikeda map for $R = 0.7$

16.2.5 $R = 0.7$

The Ikeda map with $R = 0.7$ has the inverse saddle fix point $A_0(0.3804, 02817)$ (see Fig. 16.5,a). The unstable manifold $W^u(A_0)$ of A_0 ends at the sink formed by a pair of the period 2 points $S(0.1548, 0.2030)$, $(0.6110, 0.2118)$ which is a minimal attractor. The inverse saddle point A_0 and the period 2 sink S arise from the sink A_0 while R varies from $R = 0.6$ up to $R = 0.7$. A closure of the unstable manifold $W^u(A_0)$ forms the attractor $A_1 = W^u(A_0) + S$. The Ikeda map reverse the orientation of $W^u(A_0)$ and hence the orientation of $W^s(A_0)$ is also reversed since the Ikeda map is orientation preserving. There exists the period 2 hyperbolic orbit $H_1(0.5772, 0.6788)$, $(0.3102, -0.7009)$ with transverse intersection of the stable $W^u(H_1)$ and unstable $W^u(H_1)$ manifolds forming the chaotic attractor $A_2 = W^u(H_1)$ (see Fig. 16.5,b). The attractor

A_2 has two connected components derived from components of the unstable manifold $W^u(H_1)$ for points of the orbit H_1. The attractor A_2 can be viewed as a two-periodic attractor since the Ikeda map takes one connected component of A_2 onto the other one. The unstable manifold $W^u(H)$ of the period 2 hyperbolic orbit $H(-0.1364, -0.3495)$, $(0.9931, -0.1676)$ is formed by two separatrices $W^u(H)_1$ and $W^u(H)_2$, which ends at the attractors A_1 and A_2, respectively. Thus, the closure of $W^u(H)$ makes up the attractor $A = A_1 + W^u(H) + A_2$ of the form $A = S + W^u(A_0) + W^u(H) + W^u(H_1)$ (see Fig. 16.5,c).

The stable manifold $W^s(H_0)$ of the hyperbolic fixed point $H_0(1.5062, -2.5002)$ separates the basins of attraction of the attractor A and the sink $S_0(3.1580, 3.2738)$. The unstable manifold $W^u(H_0)$ of H_0 is formed by two separatrices, the left one ends at the attractor A while the right one ends at the sink S_0. The closure of $W^u(H_0)$ generates the global attractor $A_g = A + W^u(H_0) + S_0$ (see Fig. 16.5,d).

16.2.6 $R = 0.8$

The Ikeda map has the inverse saddle A_0 at approximately $(0.4311, 0.2761)$ (see Fig. 16.6,a). Two unstable separatrices $W^u(H)_S$ of the period 2 orbit $H(0.9429, -0.1339)$, $(-0.0296, -0.2155)$ end at the period 2 sink $S(0.0387, -0.0345)$, $(0.8467, -0.0013)$ while two other ones $W^u(H)$ intersect the stable manifolds $W^s(A_0)$ and $W^s(H_1)$ (colored light) of the saddle A_0 and the period 2 hyperbolic orbit $H_1(0.3844, -0.6761)$, $(0.5798, 0.6644)$. The unstable manifolds $W^u(A_0)$ and $W^u(H_1)$ (colored dark) intersect in turn the stable manifold $W^s(H)$, forming the heteroclinic cycle $A_0 \rightarrow H_1 \rightarrow H \rightarrow A_0$ (see Fig. 16.6,a). The closure of unstable manifolds of the cycle generates the attractor A (see Fig. 16.6,b).

The attractor A contains the sink S and, hence, is not a minimal attractor. The basin of attraction $W^s(A)$ of A is bounded by the stable manifold $W^s(H_0)$ of the saddle fixed point $H_0(1.3219, -2.4527)$, the complement to the closure of $W^s(A)$ is the basin of attraction of the focus $S_0(3.0614, 1.6110)$. As above, the left unstable separatrix $W^u(H_0)_l$ of H_0 ends at the attractor A while the right one $W^u(H_0)_r$ ends at the sink S_0. The global attractor A_g is the closure of the unstable manifold $W^u(H_0)$ of the saddle $H_0 : A_g = W^u(H_0) + A + S_0$. We notice that at $R = 0.7$ the unstable manifold $W^u(A_0)$ ends at the sink S, whereas at $R = 0.8$ the sink S is the limit of the unstable separatrix $W^u(H)$, i.e. a bifurcation occurs.

16.2.7 $R = 0.9$

The Ikeda mapping with $R = 0.9$ has a chaotic minimal attractor named the Ikeda attractor. As R increases from $R = 0.8$ to $R = 0.9$, the following bifurcation occurs: the period 2 sink S and the period 2 hyperbolic orbit H disappear.

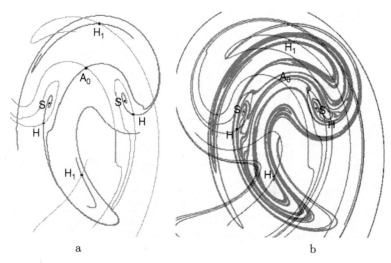

Fig. 16.6. Ikeda map for $R = 0.8$

The attractor A contains the inverse saddle $A_0(0.4819, 0.2645)$ and the period 2 hyperbolic orbit $H_1(0.5964, 0.6394)$, $(0.4497, -0.6453)$. The stable $W^s(A_0)$ and $W^s(H_1)$ and unstable $W^u(A_0)$ and $W^u(H_1)$ manifolds (separatrices) of these saddles intersect and form the heteroclinic cycle $A_0 \rightarrow H_1 \rightarrow A_0$ (see. Fig. 16.7,a) generating the chaotic attractor A which is the closure of the unstable manifolds $W^u(A_0)$ or $W^u(H_1)$. There exists a pair of the period 3 hyperbolic orbits $P_3(0.8091, 0.7834)$, $(0.9960, -1.0090)$, $(-0.0280, -0.8758)$ and $Q_3(1.3512, -0.0707)$, $(0.6568, -1.1932)$, $(-0.2418, -0.4462)$ (see Fig. 16.7,b). The stable and unstable manifolds of orbits P_3 and Q_3 intersect forming the heteroclinic cycle which also generates the attractor A. The closure of the unstable manifold of any one of the orbits A_0, H_1, P_3 or Q_3 is the attractor A (see Fig. 16.7,c). Outside the attractor A there is the saddle $H_0(1.1987, -2.3769)$ whose left separatrix $W^u(H_0)_l$ ends at the attractor A. The right unstable separatrix $W^u(H_0)_r$ ends at the sink $S_0(3.0027, 3.8945)$ (see Fig. 16.7,d). The stable manifold $W^s(H_0)$ of the saddle H_0 separates the basin of attraction $W^s(A)$ of the attractor A and the basin of attraction $W^s(S_0)$ of the sink S_0. The closure of the unstable manifold $W^u(H_0)$ generates the global attractor $A_g = A + W^u(H_0) + S_0$. The map has no other period 2 and period 3 orbits.

16.2.8 $R = 1.0$

When R goes from 0.9 to $R = 1.0$ the period 1, period 2, and period 3 orbits survive, except that their coordinates vary: when $R = 1.0$ (see Fig. 16.8,a) the inverse saddle A_0 is approximately $(0.5228, 0.2469)$, the period 2 hyperbolic orbit H_1 is approximately $(0.6216, 0.6059)$, $(0.5098, -0.6084)$, and the period 3 hyperbolic orbits P_3 and Q_3 are approximately $(0.7795, 0.7672)$, $(1.0140, -0.9832)$, $(0.0858, -0.8832)$ and $(0.6583, -1.1541)$, $(1.3297, -0.1427)$,

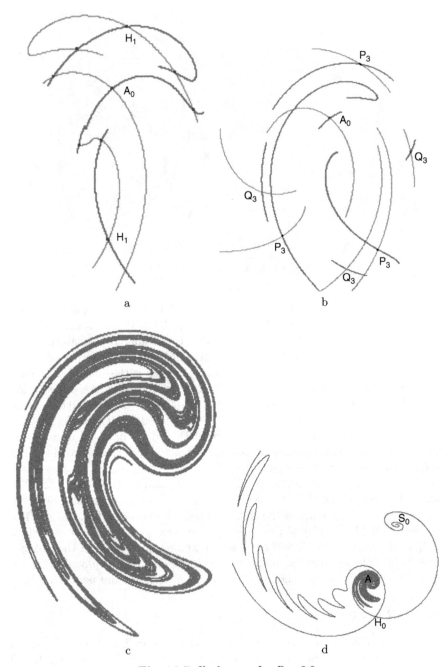

Fig. 16.7. Ikeda map for $R = 0.9$

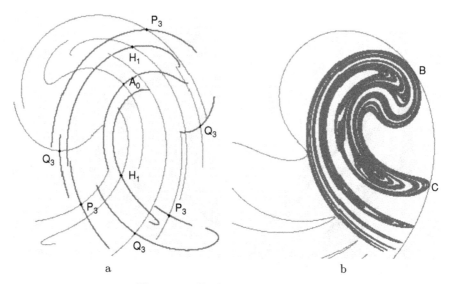

Fig. 16.8. Ikeda map for $R = 1.0$

$(-0.1353, -0.3756)$, respectively. The closure of unstable manifold of any orbit A_0, H_1, P_3 or Q_3 is an attractor A (see Fig. 16.8,b). The basin of attraction of A is bounded by the stable manifold $W^s(H_0)$ of the hyperbolic fixed point $H_0(1.1142, -2.2857)$ which is nearly tangent to A (see Fig. 16.8,b). The enlarged scale phase portraits (Figs. 16.9,c and 16.9,d) show that the distance between A and $W^s(H_0)$ near points B and C is yet positive. The stable manifold $W^s(H_0)$ is a common boundary of the basins of attraction of the attractor A and the sink $S_0(2.9721, 4.1459)$. The stable and unstable manifolds of H_0 are nearly tangent forming a sufficiently fine domain of attraction near points of "nearly tangency". The right separatrix $W^u(H_0)_r$ ends at the sink $S_0(2.9721, 4.1459)$ and the left one $W_u(H_0)_l$ approaches the chaotic attractor A.

16.2.9 $R = 1.1$

The mapping I has the following orbits with periods 1, 2, and 3: the inverse saddle $A_0(0.5837, 0.2232)$, the period 2 orbit $H_2(0.6525, 0.5641)$, $(0.5670, -0.5643)$, and the period 3 orbits $P_3(0.1906, -0.8730)$, $(1.0240, -0.9557)$, $(0.7718, 0.7342)$ and $Q_3(0.6660, -1.0738)$, $(-0.0110, -0.2430)$, $(1.2810, -0.1232)$. The relative positions of these orbits are similar to the case $R = 1.0$. The stable and unstable manifolds of the hyperbolic fixed point $H_0(1.05926, -2.1850)$ intersect transversally generating a homoclinic orbit (Fig. 16.10,a).

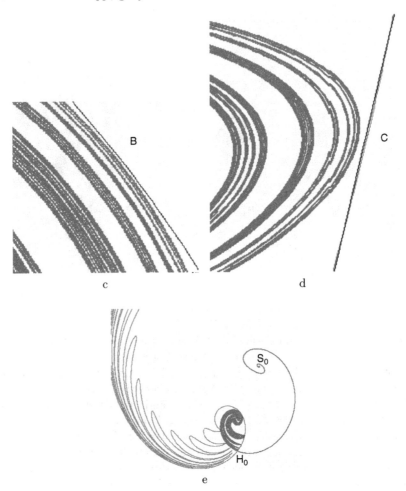

Fig. 16.9. Ikeda map for $R = 1.0$

Fig. 16.10,b displays the manner in which the manifolds $W^s(H_0)$ and $W^u(H_0)$ intersect near H_0. Furthermore, Fig. 16.10,a shows that the stable and unstable manifolds of A_0 and H_0 intersect generating a heteroclinic cycle. Thus, the attractor A fails when R goes from 1.0 to 1.1. The global attractor A_g is the closure of the unstable manifolds of H_0 or A_0. The right unstable separatrix $W^u(H_0)_r$ ends at the focus $S_0(2.9630, 4.3773)$. Moreover, all other unstable manifolds stretching along $W^u(H_0)_r$ approach S_0 as well. The set of chain recurrent points except for S_0 is the closure of points of intersection of $W^s(H_0)$ and $W^u(H_0)$, Fig. 16.11,d displays a neighborhood of the chain recurrent set.

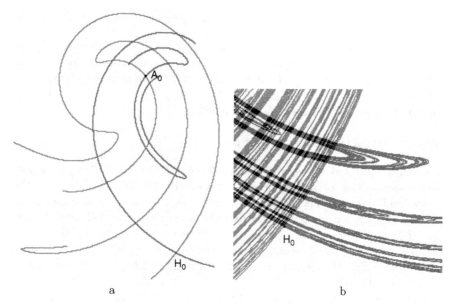

a   b

Fig. 16.10. Ikeda map for $R = 1.1$

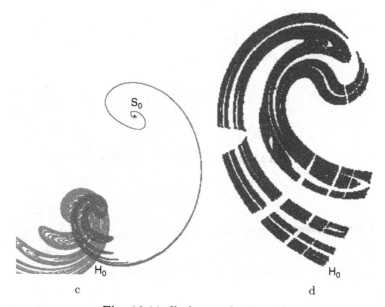

c   d

Fig. 16.11. Ikeda map for $R = 1.1$

16.3 Modified Ikeda Mappings

In this section we consider some possible modifications of the Ikeda mapping. With this aim in view, let us rewrite the Ikeda mapping in the form

$$J : (x,y) \mapsto (R + a(x\,\cos\tau - y\,\sin\tau), b(x\,\sin\tau + y\,\cos\tau)), \qquad (16.3)$$

where $\tau = 0.4 - 6/(1+x^2+y^2)$. For the normal Ikeda mapping $a = b = C_2$ and $C_2 \in (0,1)$, i.e. the mapping is an orientation preserving contraction. Now we will not assume $a = b$, in particular, a and b may be of opposite signs.

16.3.1 Mappings Preserving Orientation

Inverse Attraction: $R = 3$, $a = b = -0.9$

The mapping J has the hyperbolic fixed point $H(1.6030, 0.8268)$ with non-empty intersection of stable and unstable manifolds: $W^s(H) \cap W^u(H)$. The stable and unstable manifolds are nearly tangent at a homoclinic point (Fig. 16.12,a). Since $a = b < 0$, J revises the orientation of $W^s(H)$ and $W^u(H)$ and H is an inverse saddle. There exists the period 2 sink $S(0.0320, 0.3637)$, $(3.3216, -0.0835)$, which is contained in the limit set of $W^u(H)$. The closure of $W^u(H)$ forms the global attractor A_g (Fig. 16.12,b). The global attractor involves the chain recurrent set Q, which contains the orbits H and S and the points of intersection of $W^s(H)$ and $W^u(H)$ (homoclinic points).

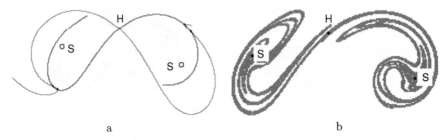

a   b

Fig. 16.12. Ikeda map for $R = 3$, $a = b = -0.9$

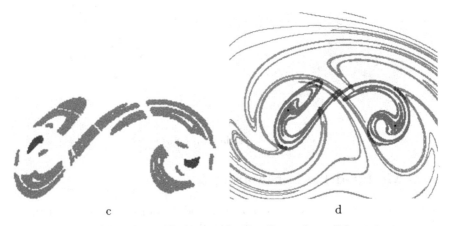

c   d

Fig. 16.13. Ikeda map for $R = 3$, $a = b = -0.9$

A neighborhood of Q obtained by the symbolic dynamics methods is shown in Fig. 16.13,c. A neighborhood of S (colored dark) is a lower bound for a basin of attraction of S. The manifolds $W^s(H)$ and $W^u(H)$ and their intersection points are presented in Fig. 16.13,d. The set of homoclinic points $W^s(H) \cap W^u(H)$ is a lower bound for the chain recurrent set Q.

Hyperbolic Mapping: $R = 1$, $a = 0.9$, $b = 1.2$

There exists hyperbolic fixed point $H(-0.1824, -2.3536)$ with nonempty intersection of the stable and unstable manifolds. The stable and unstable manifolds of H and the point $F(0.0851, 0.9643)$ homoclinic to H are shown in Fig. 16.14,a. Table presents numerical results of successive computation of points H and F.

Step	Fixed point	Homoclinic point
30	$x = -0.18235986, y = -2.35361944$	$x = -0.08509742, y = 0.96427872$
31	$x = -0.18235987, y = -2.35361803$	$x = -0.08144479, y = 0.96428413$
32	$x = -0.18235936, y = -2.35361106$	$x = -0.08519972, y = 0.96428226$

The mapping has the hyperbolic fixed point $H_1(0.5153, 0.2835)$ and the period 2 hyperbolic orbit $P(0.3708, 0.6824)$, $(0.5505, -0.7136)$. The stable and unstable manifolds of H, H_1 and P intersect generating heteroclinic cycles (see Fig. 16.14,b). Fig. 16.14,c shows how the stable and unstable manifolds of P are situated. The set of points homoclinic to H (constructed as an intersection of $W^s(H)$ and $W^u(H)$) is a lower bound of the chain-recurrent set Q and is depicted in Fig. 16.14,d. A neighborhood of Q (an upper bound) obtained by localization using symbolic dynamics methods is displayed in Fig. 16.15,e. The stable manifold $W^s(H)$ of H and stable manifolds of all other orbits from Q start from the source $S(-2.9622, 5.8918)$, see Fig. 16.15,f.

Expansion: $R = 1$, $a = b = 1.2$

The mapping J increases an area by $a^2 = 1.44$ and has a global repeller R_g. This repeller contains the hyperbolic fixed point $H(0.4368, 0.3100)$ which stable and unstable manifolds intersect generating a homoclinic contour. The fixed point H is an inverse saddle, i.e. the map J reverses orientation on $W^s(H)$ and $W^u(H)$. In addition, there exists the 2-periodic orbit $H_1(0.5132, -0.7463)$, $(0.1850, 0.7191)$ whose stable $W^s(H_1)$ and unstable $W^u(H_1)$ manifolds intersect each other and stable $W^s(H)$ and unstable $W^u(H)$ manifolds of H generating a heteroclinic contour (see Fig. 16.16,a). The closure of $W^s(H)$ (or $W^s(H_1)$) forms the repeller R. Fig. 16.16,b presents the repeller R and the manifolds $W^s(H)$ and $W^u(H)$. The set of points (colored dark) of intersection of stable and unstable manifolds of H and H_1 (a lower bound for Q) is depicted in Fig. 16.16,b. Obtained by symbolic dynamics methods, a neighborhood of the chain-recurrent set Q (an upper bound)

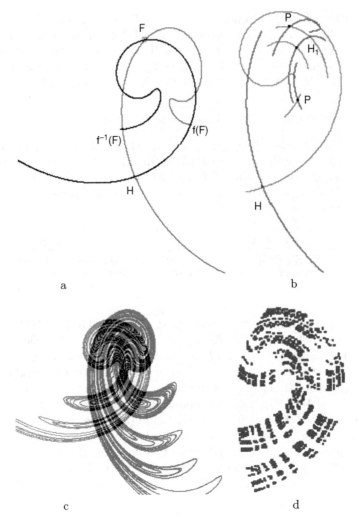

Fig. 16.14. Ikeda map for $R = 1$, $a = 0.9$, $b = 1.2$

containing R is shown in Fig. 16.17,c. It seems likely that $R = Q$. Outside R there exists a hyperbolic fixed point $H_0(-1.2588, -2.5318)$ (see Fig. 16.17,d), the left separatrix of which starts from R and the right one starts from the source $S(-3.7022, 2.3228)$.

16.3.2 Mappings Reversing Orientation

Contraction: $R = 1$, $a = 0.9$, $b = -0.9$

The map J decreases an area and has a global attractor A_g. There exist two hyperbolic fixed points $H_0(0.5726, 0.6602)$ and $H_1(0.5606, -0.5692)$

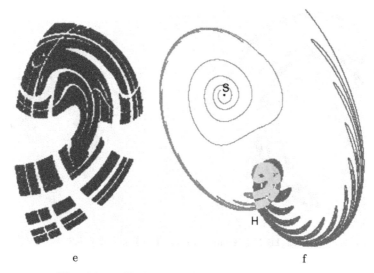

e   f

Fig. 16.15. Ikeda map for $R = 1$, $a = 0.9$, $b = 1.2$

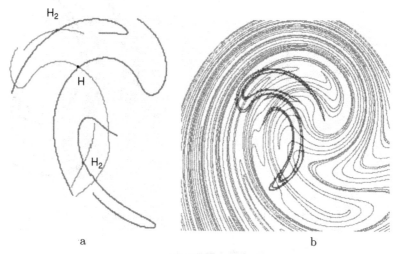

a   b

Fig. 16.16. Ikeda map for $R = 1$, $a = b = 1.2$

whose stable and unstable manifolds intersect forming a heteroclinic cycle. In addition, there is a unique 2-periodic hyperbolic orbit $P(0.9391, -0.2036)$, $(0.1539, 0.1791)$ whose stable (unstable) manifold intersects $W^u(H_0)$ and $W^u(H_1)$ ($W^s(H_0)$ and $W^s(H_0)$) forming a heteroclinic cycle (Fig. 16.18,a). Points of intersection of stable and unstable manifolds of these orbits (colored dark in Fig. 16.18,b) yield a lower bound for the chain-recurrent set Q. An upper bound for Q obtained by symbolic dynamics methods is depicted in

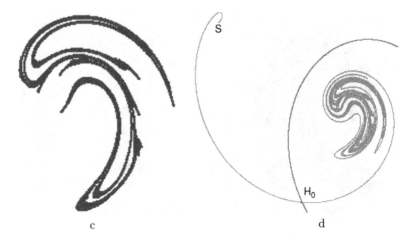

Fig. 16.17. Ikeda map for $R = 1$, $a = b = 1.2$

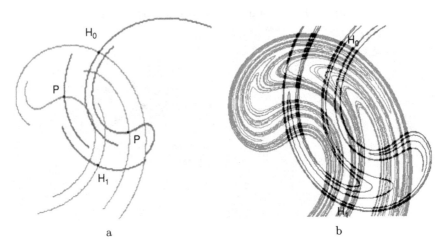

Fig. 16.18. Ikeda map for $R = 1$, $a = 0.9$, $b = -0.9$

Fig. 16.19,c. Near H_0 the manifold $W^s(H_0)$ bounds Q, with the left separatrix $W^u(H_0)_l$ involved in Q and the right one $W^u(H_0)_r$ going to the right (Figs. 16.19,a,b and d). Near H_1 the manifold $W^u(H_1)$ bounds Q, with the right separatrix $W^u(H_1)_r$ involved in Q and the left one $W^u(H_1)_l$ going to infinity (Figs. 16.18,a,b). Stretching along the right separatrix $W^u(H_0)_r$, unstable manifolds start from Q and end at the sink $S(9.7301, -1.5751)$. Stable manifolds start from Q and along the left separatrix $W^s(H_0)_l$ reach infinity in the form of "rabbit ears" (Fig. 16.19,d and Fig. 16.20). The global attractor A_g is the closure of $W^u(H_0)$ (Fig. 16.20).

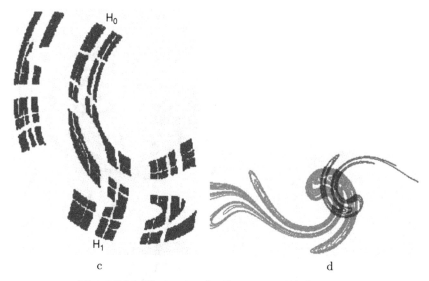

Fig. 16.19. Ikeda map for $R = 1$, $a = 0.9$, $b = -0.9$

Fig. 16.20. Ikeda map for $R = 1$, $a = 0.9$, $b = -0.9$

Contraction: $R = 2$, $a = -0.9$, $b = 0.9$.

The map J decreases an area and the global attractor A_g. There exists the unique hyperbolic fixed point $H(1.3815, -2.4746)$ (Fig. 16.21,a) whose stable and unstable manifolds $W^s(H)$ and $W^u(H)$ intersect (Fig. 16.21,a and b). In addition, there is the unique periodic orbit $P_2(0.2378, -0.7031)$, $(1.9995, 0.6681)$ stable and unstable manifolds of which intersect $W^s(H)$ and $W^u(H)$ forming a heteroclinic cycle (Fig. 16.21,a). The global attractor A_g is a closure of $W^u(H)$ or $W^u(P)$ (Fig. 16.21,b). The set $W^s(H) \cap W^u(H)$ is a lower bound for the chain-recurrent set Q. Fig. 16.21,c presents a neighborhood of Q constructed by symbolic dynamics methods. Since A_g contains all limit points, stable manifolds of orbits from A_g cover the plane R_2. Using symbolic dynamics methods we obtain the 6-period hyperbolic orbit P_6 $(1.0847, -1.0732)$, $(2.7889, -1.1242)$, $(-0.2626, -1.4846)$, $(3.3560, 0.0508)$, $(-1.0124, -0.2235)$, $(1.3964, 0.7116)$. Its Lyapunov exponents are calculated by $\lambda = \frac{1}{6} \cdot \ln |\gamma|$, where by are the eigenvalues of the differential of the Ikeda mapping along the orbit P_6. We obtain: $\gamma_1 = -23.098$, $\gamma_2 = -0.012$ and $\lambda_1 = 0.523$ and $\lambda_2 = -0.734$. The attractor has the 2-periodic orbit P_2 $(0.2385, -0.7024)$, $(1.9989, 0.6691)$. The eigenvalues of the differential

Fig. 16.21. Ikeda map for $R = 2$, $a = -0.9$, $b = 0.9$

along P_2 are $\lambda_1 = -0.134$, $\gamma_2 = -4.888$, and the Lyapunov exponents $\lambda = \frac{1}{2} \cdot \ln|\gamma|$ are $\lambda_1 = -1.004$, $\lambda_2 = 0.793$. There exists the 4-periodic orbit $P_4(-0.6836, -0.6319)$, $(0.7312, -0.9389)$, $(1.6003, 0.72792)$, $(3.0613, -0.1713)$ with the Lyapunov exponents $\lambda_1 = -0.843$, $\lambda_2 = 0.633$.

Hyperbolic Mapping: $R = 1$, $a = -0.9$, $b = 1.2$

The map J has the hyperbolic fixed point $H(-0.0950, 2.1937)$ stable manifold $W^s(H)$ of which can be bijectively projected on the x-axis. The map J reverses orientation on $W^s(H)$. The unstable manifold $W^u(H)$ can be bijectively projected on the y-axis near H, however, the lower part of $W^u(H)$ offers a complicated structure (Fig. 16.22,a). Such a behavior of $W^u(H)$ results from the fact that $W^u(H)$ intersects the stable manifold $W^s(Q_2)$ of the 2-periodic hyperbolic orbit $Q_2(-1.5584, -1.9046)$, $(3.0088, -1.2438)$, which in turn has a homoclinic point of transverse intersection of stable and unstable manifolds

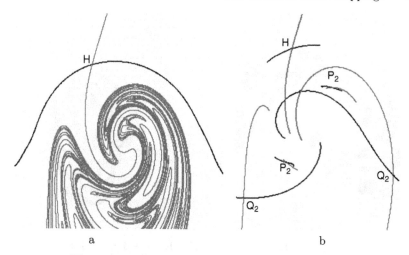

Fig. 16.22. Ikeda map for $R = 1$, $a = -0.9$, $b = 1.2$

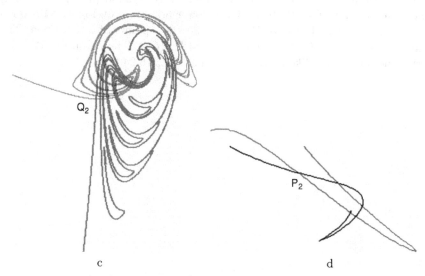

Fig. 16.23. Ikeda map for $R = 1$, $a = -0.9$, $b = 1.2$

$W^u(Q_2)$ and $W^s(Q_2)$ (Fig. 16.22,b). Fig. 16.23,c shows the manner in which $W^u(Q_2)$ and $W^s(Q_2)$ intersect near $Q_2(-1.5584, -1.9046)$. Besides Q_2 there is another 2-periodic hyperbolic orbit $P_2(-0.2554, -0.9207)$, $(1.1152, 1.1362)$ with homoclinic intersection of its stable and unstable manifolds $W^u(P_2)$ and $W^s(P_2)$. Fig. 16.23,d shows the manner in which $W^u(P_2)$ and $W^s(P_2)$ intersect near $P_2(-0.2554, -0.9207)$. Stable and unstable manifolds of orbits Q_2 and P_2 intersect forming a heteroclinic cycle. This leads to the chaotic

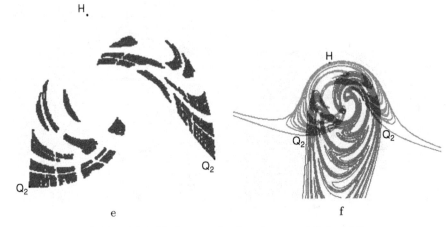

Fig. 16.24. Ikeda map for $R = 1$, $a = -0.9$, $b = 1.2$

chain-recurrent set Q. Fig. 16.24,e depicts a neighborhood (an upper bound) of Q. The set $W^u(Q_2) \cap W^s(Q_2)$ gives a lower bound for Q. Fig. 16.24,f shows the displacement of $W^u(Q_2)$, $W^s(Q_2)$, and their points of intersection. The stable manifold $W^s(H)$ is in the closure of $W^s(Q_2)$. The closure of $W^s(Q_2)$ forms the set looking like a "Napoleon" hat (Fig. 16.24,f).

A Dynamical System of Mathematical Biology

17.1 Analytical Results

The 3-dimensional dynamical system describes a discrete food chain model. Lindström [73] proposed the model that displays a lot of properties commonly known for continuous food-chains [47, 122].

The discrete food chain model is defined by the mapping f of the form

$$
\begin{aligned}
X_{t+1} &= \frac{M_0 X_t \exp(-Y_t)}{1 + X_t \max(\exp(-Y_t), K(Z_T)K(Y_t))} \\
Y_{t+1} &= M_1 X_t Y_t \exp(-Z_t) K(Y_t) \cdot K(M_3 Y_t Z_t) \\
Z_{t+1} &= M_2 Y_t Z_t,
\end{aligned}
\tag{17.1}
$$

where

$$
K(\gamma) = \begin{cases} \frac{1 - \exp(-\gamma)}{\gamma}, & \text{if } \gamma \neq 0 \\ 1, & \text{if } \gamma = 0 \end{cases},
\tag{17.2}
$$

The detailed description of the model was given in [73]. The variables are related to the different trophic levels of the system, so X is proportional to vegetation abundance whereas Z is proportional to carnivore abundance. Since the relation between herbivores and Y is nonlinear, a more complicated relation describes the situation here. However, such relationships do not change the topological properties of the system under investigation. So, for convenience, we will refer to vegetation, herbivore, and carnivore levels in the sequel.

It should be noted that $M_2 = M_3$ in the original equation. The fourth parameter M_3 is introduced in order to generate additional cases and obtain the complete analysis of system characteristics. So, Example 186 shows the existence of an invariant Möbius band at the parameter position $M_0 = 4.0$, $M_1 = 1.0$, $M_2 = 3.0$, $M_3 = 4.0$. However in this chapter we will consider the original model with $M_2 = M_3$.

It should be marked that the solutions of (17.1) remain positive and bounded. Repeating the arguments given in [73] we can show that all solutions starting in the positive cone enter the box $0 < X_t < M_0$, $0 < Y_t < M_0 M_1$, $0 < Z_t < M_0 M_1 M_2^2 / M_3$ within three iterations.

The system (17.1) has at most four equilibria [73] which are given by:

$$E_0 = (0, 0, 0),$$
$$E_1 = (M_0 - 1, 0, 0),$$
$$E_2 = \left(\frac{M_0 \log\left(\frac{M_1 M_0}{1 + M_1}\right)}{(M_0 - 1) M_1 - 1}, \log\left(\frac{M_1 M_0}{1 + M_1}\right), 0 \right),$$

and $E_3(X, Y, Z)$ is given by

$$\left(\frac{M_0 \exp(-\frac{1}{M_2}) - 1}{K\left(\frac{1}{M_2}\right) K\left(\log M_1 \left(M_0 \exp(-\frac{1}{M_2}) - 1\right)\right)}, \right.$$
$$\left. \frac{1}{M_2}, \log M_1 \left(M_0 \exp\left(-\frac{1}{M_2}\right) - 1\right) \right),$$

if $M_2 = M_3$ and $\max(\exp(-Y), K(Y)K(Z)) = K(Y)K(Z)$ at E_3.

Some general features of the system can be described by the Morse spectrum of the determinant $\det Df$ which is the rate of change of phase volume.

If $\xi = \{v_0, ..., v_p = v_0\}$ is a p-periodic ε-orbit, then the determinant exponent is defined by

$$\lambda(\det Df, \xi) = \frac{1}{p} \sum_{i=0}^{p-1} \ln |\det Df(v_i)|.$$

The Morse spectrum of the determinant is defined as the following

$$\Sigma(\det Df) = \{\lambda \in R: \ there \ are \ \varepsilon_k \to 0 \ and \ periodic \ \varepsilon_k - orbits \ \xi_k$$
$$with \ \lambda(\det Df, \xi_k) \to \lambda \ as \ k \to \infty\}.$$

It is well known [145] that if λ_1, λ_2, and λ_3 are Lyapunov exponents of a periodic orbit ξ and $\lambda(\det Df, \xi)$ is its determinant exponent then

$$\lambda_1 + \lambda_2 + \lambda_3 = \lambda(\det Df, \xi).$$

Our computing experiments show that in the selected area the considered system has the negative Morse spectrum of $\det Df$. It follows from Corollary 143 of Chapter 12 that the volume tends to zero with negative exponent along each chain recurrent orbit, being its determinant exponent has at least one negative Lyapunov exponent.

17.2 Numerical Results

We limit our discussion to the parameter range $M_0 \in [3.00; 3.65]$ and fix $M_1 = 1.0, M_2 = M_3 = 4.0$. We commence an overview about some general features and bifurcations. The selected area is located along the route to chaos.

At $M_0 \approx 2.93$ a Neimark-Sacker bifurcation occurs. The fixed point E_3 loses its stability and an invariant circle appears in its vicinity. This curve becomes the minimal attractor of the system. As the parameter M_0 increases, the attractor is alternatingly quasi-periodic and periodic, like the dynamics of circle maps [37, 144]. This holds as long as the parameter value stays moderately far from the bifurcation value.

17.2.1 $M_0 = 3.000$

The Morse spectrum of det(Df) is estimated as $[-0.403970, -0.322668]$. The system has the fixed point E_2 with the coordinates $(1.2164, 0.4055, 0)$ and the Lyapunov exponents $L_1 = 0.4837; L_{2,3} = -0.1667$. The (xy)-plane is invariant for the differential at E_2 and the exponents $L_{2,3}$ correspond to a focus on this plane. Thus, the unstable manifold $W^u(E_2)$ is a curve transversal to the (xy)-plane. The system has the fixed point E_3 with the coordinates $(1.7160, 0.2500, 0.2806)$ and the Lyapunov exponents $L_{1,2} = 0.0630; L_3 = -0.6430$. The unstable manifold $W^u(E_3)$ is a 2-dimensional surface with an unstable focus. Our computing results show that the closure of $W^u(E_3)$ is a global attractor in the positive corner $\{x > 0, \, y > 0, \, z > 0\}$. In particular, $W^u(E_2)$ tends to the closure of $W^u(E_3)$, see Fig. 17.1. Such a dynamics may be observed when the parameter M_0 changes from 3 to 4.

The closure of the unstable manifold $W^u(E_3)$ is diffeomorphic to a standard 2-dimensional closed disc, see Fig. 17.1. The boundary C of the unstable

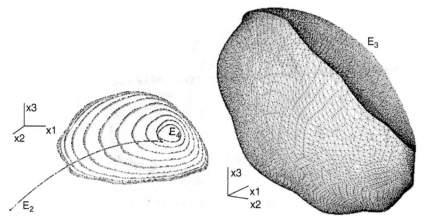

Fig. 17.1. Unstable manifolds $W^u(E_2)$ and $W^u(E_3)$, $M_0 = 3.000$

manifold $W^u(E_3)$ is homeomorphic to circle S^1. The stable invariant curve C appears at the Neimark-Sacker bifurcation at $M_0 \approx 2.93$ and looses its stability at $M_0 \approx 3.366$. As $M_0 = 3.0$ we observe a quasiperiodic behavior on C. The first approximation of this rotation looks like as 9-periodic. However, Danny Fundinger found the coordinates of a point X_0 on C whose 50,000 iterations form a line, being the iterations from 40,000 to 50,000 form a line as well. We conclude from this that the movement on the cycle is not periodic. The coordinates of the iterations are $X_0 = (1.336740, 0.379555, 0.253432)$, $X_8 = (1.471666, 0.417361, 0.145821)$, $X_9 = (1.372226, 0.385229, 0.243440)$, $X_{18} = (1.378605, 0.390052, 0.234376)$, $X_{36} = (1.392378, 0.397436, 0.218947)$. These results show that if we start from a point X_0, then X_9 is shifted a little bit from the position of X_0, X_{18} a little bit further and so on. The chain recurrent sets E_2, E_3 and C are localized by the symbolic method, the unstable manifolds $W^u(E_2)$ and $W^u(E_3)$ are constructed by the iterations of broken lines and polytopes, respectively.

When parameter M_0 changes from 3 to 3.3, both the topological structure of trajectories and the manifolds $W^u(E_2)$, $W^u(E_3)$ persist.

17.2.2 $M_0 = 3.300$

The Morse spectrum of $\det(Df)$ is estimated as $[-0.431418, -0.289534]$. The fixed points E_2 and E_3 have the coordinates $(1.27119, 0.50077, 0)$ and $(2.01597, 0.25, 0.39002)$ respectively. As in previous case they have the same type of stability. Moreover, there is a minimal attractor C with quasiperiodic motion. However, as $M_0 = 3.3$ the manifold $W^u(E_2)$ tends to the invariant curve C by winding around C, see Fig. 17.2. Hence C should not be considered as a boundary of the smooth manifold $W^u(E_2)$, but as its limit set.

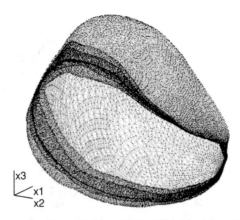

Fig. 17.2. Unstable manifolds $W^u(E_3)$, $M_0 = 3.300$

17.2.3 $M_0 = 3.3701$

The Morse spectrum of $\det(Df)$ is estimated as $[-0.526124, -0.194662]$. The stable invariant curve C looses its stability at $M_0 \approx 3.366$. When the parameter M_0 becomes 3.3701, this bifurcation results in the appearance of a Möbius band MB(3.3701). Later we prove that the invariant manifold MB(3.3701) is non-oriented. The bifurcation is like period doubling bifurcation which is usually observed in continuous dynamical systems. However, we can not speak of period doubling bifurcations here in the same sense as is usually meant for discrete systems. More precisely, in the discrete system we observe a pattern which is typical for continuous systems [40, 48]. So we can say about a "Feigenbaum-like bifurcation". As we will see later, several Feigenbaum-like bifurcations of the same kind happen close to the transition to chaos. The manifold MB(3.3701) is the limit set of the unstable manifold $W^u(E_3)$. Construct the unstable manifold $W^u(E_2)$ of the fixed point E_2 as above. The boundary $L = \partial MB(3.3701)$ is a limit set of the unstable manifold $W^u(E_2)$, see Fig. 17.3. Center line C of $MB(3.3701)$ is an unstable invariant curve (on $MB(3.3701)$) with quasiperiodic motion, see Fig. 17.4. Note that L is a stable invariant curve homeomorphic to circle which is two times longer than the center line C. On L there are the stable 55-periodic orbit $\{P, \dots\}$ and the hyperbolic 55-perodic orbit $\{H, \dots\}$ which is stable on L. The rotation number [48] of the system on L is $3/55$. The points $R(1.631969, 0.105806, 0.778837)$, $Q(1.456810, 0.157710, 0.718353)$ lies on the orbit of $P(1.519275, 0.140199, 0.847081)$, and H has the coordinates $(1.582335, 0.118405, 0.815238)$, see Fig. 17.4. The eigenvalues of the differential Df^{55} at P are approximately equal to $\lambda_{1,2} = 0.8584 \pm 0.2398i$ and $\lambda_3 = 5 \cdot 10^{-8}$. Hence the point P is focus for f^{55}. To prove non-orientability of the band MB(3.3701) we consider the direction of the rotation on the orbit of the point P. Comparing the rotation at points along the center line, i.e. at P, R, etc. (see Fig. 17.4), we see that the direction of the rotation persists. But the directions of the rotation at

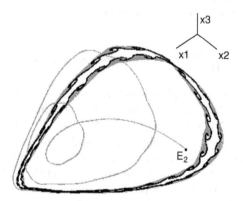

Fig. 17.3. Unstable manifold $W^u(E_2)$ and the Möbius band MB(3.3701)

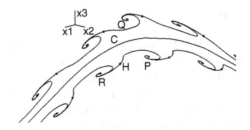

Fig. 17.4. Detail of the Möbius band MB(3.3701)

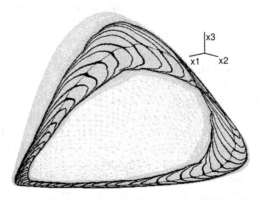

Fig. 17.5. The unstable manifold $W^u(E_3)$ and the Möbius band $MB(3.4001)$

the points P and Q are opposites. So, if we go from P to Q along the center line, we save the orientation and if we go transversal to C we get opposite orientation. This is possible if and only if the band is non-oriented. There is only single 2-dimensional non-oriented strip - Möbius band.

Thus, the global dynamics of the system in the positive corner $\{x > 0, y > 0, z > 0\}$ is the following. The stable 55-periodic orbit of the point P is a single attractor A minimal by inclusion. Other trajectories, except the fixed points E_3, the center line C, and the hyperbolic 55-periodic orbit of the point H, tend to A.

17.2.4 $M_0 = 3.4001$

The Morse spectrum of det(Df) is estimated as $[-0.580550, -0.186926]$. By using symbolic methods we determined the coordinates of the fixed points E_2 (1.288505, 0.5309395, 0.0), E_3 (2.11595, 0.24995, 0.42300) and two chain recurrent curves C and L. The eigenvalues at E_2 and E_3 are $\lambda_1(E_2) = 2.124$, $\lambda_{2,3}(E_2) = 0.628869 \pm 0.627i$, and $\lambda_{1,2}(E_3) = 0.806993 \pm 0.812688i$, $\lambda_3(E_3) = 0.447195$. Both C and L are homeomorphic to a circle, being L is two times longer than C. The curve L becomes the minimal attractor of the system. These two curves belong to a 2-dimensional invariant manifold $MB(3.4001)$ which is homeomorphic to an Möbius band, Fig. 17.5. On the picture you can

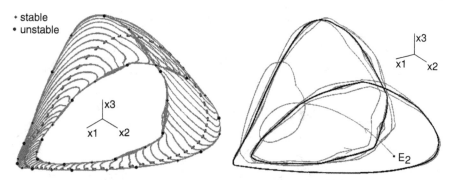

Fig. 17.6. The boundary L of $MB(3.48)$ is a limit set of $W^u(E_2)$

see a collection of curves being ustable manifold of an orbit on the line C. Geometrically, the curves resemble the shape of a Möbius strip. The stable invariant curve can be imagined as the edges of the strip, and the unstable curve as its center line. Numerical studies of forward and backward iterations so far indicate that the curve at the center line has saddle type in the space and unstable type on $MB(3.4001)$. We constructed the unstable manifold $W^u(E_3)$ and the Möbius band $MB(3.4001)$, see Fig. 17.5. Appendix B by Danny Fundinger contains full-length information about dynamics of the given system.

17.2.5 $M_0 = 3.480$

The Morse spectrum of $\det(Df)$ is estimated as $[-0.552695, -0.181412]$. The Möbius band $MB(3.48)$ persists. It is a global attractor in the positive corner $\{x > 0,\ y > 0,\ z > 0\}$. The center line C of $MB(3.48)$ is an unstable invariant curve on $MB(3.48)$. The dynamics on C is periodic with the rotation number $7/64$. There is a stable 64-perodic orbit of the point $P \approx (1.231941, 0.800882, 0.137928)$ on C. The limit set of $MB(3.48)$ is a stable invariant curve L with periodic dynamics of the rotation number $1/18$. There is a hyperbolic 18-periodic orbit of the point $U \approx (2.194811, 0.356991, 0.065345)$ on L. The differential $Df^{18}(U)$ has the eigenvalues $\lambda_1 \approx 1.21$, $\lambda_2 \approx -0.28$, $\lambda_3 \approx -0.002$. The first eigenvalue corresponds to L so the orbit of U is unstable on L. There is a stable 18-periodic orbit of the point $S \approx (1.652079, 0.131587, 0.502229)$ on L. The differential $Df^{18}(S)$ has the eigenvalues $\lambda_1 \approx 0.81$, $\lambda_2 \approx -0.40$, $\lambda_3 \approx -0.003$. Since the pair of the eigenvalues at U and S are negative, the Möbius band $MB(3.48)$ tends to L by winding around one. The periodic orbit of S is a single attractor minimal by inclusion. Thus, almost all trajectories from positive corner tend to this orbit.

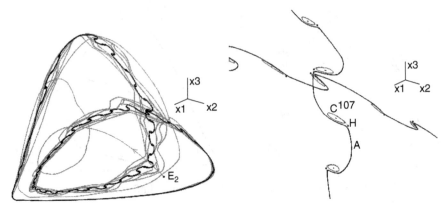

Fig. 17.7. Unstable manifold $W^u(E_2)$ and its limit set $LimW^u(E_2)$, $M_0 = 3.532$

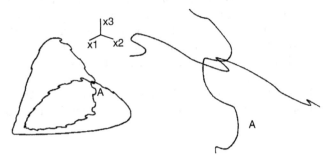

Fig. 17.8. The attractor A and its details, $M_0 = 3.532$

17.2.6 $M_0 = 3.532$

The Morse spectrum of $\det(Df)$ is estimated as $[-0.572940, -0.162688]$. The coordinates of the fixed point E_2 are $(1.305196, 0.561747, 0)$. We constructed the unstable manifold $W^u(E_2)$ and its limit set $LimW^u(E_2)$. It turns out that $LimW^u(E_2)$ has nontrivial structure, see Fig. 17.7. The limit set consists of the 107 circles (C^{107}), the hyperbolic 107-periodic orbit H, its unstable manifold $W^u(H)$, and the attractor A, see Fig. 17.8. The unstable 107-periodic orbit U is inside of the set C^{107}. The limit set of the Möbius band $MB(3.532)$ coincides with $LimW^u(E_2)$. The attractors A and C^{107} are minimal by inclusion, so we have a non-ordinary phenomenon — the existence of two minimal attractors in a biology system. It should be noted that the set C^{107} exists in a short interval for M_0. The circles, the orbits U and H disappear when $M_0 \approx 3.536$, whereas the attractor A persists. The paper [44] deals with the appearance of multiple attractors in the chain food dynamics and contains detailed information about the considered system.

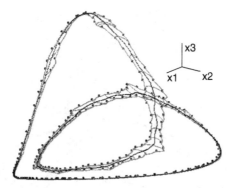

Fig. 17.9. The second Möbius band $MB_2(3.54)$, its unstable center line C_2, and the stable 283-periodic orbit S on L_2

17.2.7 $M_0 = 3.540$

The Morse spectrum of $\det(Df)$ is estimated as $[-0.578320, -0.173867]$. The attractor A is the limit set of the Möbius band MB from $M_0 \approx 3.536$ to $M_0 \approx 3.538$, but for all that the invariant curve A looses its stability for the last parameter value. When the parameter M_0 becomes 3.54, the bifurcation results in the appearance of the second Möbius band $MB_2(3.54)$, see Fig. 17.9. Hence we have "Feigenbaum-like bifurcation".

The second Möbius band $MB_2(3.54)$ appears as a limit set of the first Möbius band $MB_1(3.54)$. The center line of the second Möbius band MB_2 (3.54) is unstable invariant curve C_2 with quasiperiodic motion. The boundary L_2 of $MB_2(3.54)$ is a stable invariant circle which is two times longer than the center line C_2. On the boundary L_2 there are two 283-periodic orbits: stable $S \approx \{(1.354245, 0.467325, 0.679182), \dots\}$ and unstable $U \approx \{(1.354245, 0.467325, 0.679182), \dots\}$. The points of these orbits alternate.

17.2.8 $M_0 = 3.570$

The Morse Spectrum of $\det(Df)$ is estimated as $[-0.565295, -0.161637]$. When the parameter M_0 is greater than 3.538, the second Möbius band becomes a limit set of the first Möbius band. More precisely, the trajectories on MB_1 tend to MB_2 by winding around it. The set $MB_1(3.57)$, the unstable malifold of an orbit on the center line C_1 and its behavior near the limit set MB_2 are shown on the Fig. 17.10.

Now we observe the next "Feigenbaum-like bifurcation" near the boundary of the second Möbius band MB_2. Here the bifurcation is the following. The invariant curve $L_2 = \partial MB_2$ loses its stability and a strange invariant set MB_3 appears, see Fig. 17.11. The set $MB_3(3.570)$ consists of 71 pieces, each piece has both 2- and 1-dimensional parts. This decomposition is invariant, i.e. an image of a piece part is a piece part. The 2-dimensional part is the

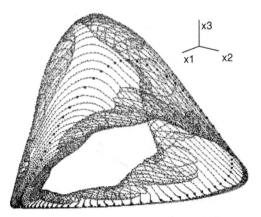

Fig. 17.10. The trajectories of the first Möbius band tend to the second Möbius band by winding around it, $M_0 = 3.570$

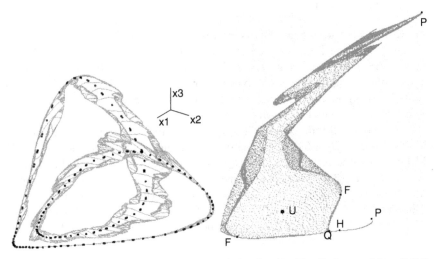

Fig. 17.11. The second Möbius band and the detail of its limit set, $M_0 = 3.570$

closure of the unstable manifold $W^u(U)$ of the hyperbolic 71-periodic orbit $U \approx \{(1.323374, 0.186463, 0.646940), \dots\}$. The eigenvalues of the differential $Df^{71}(U)$ are $\lambda_1 \approx 1.374292$, $\lambda_2 \approx -1.501502$, $\lambda_3 \approx -0.794848$. The eigenvalues λ_1 and λ_2 correspond to the unstable manifold $W^u(U)$. The system behavior on $W^u(U)$ is similar to its dynamics on a Möbius band. As $\lambda_1\lambda_2 < 0$, the 71-th iteration f^{71} inverts the orientation on $W^u(U)$. The 1-dimensional part is formed by the unstable manifold $W^u(H)$ of the hyperbolic 71-periodic orbit $H \approx \{(1.576668, 0.44846, 0.72410), \dots\}$. The manifold $W^u(H)$ ends at the stable 71-periodic cycles $Q \approx \{(1.913398, 0.677286, 0.093128), \dots\}$ and $P \approx \{(2.489430, 0.086803, 0.027202), \dots\}$, see Fig. 17.11. The limit set of

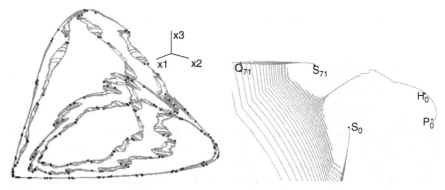

Fig. 17.12. The invariant set of $MB_3(3.571)$ and the unstable manifold $W^u(H_0)$ on it

the 2-dimensional manifold $W^u(U)$ is a stable invariant curve which forms by the 1-dimensional unstable invariant manifold $W^u(F)$ of the hyperbolic 142-periodic orbit F. The structure of the both Möbius bands persists near $M_0 \approx 3.5708$. Moreover, the structure of the invariant set MB_3 persists as well. It should be noted that in this case we detect a new phenomenon. When $M_0 = 3.5708$ the set MB_3 contains at least three stable periodic orbits: the known 71-periodic orbits $Q \approx \{(1.560941, 0.447309, 0.731145), \dots\}$ and $P \approx \{(1.601819, 0.436388, 0.74072), \dots\}$ and a new stable 142-periodic orbit $S \approx \{(1.547730, 0.450719, 0.727846), \dots\}$. Thus, we obtain three minimal attractors in the biology system.

17.2.9 $M_0 = 3.571$

In this case the Morse spectrum of the det(Df) is estimated as $[-0.596287, -0.163302]$. The structure of $MB_3(3.571)$ is the following. There are two attractors minimal by inclusion. One of them — P — is a stable 71-periodic cycle generated by the point $P_0 \approx (1.6014, 0.4362, 0.7415)$, see Fig. 17.12. The second attractor is a stable cycle S formed by 142-periodic orbit of the point $S_0 \approx (1.93364, 0.60068, 0.0002)$. The eigenvalues of these cycles are following:

$$\lambda_{1,2}(P) \approx 0.6477 \pm 0.3656i, \quad |\lambda_{1,2}(P)| = 0.7438 < 1, \quad \lambda_3(S^{71}) \approx 0$$

and

$$\lambda_{1,2}(S) \approx 0.5079 \pm 0.3626i, \quad |\lambda_{1,2}(S)| = 0.6241 < 1, \quad \lambda_3(S) \approx 0.$$

Thus, the both cycles are of focus type and obviously, they are attractors minimal by inclusion. Additionally, there are two unstable periodic cycles. One of them H is 71-periodic orbit of the point $H_0 \approx (1.5889, 0.4438, 0.7305)$ and the other Q is 142-periodic orbit of the point $Q_0 \approx (1.578620, 0, 410556, 0.798312)$,

$f^{71}(Q_0) = Q_{71} \approx (1.649636, 0.279537, 1.066632)$. Again, we analyze the eigenvalues of these cycles:

$$\lambda_1(H) \approx 1.4817, \quad \lambda_2(H) \approx 0.5603, \quad \lambda_3(H) \approx 0;$$

and

$$\lambda_1(Q) \approx 1.8779, \quad \lambda_2(Q) \approx -0.0669, \quad \lambda_3(Q) \approx -0.$$

Moreover, we estimate the corresponding eigenvectors and eigenspaces. So the periodic cycles H and Q are of hyperbolic type with 1-dimensional unstable manifolds. The unstable manifold $W^u(H)$ are constructed and we see (Fig. 17.12) that the right part of $W^u(H_0)$ has a simple behavior and ends at the point P_0 of the orbit P. The left part of $W^u(H_0)$ has more complex oscillated behavior. The limit set of $W^u(H_0)$ contains the points S_0 and $S_{71} = f^{71}(S_0)$ of the stable orbit S. The limit set of $W^u(H_0)$ contains the points Q_0 and Q_{71} of the hyperbolic orbit Q, see Fig. 17.13. We estimate the angle between the stable subspace of the orbit Q at Q_0 and $W^u(H_0)$ and verify that their intersection is transversal near Q_0. Thus, we have the heteroclinic transversal connection $H \to Q$.

The next step is the construction of the unstable manifold $W^u(Q_0)$ at the point Q_0. It turn out that here we have a similar behavior: one part of

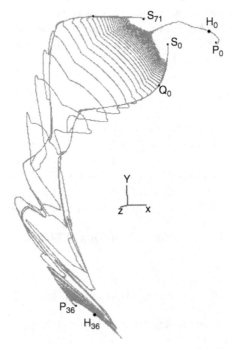

Fig. 17.13. The closure of the unstable manifold $W^u(H)$, the stable orbits S and F

$W^u(Q_0)$ ends at S_0 of the stable orbit S. The other part of $W^u(Q_0)$ has more complex oscillated behavior such that its limit set contains the point H_{36} of the hyperbolic orbit H and the point P_{36} of the stable orbit P. Here we also obtain the heteroclinic transversal connection $Q \to H$. From this it follows that there is the homoclinic connections $Q \to Q$ and $H \to H$. To check this conclusion we construct the global unstable manifold $W^u(H)$ of the hyperbolic orbit H. The result of our computing displayed on Fig. 17.13 shows that the limit set of $W^u(H_0)$ contains $H_{36} = f^{36}(H_0)$. Thus, there exists the homoclinic orbit $H \to H$, which usually leads to chaotic dynamics near this orbit [45, 48, 132, 133, 136, 137]. Our numerical investigation shows that a chaotic dynamics is located inside of the closure of $W^u(H)$. This closure is not minimal by inclusion because it contains the stable orbits S and P, see Fig. 17.12.

17.2.10 Chaos

Consider the system behaviour for $M_0 = 3.573$, 3.580 and 3.650. The Morse spectrum of the differential was obtained by the symbolic analysis methods. For $M_0 = 3.573$ the Morse spectrum of $\det(Df)$ is estimated as $[-0.601288, -0.167536]$. In the case $M_0 = 3.58$ the spectrum is estimated as $[-0.582492, -0.156852]$ and for $M_0 = 3.65$ the estimate is $[-0.601972, -0.136850]$. Thus, we have negative spectrum in all cases. It results in zero volume of the chain recurrent set and, as a consequence, in zero volume of the chaotic attractor. It was mentioned above that the chaotic dynamics appears when $M_0 \approx 3.571$ and is located in very small domain. When $M_0 \in [3.573, 3.650]$ several local and global bifurcations lead to a number of subsequent changes in the dynamics. This involves the appearance of chaotic attractor which grows as M_0 increases. Our construction of a chaotic attractor is trivial: we pick up a point from its domain of attraction and consider its iterations. If the iteration number is huge, we obtain the desired attractor. In our cases the chaotic attractor is minimal by inclusion, so almost all points from the positive corner are in the basin of the desired attractor. When $M_0 = 3.573$ we can observe (see Fig. 17.14) that the chaotic attractor coincides with the invariant set $MB_3(3.573)$. If the parameter M_0 increases and reaches value $M_0 = 3.58$, the chaos occupies the second Möbius band $MB_2(3.58)$, see Fig. 17.14. After all, as $M_0 = 3.650$ the chaos covers the first Möbius band $MB_1(3.65)$, see Fig. 17.15.

17.3 Conclusion

In this chapter we have proved that a real discrete system with biological origin possesses a non-oriented invariant manifold — Möbius band. The obtained results demonstrate the existence of multiple attractors in food-chains models. We have found several parameter regions where attractors of this kind exist.

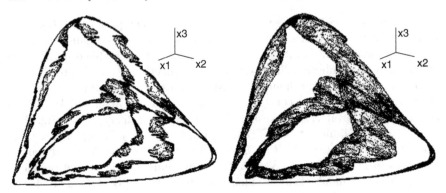

Fig. 17.14. The chaotic attractors for $M_0 = 3.573$ and $M_0 = 3.58$

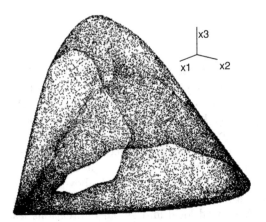

Fig. 17.15. The chaotic attractor for $M_0 = 3.65$

Moreover, the parameter region with three coexisting and closely-spaced attractors was found. It should be noted that such a proximity does not exclude the possibility that a complicated situation may appear, which may lead to more intriguing biological consequences in the system under study or similar systems.

We have shown the route to chaos in the food-chain dynamics. The initial system ($M_0 < 2.9$) has a single stable fixed point, when the parameter M_0 increases the systems passes through non-trivial cascad of the bifurcations which, when $M_0 = 3.65$, results in the appearance of a minimal chaotic attractor covering a Möbius band.

Acknowledgement.

This research was supported by the Royal Swedish Academy of Science.

References

1. A.V. Aho and J.E. Hopcroft and J.D. Ullman, *Data Structures and Algorithms*, Addison-Wesley, 1987
2. G. Alefeld and J. Herzberger, *Introduction to Interval Computations*, Academic Press, 1983
3. V.M. Alekseev, Quasi-random oscillations and qualitative problems of celestial mechanics. In: *9th Mathematical school*, Kiev, 1972 (in Russian)
4. V.M. Alekseev, *Symbolic Dynamics, 11th Mathematical School*, Kiev, 1976 (in Russian)
5. A.A. Andronov and L.S. Pontryagin, Rough Systems, *Doklady Academy Nauk SSSR*, v.14, no.5, 247–250 (1937) (in Russian)
6. D. V. Anosov, *Geodesic flow on closed Riemannian manifold of negative curvature*, Trudy Math. Steclov Institute, v.90, 1967, 210 p., (in Russian)
7. N. Ampilova, K. Mot'kina, An algorithm for construction of the homotopic paths on a symbolic image, *The Fourth International Conference "Tools for Mathematical Modelling"*, St. Petersburg, 2003, book of abstracts, p. 160
8. J. Argyris, G. Faust, and M. Haase, *An Exploration of Chaos: An Introduction for Natural Scientists and Engineers*, Elsevier Science Ltd., 1994
9. L. Auslender and R. MacKenzie, *Introduction to Differentiable Manifolds*, N.Y., 1963
10. V. Avrutin, R. Lammert, M. Schanz, G. Wackenhut, and G. Osipenko, On the software package AnT 4.669 for the investigation of dynamical systems, *The Fourth International Conference "Tools for Mathematical Modelling"*, Mathematical Reaseach, v.9 (2003), 24–35, St. Petersburg State Polytechnic University, Russia
11. V. Avrutin and M. Schanz, Border-collision period-doubling scenario, *Phys. Rev. E*, v.70 (2004), 026222/1–11
12. N.P. Bhatia, G.P. Szego, *Stability theory of dynamical systems*, New York, Springer, 1970
13. G.D. Birkhoff, Nouvelles recherches sur les systemes dynamique, *Memoriae Pont. Acad. Novi Lincaei* v.3, no. 1(1935), 85–216
14. R. Bowen, *Equilibrium States and the Ergodic Theory of Anosov Diffeomorphisms, Lectures Note in Mathematics*, 470, Springer-Verlagh, N.Y., 1982
15. R. Bowen, *Symbolic Dynamics*, Ann. Math. Soc. Providence, R.I., vol.8, 1982

234   References

16. M. Brin, On incusion of diffeomorphism into flow, *Izvestiya VUZov*, no. 8(123), 1972, 19–25 (in Russian)

17. H. Broer, C. Simo, Hill's equation with quasi-periodic forcing: resonance tongues, instability pockets and global phenomena, *Bul. Soc. Bras. Mat.*, v.29 (1998), 253–293

18. I.U. Bronshtein, Theorem on structural stability of smooth extension of cascade, in *Algebraic invariants of dynamical systems, Mat. Issledovaniya*, v.67, Kishinev, Shtinisa, 12–29 (1980) (in Russian)

19. I.U. Bronshtein, *Nonautonomous dynamical system*, Kishinev, Shtinisa, 1984 (in Russian)

20. B.F. Bylov, R.E. Vinograd, D.M. Grobman, V.V. Nemytskii, *Theory of Lyapunov exponents*, Moscow, Nauka, 1966 (in Russian)

21. M.L. Cartwright and J.E. Littlewood, On non-linear differential equations of the second order, III, IV, *Acta Math.*, v.97 (1957), no. 3–4, 267–308; v.98 (1957), no 1–2, 1–110

22. T. Cormen, C. Leiserson, and R. Rivest, *Introduction to algorithms*, MIT Press, 2000

23. P. Cvitanović, Periodic orbits as the skeleton of classical and quantum chaos, *Physica D*, 51, 1991

24. P. Cvitanović. Focus issue on periodic orbit theory. *Chaos*, 2, 1992

25. F. Colonius and W. Kliemann, *The Morse spectrum of linear flows on vector bundles*, Trans. Amer. Math. Soc., 348, 4355–4388 (1996)

26. F. Colonius and W. Kliemann, *The Lyapunov spectrum of families of time varying matrices*, report 504, Inst. of Math. Augsburg Univ., 1994

27. F. Colonius and W. Kliemann, *The Dynamics of Control*, Burkhauser, 2000

28. C. Conley, Isolated Invariant set and the Morse Index, *CBMS Regional Conference Series*, v.38, Amer. Math. Soc., Providence, 1978

29. B. Coomes, H. Kocak, K. Palmer, Periodic shadowing, chaotic numerics, *Contemporary Mathematics*, 172, 1994

30. B. Coomes, H. Kocak, K. Palmer, Computation of long period orbits in chaotic dynamical systems, *Aust. Math. Soc. Gaz.*, v.24, no.5, 183–190 (1997)

31. M. Dellnitz, A. Hohmann, The computation of unstable manifolds using subdivision and continuation, in *Nonlinear Dynamical Systems and Chaos*, by Broer, H.W.; Gils, S.A. van; Hoveijn, I.; Takens, F. (eds.) PNLDE 19, Birkhauser, 449–459, 1996

32. M. Dellnitz, A. Hohmann, A subdivision algorithm for the computation of unstable manifolds and global attractors, *Numerische Mathematik* 75, 293–317, 1997

33. M. Dellnitz, O. Junge, An adaptive subdivision technique for the approximation of attractors and invariant measures, *Comput. Visual. Sci.* 1, 63–68, (1998)

34. M. Dellnitz, O. Junge, Almost invariant sets in Chuas circuit, *Int. J. Bif. and Chaos*, 7(11), 2475–2485, 1997

35. M. Dellnitz, G. Froyland, O. Junge, The algorithms behind GAIO – Set oriented numerical methods for dynamical systems, in *Ergodic Theories, Analysis, and Efficient Simulation of Dynamical Systems*, Springer, 2001, 145–175

36. M. Dellnitz and O. Junge, Set Oriented Numerical Methods for Dynamical Systems, in *Handbook of Dynamical Systems II: Towards Applications*, World Scientific, 2002, 221–264

37. R.L. Devaney, *An Introduction to Chaotic Dynamical Systems*, Addison-Wesley pub. com., 1990

38. P. Diamond, P. Kloeden, A. Pokrovskii, Cycles of spatial discretizations of shadowing dynamical systems, *Math. Nachr.*, v.171 (1995), 95–110

39. W. Dijkstra, A Note on two problems in connection with graphs, *Numerische Math.*, v.1 (1959), 269–271

40. M.J. Feigenbaum, The Universal Metric Properties of Nonlinear Transformations, *J. Stat. Phys.*, v.21, no. 6 (1979), 669–707

41. M.J. Feigenbaum, Quantitative Universality for a Class of Nonlinear Transformations, *J. Stat. Phys.*, v.19, no. 1 (1978), 25–53

42. J. Franke and J. Selgrade, Hyperbolicity and chain recurrence, *J. Differential Equations*, 26 (1977), 27–36

43. G. Froyland, O. Junge, G. Ochs, Rigorous computation of topological entropy with respect to finite partition, Web-article, URL: *http://www.math.uni-paderborn.de/agdellnitz/papers/topen15.ps.gz*

44. D. Fundinger, T. Lindström, G. Osipenko, On the appearance of multiple attractors in discrete food-chains, submitted in *Applied Mathematics and Computation*, 2006

45. N.K. Gavrilov and L.P. Silnikov, On three-dimansional dynamical systems close to systems with a structurally unstable homoclinic curve, *Math. USSR Sb.*, v.88 (1972), 467–485; v.90 (1973), 139–156

46. G. Golub, C. Van Loan, *Matrix computations*, Mir, Moscow, 1999

47. A. Gragnami, O. De Feo, and S. Rinaldi, Food chains in the chemostat: Relationships between mean yield and complex dynamics, *Bulletin of Math. Biology*, v.60, 703–719, 1998

48. J. Guckenheimer, P. Holmes, *Nonlinear Oscilations, Dynamical Systems and Bifurcations of Vector Fields*, Springer-Verlag, N.Y., 1983

49. B. Hao and W. Zheng, *Applied Symbolic Dynamics and Chaos*, World Scientific, Singapore, 1998

50. J. Hadamard, Les surfaces a courbures opposees et leurs lignes geodesiques, *Journal de Matematiques Pure et Applique*, v.4 (1898), 27–73

51. P. Hartman, *Ordinary Differential Equations*, N.Y., 1964

52. M.E. Henderson, Computing invariant manifolds by integrating fat trajectories, *IBM Research, Technical Report RC22944*, 2003

53. M. Henon, A two-dimensional mapping with strange attractor. *Copmm. Math. Phys.*, 50, 1976, 69–77

54. M. Hirsch and S. Smale, *Differential Equations, Dynamical Systems and Linear Algebra*, N.Y. 1970

55. D. Hobson, An efficient method for computing invariant manifolds, *J. Comput. Phys.*, v.104 (1991), 14–22

56. S.L. Hruska, Constructing an expanding metric for dynamical systems in one complex variable, *Nonlinearity*, v.18 (2005), 81–100

57. C.S. Hsu, *Cell-to-Cell Mapping*. Springer-Verlag, N.Y., 1987

58. F. Hunt, Unique ergodicity and the approximation of attractors and their invariant measures using Ulam's method, *Nonlinearity*, v.11, no. 2, 307–317 (1998)

59. M. Hurley, Chain recurrence and attraction in non-compact spaces, *Ergodic Theory and Dynamical Systems*, 11 (1991), 709–729

60. K. Ikeda, Multiple-valued stationary state and its instability of the transmitted light by a ring cavity system, *Opt. Commun.*, 30, 257–261, 1979

61. A. Jorba, and C. Simo, On quasiperiodic perturbation of elliptic equilibrium points, *SIAM J. of Math. Anal.*, 27 (1996), 1704–1737

62. O. Junge, Rigorous discretization of subdivision techniques, in *Proceedings of Equadiff '99*, Berlin, 2000

63. T. Kaczynsky, K. Mishaikow, M. Mrozek, *Computation homology*, Applied mathematic sciences, 157, Springer, N-Y, 2004

64. K. Kaneko (editor), *Theory and Applications of Coupled Map Lattices*, Wiley, New York, 1993

65. S. Khryashchev, *On local controllability along trajectories*, VINITI, 868-B 97, 1997, 1–14

66. S.Yu. Kobjakov, D.Yu. Matiassevich, G.S. Osipenko, Location of the invariant sets. *Proceedings of the Fourth International Conference Tools for Mathematical Modelling*, Saint-Petersburg State Polytechnic University, 2003, 300–3007

67. A.N. Kolmogorov, S.W. Fomin, *An elements of the theory of functions and the functional analysis*, Science, Moscow, 1968 (in Russian)

68. A.N. Kolmogorov, New mesure invariant for transitive dynamical systems and endomorphisms of Lebesgue spaces, *Doklady AN USSR*, 115, no. 5 (1958), 861–864

69. B. Krauskopf, H.M. Osinga, Globalizing two-dimensional unstable manifolds of maps, *Int. J. Bifurcation and Chaos*, v.8, no. 3 (1998), 483–503

70. B. Krauskopf, H.M. Osinga, E.J. Doedel, M.E. Henderson, J. Guckenheimer, A. Vladimirsky, M. Dellnitz, O. Junge, A survey of methods for computing (un)stable manifolds of vector fields, *Int. J. Bifurcation and Chaos*, v.15, no. 3 (2005)

71. B.Yu. Levit and V.N. Livshits, *Nonlinear Network Transport Problems* Moscow, Transport, 1972 (in Russian)

72. D. Lind, B. Marcus, *An introduction to symbolic dynamics and coding*, New York, 1995

73. T. Lindström, On the dynamics of discrete food chains: Low- and high-frequency behavior and optimality of chaos, *Journal of Mathematical Biology*, v.45 (2002), 396–418

74. E.N. Lorenz, Deterministic nonperiodic flow, *J. Atmos. Sci.*, 20:130–141, 1963

75. A.M. Lyapunov, *Probleme General de la Stabilite an Mouvement*, Kharkov, 1892

76. G.G. Malinetskii, A.V. Potapov, A.I. Rakhmatov, E.B. Rodichev, Limination of delay reconstruction for chaotic systems with board spectrum, *Phys. Lett. A.*, v.179, 15, 1993

77. G.G. Malinetskii, A.V. Potapov, A.I. Rakhmatov, Limination of delay reconstruction for chaotic systems with board spectrum, *Phys. Rev. E.*, v.48, 904–912, 1993

78. G.G. Malinetskii, A.V. Potapov, *Modern problems in nonlinear dynamics*, Moscow, 2000 (in Russian)

79. R. Mane, Characterizations of AS diffeomorphisms, *Lect. Notes Math.*, v.597, 389–394 (1977)

80. R. Mane, A proof of the C^1 stability conjecture, *Publ. Math., Inst. Hautes Etud. Sci.*, 66, 161–210 (1988)

81. D.Yu. Matiyasevich, Localization of Invariant Sets of Dynamical Systems, *Journal of Mathematical Sciences*, v.124 (2004), no. 3, 4990–5000

82. D.Yu. Matiyasevich, E.I. Petrenko, Algorithms for the construction of isolated invariant subsets of the symbolic image, *Proceedimgs of XXXVI conference "Control Processes and Stability"*, St.Petersburg, 2005, 341–347

83. J.R. Miller and J.A. Yorke, Finding all periodic orbits of maps using newton methods: Sizes of Basins, *Physica D*, 135: 195–211, 2000

84. K. Mischaikow, Topological techniques for efficient rigorous computations in dynamics, *Acta Numerica*, 2002

85. D.A. Mizin, *Algorithms based on applied symbolic dynamics*, PhD dissertation, St. Petersburg State Polytechnic University, 2002 (in Russian)

86. D.A. Mizin, G. Osipenko, S. Kobyakov, The estimats for the topological entropy of the dynamic system, *Proceedings of the third international conference "Tools for mathematical modelling"*, 2001, 85–105

87. A.A. Moiseev, Symbolic image od dynamical system and algorithms for investigation, *Differential Equations and applications,* abstracts of the International Conference, St. Petersburg, 1996, 152–153

88. R. Montgomery, The N-body problem, the briad group, and action-minimizing periodic solutions, *Nonlinearity*, v.11 (1998), 363–376

89. M. Morse, A one-to-one representation of geodesics on a surface of negative curvature, *Amer. J. Math.*, v.43 (1921), no. 1, p.33–51

90. M. Morse, G.A. Hedlund, Symbolic dynamics I, II *Amer. J. Math.*, v.60 (1938), no. 4, p.815–866; v.62 (1940), no 1, p.1–42

91. S.E. Newhouse, Diffeomorphisms with infinitely many sinks, *Topology*, v.13, 1974, 9–18

92. Z. Nitecki, *Differentiable Dynamics*. London, 1971

93. Z. Nitecki and M. Shub, Filtrations, decompositions, and explosions, *Amer. J. of Math.*, v.97 (1975), no.4, 1029–1047

94. H.E. Nusse, J.A. Yorke, *Dynamics: Numerical Explorations,* Springer-Verlag, 1997

95. G.S. Osipenko, On a symbolic image of dynamical system, in *Boundary value problems*, Perm, 1983, 101–105 (in Russian)

96. G.S. Osipenko, Verification of the transversality condition by the symbolic-dynamical methods, *Differential Equations*, v.26, N9, 1126–1132; translated from *Differentsial'nye Uravneniya*, v.26, N9, 1528–1536, 1990

97. G.S. Osipenko, The periodic points and symbolic dynamics, in *Seminar on Dynamical Systems. Euler International Mathematical Institute, St. Petersburg, Russia, October and November, 1991*, Birkhauser Verlag, Basel, Prog. Nonlinear Differ. Equ. Appl. 12, 261–267 (1993)

98. G.S. Osipenko, Localization of the chain recurrent set by symbolic dynamics methods, *Proceedings of Dynamics Systems and Applications*, v.1, (1994), 227–282, Dynamic Publishers Inc

99. G. Osipenko and I. Komarchev, Applied symbolic dynamics: construction of periodic trajectories, *WSSIAA*, 4(1995), 573–587

100. G. Osipenko and I.Il'in, Methods of Applied Symbolic Dynamics, *Proceedings of Dynamics Systems and Applications*, v.2 (1996), 451–160

101. G. Osipenko, Indestructibility of invariant locally non-unique manifolds, *Discrete and Continuous Dynamical Systems*, v.2, no.2 (1996), 203–220

102. G. Osipenko, E. Ershov and J. Kim, *Lectures on Invariant Manifolds of Perturbed Differential Equations and Linearization*, St. Petersburg Polytech. University, 1996

238   References

103. G. Osipenko and Eu. Ershov, Perturbation of invariant manifolds of ordinary differential equations, in *Six lectures on dynamical systems,* editors B. Aulbach and F. Colonius, Wold Scientific, Singapore, 1996

104. G. Osipenko, Morse Spectrum of Dynamical Systems and Symbolic Dynamics, *Proceedings of 15th IMACS World Congress,* v.1 (1997), 25–30

105. G. Osipenko, Linearization near a locally non-unique invariant manifold, *Discrete and Continuous Dynamical Systems,* v.3, no.2 (1997), 189–205

106. G. Osipenko and S. Campbell, Applied Symbolic Dynamics: Attractors and Filtrations, *Discrete and Continuous Dynamical Systems,* v.5, no.1&2, 43–60 (1999)

107. G. Osipenko, Spectrum of a Dynamical System and Applied Symbolic Dynamics, *Journal of Mathematical Analysis and Applications,* v.252, no. 2, 587–616 (2000)

108. G.S. Osipenko, J.V. Romanovsky, N.B. Ampilova, E.I. Petrenko, Computation of the Morse Spectrum, *Journal of Mathematical Sciences,* v.120, no. 2 (2004), 1155–1166

109. G.S. Osipenko, *Lectures on symbolic analysis of dynamical systems,* St. Petersburg State Polytechnic University, 2004

110. J. Palis, J. Melo, *Geometric Theory of Dynamical Systems,* Springer-Verlag, New York (1982)

111. P. Pilarczyk, Computer assisted method for proving existence of periodic orbits. *TMNA,* 13(2): 365–377, 1999

112. P. Pilarczyk, Homology software, in *Computational homology program,* http:www.gatech.edu/chom

113. S.Yu. Pilyugin, *The space of Dynamical Systems with C^0-Topology,* Springer-Verlag, 1994, Lec. Notes in Math., 1571, 180p

114. S. Pissanetzky *Sparse Matrix Technology,* New York, Academic Press, 1984

115. V.A. Pliss, *Integral Sets of Periodic System of Differential Equations,* Moscow, 1977 (in Russian)

116. H. Poincare, Sur le probleme des trois corps et les equations de la dynamique, *Acta Mathematica* v.13 (1890), 1–271

117. H. Poincare, *Les methodes nouvelles de la mecanique celeste,* I–III (1892–1899), Gauthiers-Villars

118. J. Robbin, A structural stability theorem, *Annals of Math.,* v.94, no.3, 447–493 (1971)

119. C. Robinson, Structural stability of C^1-diffeomorphism, *J. Diff. Equat.,* v.22, no.1, 28–73 (1976)

120. C. Robinson, *Dynamical Systems: Stability, Symbolic Dynamics and Chaos,* 1995

121. Yu.V. Romanovsky, Optimization and stationary control of discrete deterministic process in dynamic programming *Kibernetika* **2**, 66–78 (in Russian); Engl. transl. *Cybernetics* **3**, 1967

122. M.L. Rosenzweig, Exploitation in three throphic levels, *The American Naturalist,* 107 (954), 275–294, 1973

123. R. Sacker and G. Sell, Existence of dichotomies and invariant splitting for linear differential systems I-III, *J. Diff. Eq.* v.15, no3 (1974), 429–458, v.22, no.2 (1976) 476–522

124. R. Sacker and G. Sell, A spectral theory for liner differential systems, *J. Diff. Eq.,* v.27, no. 3, (1978), 320–358

125. D. Salamon and E. Zehnder, Flows on vector bundles and hyperbolic sets, *Trans. Amer. Math. Soc.*, v.306, no. 2 (1988), 623–649

126. J. Selgrade, Isolated invariant sets for flows on vector bundles, *Trans. Amer. Math. Soc.*, v.203 (1975), 359–390

127. G. Sell, Nonautonomous differential equations and topological dynamics, *Trans. AMS*, 127 (1967), 241–283

128. R. Sedgewick, *Algorithms in Modula 3*, Addison-Wesley, Massachusetts, 1993

129. A.N. Sharkovsky, Structure theory of differentiable dynamical systems and weak nonwandering points, *Abh. Akad. Wiss. DDR. Abt. Math. Naturwiss. Techn.*, v.4 (1977), 193–200

130. A.N. Sharkovsky, Yu.L. Maistrenko, E.Yu. Romanenko, *Diffenrence equations and applications,* Naukova dumka, Kyev, 1986 (in Russian)

131. M. Shub, Stabilite globale de systems denamiques, *Asterisque*, v.56 (1978), 1–21

132. L.P. Silnikov, A case of the existence of a denumerable set of periodic motions, *Sov. Math. Dokl.,* v.6 (1965), 163–166

133. L.P. Silnikov, A contribution to the problem of the structure of an extended neighborhood of a rough state of saddle-focus type, *Math. USSR Sb.,* v.10 (1970), 91–102

134. C. Simo, Effective computations in hamiltonian dynamics, in *Cent ans apes les Methodes Nouvelles de H. Poincare*, 1–23, 1996

135. C. Simo, Invariant curves of analytic perturbed nontwist area preserving maps, *Regular and Chaotic Dynamics*, v.3 (1998), no. 3, 180–195

136. S. Smale, Diffeomorphisms with many periodic points, in *Differential and Combinatorial Topology*, Princeton Univ., 1965, 63–80

137. S. Smale, Differentiable dynamical systems, *Bull. Amer. Math. Soc.,* 73, 1967

138. S. Smale, The $\Omega-$ stability theorem, in *Global Analysis, Proc. Symp. in Pure Math.*, v.XIV, Amer. Math. Soc. 1970

139. C. Sparrow, *The Lorenz Equations: Bifurcations, Chaos, and Strange Attractors*, Springer, N.Y., 1973

140. H.J. Sussmann, Some properties of vector fields systems witch are not altered by small perturbations, *J. Differential Equations*, v.20 (1976), no. 2, 292–315

141. R. Tarjan, Depth-First Search and Linear Graph Algorithms, *SIAM J. Comput,* v.1 (1972), 146–160

142. R. Tarjan, *Data structures and network algorithms*, Philadelphia, Pa., 1991

143. H. Whitney, Diferentiable Manifolds, *Ann. Math.*, 37 (1936), pp. 645–680

144. S. Wiggins, *Introduction to Applied non linear Dynamical Systems and Chaos*, 1990

145. A. Wolf, J. Swift, H. Swinney and J. Vastano, Determining Lyapunov exponents from a time series, *Physica D,* 16 (1985), 285–317

146. Z. You, E.J. Kostelich, J.A.Yorke, Calculating stable and unstable manifolds, *Internat. J. Bifur. Chaos Appl. Sci. Engrg.*, v.1, no. 3, (1991), 605–623

147. Zh.T. Zhusubaliyev and E. Mosekilde, Bifurcations and chaos in piecewise-smooth dynamical systems, v.44, *Nonlinear Science A*, New Jersey, 2003

A

Double Logistic Map

Ampilova N.B.

St. Petersburg State University, Dept. Comp. Science
Russia

nataly@is1483.spb.edu

A.1 Introduction

Consider the map $T = T_{\lambda,\mu}$ of extended plane $\tilde{\mathbb{R}}^2 = \mathbb{R}^2 \cup \infty$ into itself: $z_{n+1} = T(z_n), z = (x,y)$, where T has the form

$$T : \quad \begin{aligned} x_1 &= (1-\lambda)\,x + \lambda\mu y(1-y), \\ y_1 &= (1-\lambda)\,y + \lambda\mu x(1-x), \end{aligned} \tag{A.1}$$

and λ, μ — are real parameters.

The topology on $\tilde{\mathbb{R}}^2$ is supposed to be induced by the Riemann sphere topology, namely, various neighbourhoods of the point at infinity are the complements of all possible compacts. We assume that ∞ is fixed point. The dynamical system generated by the map (A.1) was studied in [1,6]. The estimate of the domain of attraction to the point at infinity was obtained. Among other things it was proved that the outside of the disk

$$(x-r)^2 + (y-r)^2 \le 2r^2, \tag{A.2}$$

where $r = 0.5(1-\lambda+\lambda\mu)/\lambda\mu)$, is a neighbourhood of that point.

The map (A.1) has 4 fixed points: $O_0(0,0)$, $O_1(1-\frac{1}{\mu}, 1-\frac{1}{\mu})$ lying on the diagonal $y = x$, and $O_2(t,u)$, $O_3(u,t)$, where

$$t = \frac{1+\mu-\sqrt{(\mu-1)^2-4}}{2\mu}, \quad u = \frac{1+\mu+\sqrt{(\mu-1)^2-4}}{2\mu}. \tag{A.3}$$

The point O_0 is saddle for $\lambda \in (0, \frac{2}{\mu+1})$ and unstable node for $\lambda \geq \frac{2}{\mu+1}$.
Similarly, O_1 is saddle for $\lambda \in (0, \frac{2}{\mu-1})$ and unstable node for $\lambda \geq \frac{2}{\mu-1}$.
The fixed points O_2 and O_3 have eigenvalues s_1, s_2 equal to $1 - \lambda \pm \lambda i \sqrt{(\mu-1)^2 - 5}$, so they are focuses for $\mu \geq 1 + \sqrt{5}$ or $\mu \leq 1 - \sqrt{5}$ and nodes elsewise.

We explore the map in the focus O_2 neighbourhood and consider the conditions of appearance an invariant curve and periodic orbits in its vicinity.

A.2 Hopf Bifurcation

It is known that under some assumptions a bifurcation of a fixed point when it is focus leads to the occurence of an invariant circle which is smooth for parameters values in a neighborhood of the bifurcation point. The fixed point changes its stability type when the absolute value of the eigenvalue of Jacobi matrix in this point crosses a unit circle. Let the point on the unit circle be $s_0 = e^{i2\pi p/q}$, with (p, q) in lowest terms. It is said that p/q resonance occurs on the invariant circle if there is a pair of periodic orbits, one consisting of saddles and the other of sinks, which rotation number is p/q.

The case $q \geq 5$ is called weak resonance. An infinite number of periodic orbits are created and destroyed on the circle immediately after the Hopf bifurcation. Given p/q, a pair of periodic orbits exists on the invariant circle for parameters values lying in the narrow cusped region ("resonance horn" or "Arnold tongue") [5]. By Arnold's method this horn can be constructed by analytical methods. Unfortunately, such an approach is valued only for parameter when the invariant circle is smooth. In other case we need a computer simulation.

The strong resonance, $q = 1, 2, 3, 4$ exhibits rather different behaviour, the details of which are not yet fully understood. In this situation Arnold's method is not applicable as well. The articles [5, 7] are devoted to detailed investigation of Hopf bifurcation for two-parametric families of plane maps. We applied a combined approach to the map (A.1). It has been shown that a transition to chaotic regime in the case of strong resonance is a result of recurring Hopf bifurcation.

Arnold's method for two-parameter families of plane maps.
V. Arnold suggested using the following scheme to investigate such families. A map $\mathbb{R}^2 \to \mathbb{R}^2$ can be written as a function of one complex variable, and two real parameters can be written as a single complex parameter. Consider a map

$$z \to f_\mu(z) = \mu z + O(|z|^2), \tag{A.4}$$

where μ is a complex parameter.

Note that $z = 0$ is a fixed point of f_μ for all μ. The eigenvalues at this fixed point are μ and $\bar{\mu}$.

It should be noted that almost any two-parameter family of maps having a fixed point with complex eigenvalues can be transformed to the form (A.4) as follows. First translate the fixed point to the origin. Then make a complex linear change of coordinates which diagonalizes the linear part of the map. Finally, introduce the eigenvalue at the origin as a new complex parameter. It is necessary that the function which gives this eigenvalue in terms of the original parameters should be invertible.

Thus, the fixed point $z = 0$ is stable for $|\mu| < 1$ and unstable for $|\mu| > 1$. To study the bifurcation occuring for $|\mu| = 1$ (loss of stability), consider the point μ_0 on the unit circle. We assume that $\mu_0^k \neq 1, k = 1, 2, 3, 4$ and write f_μ in its normal form

$$f_\mu(z) = \mu z + c(\mu) z |z|^2 + O(|z|^4). \tag{A.5}$$

Let $Re(\bar{\mu}_0 c(\mu_0)) < 0$. Then under these assumptions, the Hopf bifurcation theorem [4] states, that for μ near μ_0 with $|\mu| > 1$ the map f_μ has an attracting invariant circle surrounding $z = 0$. Moreover, the areas of existence of periodic orbits ("resonance horns") may be constructed by analytical methods.

Numerical methods of investigation for two-parameter families of plane maps.
Arnold's method is valued only in the case of weak resonance for parameter values when the invariant circle is smooth. Hence to extend a resonance horn into the parameter area where the circle losses continuity we need a computer simulation. Given in [5] numerical method of construction of resonance horns is based on the following scheme.

When the parameter is outside the horn, no periodic point of rotation number p/q is present. When the parameter encounters the boundary of the horn, q saddle-nodes appear on the invariant circle. As the parameter passes into the interior of the horn, the saddle-nodes bifurcate forming saddle-sink pairs. There are now two periodic orbits on the invariant circle, one consisting of saddles and the other of sinks. The qth iterate of the map has $2q$ fixed points, alternating between saddles and sinks around the invariant circle. As the parameter continues to move across the horn, the saddles and sinks move apart. As the parameter approaches the other boundary, these points form different saddle-sink pairs. When the parameter approaches another boundary of the horn, these new saddle-sink pairs combine to form saddle-nodes.

Consider a two-parameter family of plane maps $F_{a,b}$. Given a, b and a number q, to find a q-periodic orbit of the map means to solve the equation $F^q(z) = z$, where $z = (x, y)$. Let z_0 be a solution of this equation. If the determined orbit is saddle or sink, then the point (a, b) is in the resonance horn. If there is 1 among eigenvalues of $DF^q(z_0)$, then z_0 is a periodic saddle-node and the point (a, b) lies on the boundary of the resonance horn. These boundary points are computed using Newton's method to solve the system equations: the above equation and the condition that 1 is an eigenvalue of the periodic point.

When the eigenvalue of the Jacobi matrix of the mapping F^q in the fixed point has absolute value 1, we can write it in the form

$$e^{i2\pi p/q} = cos2\pi p/q + isin2\pi p/q = u(a,b) + iv(a,b).$$

That allows to detail a point on the unit circle where a resonance horn starts to grow from.

For a case of strong resonance we apply a computer simulation and Newton's method to find periodic orbits and determine "horns".

A.2.1 The Application to Double Logistic Map

For the case of the focus O_2 these eigenvalues when crossing the unit circle can be written as $|s_1| = (1 - \lambda)^2 + \lambda^2(\mu^2 - 2\mu - 4)$ and

$$|s_1| = 1 \iff \lambda(\lambda(\mu^2 - 2\mu - 3) - 2) = 0.$$

As $\lambda > 0$, we have

$$(\mu - 1)^2 = 4 + \frac{2}{\lambda}. \tag{A.6}$$

This relation determines a line (Hopf bifurcation line) on the plane of parameters (λ, μ) such that for $1 + \sqrt{5} \leq \mu \leq 1 + \sqrt{4 + \frac{2}{\lambda}}$ the focuces O_2, O_3 are stable. When $\mu = 1 + \sqrt{4 + \frac{2}{\lambda}}$ an invariant curve appears in a O_2 (O_3) neighborhood, which is destroyed under the parameters changing.

In [2] the values of the parameter λ corresponding to an occurence of invariant curve were obtained by computer simulation methods. Let a point (λ_0, μ_0) is on the Hopf bifurcation line. Changing λ from λ_0 by a step ε, we can find the parameter λ value when an invariant curve occurs(λ_b) and is destroyed (λ_e). The obtained data are shown in Table A.1. It should be noted that the conditions $\mu \geq 1 + \sqrt{5}$ and $(\mu - 1)^2 \leq 4 + \frac{2}{\lambda}$ are satisfied simultaneously for $\lambda \leq 2$.

The following periodic orbits (weak resonance) were obtained by approximative method and specification using Newton's method:

$$\lambda = 0.512603, \mu = 4.11, p/q = 1/7; \lambda = 0.55260, \mu = 4.11, p/q = 3/22.$$

As for $p/q = 1/7$ we have $cos\frac{2\pi}{7} = 1 - \lambda$ hence $1 - \lambda \approx 0.6239$ and $\lambda \approx 0.3761$. So, 1/7-resonance occurs from the bifurcation line $\mu = 1 + \sqrt{4 + \frac{2}{\lambda}}$ when $\lambda \approx 0.3761$ and $\mu \approx 4.052$. After the destroying of the invariant curve the periodic orbit with $p/q = 2/15$ was found for $\lambda = 0.57260, \mu = 4.11$. The case of strong resonance means that $q \in \{1, 2, 3, 4\}$ and $(p, q) = 1$. Hence $p/q \in \{1, 1/2, 1/3, 1/4, 2/3, 3/4\}$. The correspondence between p/q, λ and μ is given in the following table.

Table A.1.

μ	$\lambda = \frac{2}{(\mu-1)^2-4}$	$(\lambda_b(\mu), \lambda_e(\mu))$	μ	$\lambda = \frac{2}{(\mu-1)^2-4}$	$(\lambda_b(\mu), \lambda_e(\mu))$
3.41	1.10613	1.11–1.2	3.56	0.783	0.797–0.905
3.42	1.07735	1.078–1.193	3.57	0.767	0.781–0.87
3.43	1.04992	1.061–1.15	3.58	0.75	0.79–0.86
3.44	1.02375	1.036–1.139	3.59	0.738	0.752–0.85
3.45	0.99875	1.011–1.13	3.6	0.724	0.739–0.85
3.46	0.9748	0.986–1.12	3.7	0.607	0.624–0.7
3.47	0.9519	0.966–1.08	3.8	0.52	0.536–0.667
3.48	0.93	0.945–1.05	3.9	0.4535	0.47–0.61
3.49	0.909	0.92–0.98	4.0	0.4	0.41–0.55
3.50	0.888	0.9–0.97	4.1	0.356	0.38–0.5
3.51	0.869	0.88–0.96	4.2	0.32	0.34–0.47
3.52	0.85	0.86–0.95	4.5	0.242	0.263–0.4
3.53	0.83	0.846–0.95	4.6	0.223	0.24–0.32
3.54	0.815	0.829–0.94	4.7	0.2	0.22–0.3
3.55	0.799	0.812–0.92	4.8	0.191	0.21–0.29

p/q	1	1/2	1/3	1/4	2/3	3/4
λ	0	2	3/2	1	3/2	1
μ	$\forall$	$1+\sqrt{5}$	$1+4\sqrt{3}/3$	$1+\sqrt{6}$	$1+4\sqrt{3}/3$	$1+\sqrt{6}$

Since the case of $p/q = 1(\lambda = 0)$ is singular and the fixed point O_2 is focus for $\lambda \leq 2$, we investigate the cases when $\lambda = 1$ and $\lambda = 3/2$.

$\lambda = 1$.

On the bifurcation line (A.6) it corresponds to $\mu = 1 + \sqrt{6}$. The eigenvalue is equal $\pm i$ and $e^{i\alpha} = \cos\alpha + i\sin\alpha$, so $\alpha = \pi/2$ and $e^{i\pi/2} = e^{2\pi i/4}$, i.e. $p/q = 1/4$. Nevertheless, there are no 4-periodic orbits in this case. The following periodic orbits were found:

$$\lambda = 1.14, \ \mu = 3.44, \ p/q = 1/11;$$

$$\lambda = 1.16, \ \mu = 3.44, \ p/q = 1/17;$$

$$\lambda = 1.18, \ \mu = 3.44, \ p/q = 1/7.$$

The behaviour of trajectories in a neighborhood of the invariant curve is shown in Fig. A.1–A.2. Fig. A.1 shows an appearance of the invariant curve and its deformation. Fig. A.2 illustrates recurrent Hopf bifurcation and, as a consequence, transition to chaotic regime.

$\lambda = 3/2$.

In this case the situation is more complicated. The eigenvalue is equal $-1/2 + i\sqrt{3}/2$ and $q = 3$. Computer simulation shows the existence of a 3-periodic

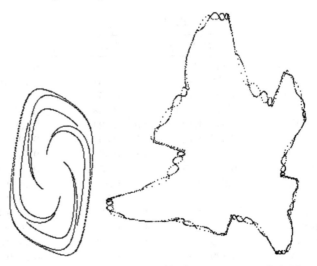

Fig. A.1. Strong resonance for the map (A.1): appearance of the invariant curve for $\lambda = 1.002$, $\mu = 3.44949$; $x \in [0.825, 0.865]$, $y \in [0.41, 0.46]$ and its deformation for $\lambda = 1.144$, $\mu = 3.44949$; $x \in [0.8, 0.9]$, $y \in [0.32, 0.56]$

Fig. A.2. Strong resonance for the map (A.1): recurrent Hopf bifurcation for $\lambda = 1.147$, $\mu = 3.44949$; $x \in [0.8, 0.9]$, $y \in [0.32, 0.56]$ and the transition to chaotic regime for $\lambda = 1.162$, $\mu = 3.44949$; $x \in [0.8, 0.9]$, $y \in [0.32, 0.56]$

orbit: Fig. A.3 demonstrates a set, arizing in a vicinity of the focus O_2 and its detailed part, Fig. A.4 also illustrates this part for different values of the parameter λ. It should be noted that this structure arises for $\lambda < 3/2 (\approx 1.49)$ and persists with some modifications up to $\lambda \approx 1.50080$.

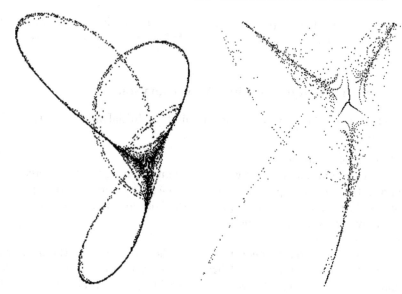

Fig. A.3. Strong resonance for the map (A.1), case $\lambda = 3/2$. The appearance of 3-periodic saddle orbit for $\lambda = 1.4938$, $\mu = 3.3094$; $x \in [0.7, 0.85]$, $y \in [0.3, 0.62]$ and the magnified middle part of the picture for $\lambda = 1.4941$, $\mu = 3.3094$; $x \in [0.8, 0.85]$, $y \in [0.4, 0.5]$

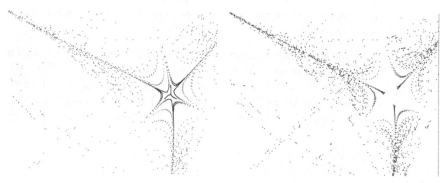

Fig. A.4. Strong resonance for the map (A.1), case $\lambda = 3/2$. The phase portrait in a vicinity of 3-periodic oprbit for $\lambda = 1.4955$, $\mu = 3.3094$; $x \in [0.81, 0.83]$, $y \in [0.44, 0.5]$ and for $\lambda = 1.4956$, $\mu = 3.44949$; $x \in [0.81, 0.83]$, $y \in [0.44, 0.5]$

A.3 Construction of Periodic Orbits

As was mentioned in the previous section, methods of computer simulation were applied to investigate the complex behaviour of the map in a vicinity of the invariant curve for Hopf bifurcation. λ, μ, for which an invariant curve occurs in the focus vicinity was obtained. In particular, it was shown that for $\lambda = 0.512603$, $\mu = 4.11$ the 7-periodic orbits appear. The numerical method of the construction of a periodic orbit is a construction of an initial approximation of the periodic orbit and its following refinement by Newton's method.

A.3.1 Construction of the First Approximation

The following methods were applied to find an initial approximation of a periodic orbit:

1. Methods of symbolic dynamics.
   For the system under investigation its symbolic image is considered. It is known that periodic paths on the symbolic image correspond to periodic orbits of the system. So, we apply a method of the search of periodic orbits with given period on a graph.
2. Lattice method.
   A lattice with a step h is constructed on the set defined by the inequality (A.2). Then the condition $||(T^k(z) - z)|| < \varepsilon$, where k is a given period and $\varepsilon \geq h$, is verified for every point z of the lattice. The points of the lattice for which the above condition is fulfilled are the points of the first approximation. Denote by P the set of such points. The value of ε is chosen to satisfy the condition $\varepsilon \geq h$. The sets P consructed for $\varepsilon = 0.01228$ and $\varepsilon = 0.06$ are shown in Fig. A.5. The circle shows a boundary for the invariant set of the system, the dark background displays the invariant set and blank squares depict the fixed points. The areas filled by the black color show the sets P.

The set P contains the first approximations to the roots of the equation $T^k(z) - z = 0$. To refine the roots we apply Newton's method, namely Theorem 25. In our problem $F(z_0) = T^k(z_0) - z_0$ and $DF(z_0) = DT^k(z_0) - I$. Denote by Cr the set of points for which Newton method is not defined, i.e. $Cr = \{z, |DT^k(z) - I| = 0\}$. Consider the set $P_1 = P \setminus (P \cap Cr)$. This is the set

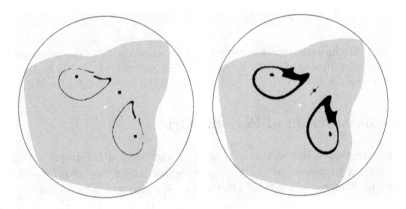

Fig. A.5. The sets P for $\varepsilon = 0.01228$ and $\varepsilon = 0.06$

of the first approximations to the roots of the equation $T^k(z) - z = 0$. As the map T specifies a two-degree polinomial, the equation has the power $2k$. Hence, it is requared to find $2k$ roots of the equation $F(z) = 0$. The results of numerical explorations show that not all points from the set P_1 are the initial approximation points for Newton's method. Assuming that a point $z_0 \in P_1$ is an initial approximation of a root of the given equation, we find its successive aproximations by Newton's method. It should be noted that even though the process converges, the inequality $KRL < 0.25$ from Theorem 25 may not hold for some $z_0 \in P_1$. In this case we take for x_0 the iteration z_l for which the inequality holds.

If Newton's process converges to a point z^* we determine its stability type calculating the eigenvalues of the matrix $DT^k(z^*) - I$. One need find only two points with different stability type which Newton's process converges to. Denote by z^*_{st} the stable point and z^*_{sd} the saddle one. Then their orbits may be considered as the first approximations to the k-periodic orbits. Actually, using this method we obtain an ε-trajectory with period k, where $\varepsilon = \rho(z^*_s, T^k(z^*_s))$ and "s" denotes either "st" or "sd".

A.3.2 Refinement of Periodic Orbits

Proposition 188. [3] *1. The derivative of the map T is Lipshitz with constant $L = 2\lambda\mu$ in U.*

2. Providing that the conditions $\lambda < 1, \mu > 0$ are fulfilled the derivative of the map $F(z) = T^k(z) - z$ is Lipshitz with constant $L = 2d(\alpha + \beta)^k$ in U, where $\alpha = 1 - \lambda$, $\beta = \lambda\mu$, $d = 2\sqrt{2}r$.

Construct the periodic orbits $SdOrb$ and $StOrb$ starting from the points z^*_{sd}, z^*_{st} (the initial approximations are equal to z^1_0 and z^2_0 respectively). According to the above proposition the derivative of the map $F(z)$ is Lipshitz. Calcucate the values R_1, R_2, where $R_1 = ||DF(z^1_0)^{-1}F(z^1_0)||, R_2 = ||DF(z^2_0)^{-1}F(z^2_0)||$. For the points from the sets $SdOrb$ and $StOrb$ find the pair (saddle,sink) such that the distance between them (denoted by ρ) is minimal. Then the size of the vicinity where a root of the equation lies may be chosen as $\min\{R_1, R_2, \rho/2\}$. Hence, the points of the orbits of the first approximation are enclosed by nonoverlapping neighbourhoods $\{V_i\}$ and we can use Theorem 26 to construct the true orbits. The 7-periodic orbits of the map T were obtained for the following parameter values: $\lambda = 0.51260309232291$; $\mu = 4.11$; $h = 1/2000$; $\varepsilon = h$.

The set of the first approximations P_1 in a neighbourhood on of the non-diagonal fixed points is given in Table A.2. The points from P_1 (belonging to the orbits of different kinds) for which Newton's method converges are given in Table A.3. For the first point the constants from Theorem 25 are the following:

$$L = 2753.87180686379, \quad K = 1.06329193217488,$$

Table A.2.

x	y	x	y
	The set of first approximation		
0.91519053028092	0.16117104015014	0.91954399154033	0.16291242465391
0.92999229856293	0.16813657816521	0.93434575982235	0.17074865492086
0.93956991333364	0.17423142392839	0.94305268234118	0.17684350068404
0.75759523269007	0.20818842175183	0.75672454043819	0.20905911400372
0.75585384818631	0.20992980625560	0.73321584963735	0.23866265056774
0.73147446513358	0.24127472732339	0.72973308062982	0.24388680407904
0.72799169612605	0.24649888083469	0.72625031162228	0.24911095759034
0.99007006394287	0.28655072442131	0.99007006394287	0.28742141667320
0.96656137314202	0.43805117624897	0.96569068089014	0.44066325300462
0.96307860413449	0.44762879101969	0.95698375837131	0.46678402056112
0.95611306611942	0.46939609731677	0.95524237386754	0.47287886632430
0.74192277215618	0.53121524720047	0.90474222325832	0.53208593945235
0.90474222325832	0.53295663170423	0.74366415665995	0.53382732395612
0.90387153100644	0.53382732395612	0.90387153100644	0.53469801620800
0.74540554116371	0.53643940071177	0.89516460848761	0.55472393800131
0.89516460848761	0.55559463025319	0.89429391623572	0.55733601475696
0.89429391623572	0.55820670700884	0.88819907047254	0.66007770047917
0.88732837822066	0.66094839273105	0.88645768596878	0.66181908498293
0.88732837822066	0.66181908498293	0.87426799444241	0.66617254624235
0.87513868669429	0.66617254624235	0.87600937894618	0.66617254624235

Table A.3.

| | | | The convergence points | |
| | initial point | | result | |
x	y	x	y	type
0.91519053028	0.16117104015	0.912391624536	0.160105567817	saddle
0.89429391623	0.55820670700	0.894754262324	0.556339195966	stable

$$R = 0.00021927999662, \quad KRL = 0.642089036344767.$$

For the second point these constants are equal

$$L = 2753.87180686379, \quad K = 0.76224726788667,$$

$$R = 0.00026423675833, \quad KRL = 0.554667639692385.$$

So, $R_1 = 0.00021927999662$, $R_2 = 0.00026423675833$.

The approximate orbits constructed for the points from Table A.3 are shown in Table A.4. As seen from the data, these orbits are (p, ε)-trajectories with $p = 7$ and $\varepsilon = 10^{-8}$.

Table A.4.

Approximation of the saddle orbit		Approximation of the stable orbit	
x	y	x	y
0.9123916245362	0.1601055678173	0.8947542623246	0.5563391959663
0.7280018183558	0.2464380117305	0.9561129376275	0.4695532838862
0.7460716663694	0.5372912311996	0.9907531588342	0.3172621741925
0.8874029104849	0.6610044350345	0.9392371359164	0.1739336957911
0.9046037737272	0.5326807061340	0.7604877264424	0.2050113035449
0.9653506381715	0.4414347801262	0.7140289360308	0.4836671372821
0.9899825159836	0.2856238007974	0.8741531580855	0.6659284970523
0.9123916199167	0.1601055624441	0.8947542806894	0.5563391822603

Table A.5.

The saddle 7-orbit		The stable 7-orbit	
x	y	x	y
0.9123916150407	0.1601055603641	0.8741529679879	0.6659283848862
0.7280018132088	0.2464380194013	0.8947542480865	0.5563397417942
0.7460716722669	0.5372912400927	0.9561128010091	0.4695535736031
0.8874029121721	0.6610044334643	0.9907555556419	0.3172131322644
0.9046037758254	0.5326807028257	0.9392374323408	0.1739339535371
0.9653506398601	0.4414347751477	0.7604882250030	0.2050108805026
0.9899825157887	0.2856237952711	0.7140286530890	0.4836663837345

The points (0.887402910484926, 0.661004435034576) and (0.874153158085516, 0.665928497052353) are the most close with $\rho/2 \approx$ 0.00706. In line with Theorem 26, the size of V_i is min $\{2R_1, 2R_2, \rho/2\} =$ 0.00044, ε is 10^{-8} and

$$a = 2.10828189573076, \quad L = 4.21359741887862, \quad K = 3.19327339144196,$$
$$\sigma = 0.00010611199917, \quad LK^2 \left(\frac{a^p - 1}{a - 1}\right)^2 \varepsilon = 0.118610200921989.$$

Table A.5 shows the refined orbits.

The obtained orbits are depicted in Fig. A.6. The periodic orbits in a vicinity of the point O_3 are shown as well, being circle signs mark saddles and filled cirles mark sinks.

It should be noted that the results of the construction of the first approximation both using symbolic dynamics methods and the lattice method with the refinement by Newton's method are practically the same. The value of ε when constructing the set P is in the interval (0.0001,0.001). Newton's method for orbits converges for 4-5 iteration. The orbit is obtained with accuracy 10^{-9}.

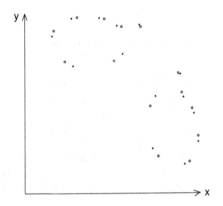

Fig. A.6. 7-periodic orbits in a vicinity of the invariant curve

References

1. N. Ampilova, A. Osipov, *Local bifurcations for Gardini map,* VINITI, 14.06.96, no. 1969-B96
2. N.B. Ampilova, Numerical investigation of invariant curves behaviour in the vicinity of a fixed point of the Gardini map, *Nonlinear dynamical systems,* v.1, Ed. G.Leonov, St. Petersburg, 1997, 5–13
3. N.B. Ampilova, Numerical methods of the construcrtion of periodic orbits near invariant curve for Hopf bifurcation, *Nonlinear dynamical systems,* v.2, Ed. G. Leonov, St. Petersburg, 2000, 71–80
4. V.I. Arnold, Loss of stability of oscillations close to resonant and versal deformation of equivariant vector fields, *Functional analysis and applications,* v.11, 1977, 1–10
5. D.G. Aronson, M.A. Chory, G.R. Hall and R.P. McGehee, Bifurcation from an invariant circle for two-parameter families of maps of the plane: a computer-assisted study, *Commun. Math. Phys.* 83, 3(1982), 303–354
6. L. Gardini, R. Abraham, R.J. Record, D Fournier-Prunaret, A double logistic map, Int. J. Bifurcation and Chaos, 1994, v.4, 145–176
7. D. Whitley, Discrete dynamical systems in dimensions one and two, *Bull. London Math. Soc.,* 15(1983), 177–217
8. J. Hale, H. Kocak, *Dynamics and Bifurcations,* Springer-Verlag, 1991

B

Implementation of the Symbolic Image

Danny Fundinger,
University of Stuttgart,
Institute of Parallel and Distributed Systems,
Universitätsstraße 38, 70569 Stuttgart, Germany,
research@danny.fundinger.de

The construction of the symbolic image is the basic task for investigations based on symbolic analysis. Once the system flow has been transformed into such a graph, all investigation tasks can be formulated as graph algorithms. Hence, an efficient implementation of this construction process is crucial for every investigation. In this chapter, we propose algorithms and adequate data structures which are appropriate to achieve this task. Additionally, we introduce some basic investigation methods on the graph in order to apply the implementation in practice. A theoretical analysis regarding the performance and accuracy of the computations is given. Afterwards, we focus on several important aspects of a practical application. Several extensions and tunings of the original concepts are proposed which extend the field of application for the method. We discuss these techniques and, if necessary, also mention some aspects of implementation and useful heuristics for an efficient usage. Numerical computations are performed for several dynamical systems in order to verify the proposed implementation. We study some reasonable parameter settings and the steps to be taken for the acquisition of the data. While doing so, the reader will be introduced to possible fields of application for symbolic analysis.

Acknowledgements

The author would like to thank V. Avrutin for discussions and for contributing some numerical computations to this work (Fig. B.2). The author also thanks G. Söderbacka for discussions about the Coupled Logistic Map.

B.1 Implementation Details

From the viewpoint of an implementation by a computer program, the principle scheme of every investigation can be considered as an iterative process which will be repeated for increasing levels of phase space discretization. At each level the calculation involves three main steps which have to be performed several times:

1. Subdivision of selected parts of the phase space into smaller parts.
2. Construction of the symbolic image for the current discretization of the phase space.
3. Application of an investigation method on the symbolic image graph. As a result, parts of the state space get selected for further subdivision and a more precise investigation.

These steps will be repeated until a termination criterion is fulfilled. This condition depends on the desired accuracy as well as on the existing computation power, see Sect. B.1.3.

In this section we describe the implementation of this basic framework and discuss those aspects which have to be taken into consideration for an efficient implementation. The basic framework as proposed here was implemented and tested within the Ant-project [3], a larger, non-commercial software package for the investigation of dynamical systems. The software is available for download, see [1].

B.1.1 Box and Cell Objects

In order to build a symbolic image for a domain $M \subset \mathbb{R}^d$ of the phase space, a finite covering $C = \{M(1), \cdots, M(n)\}$ has to be defined. In contrast to the theoretical approach, it is usually not possible to cover the complete domain $\mathcal{M}$ of the function f. Instead, in a practical approach we choose an area of investigation $M \subset \mathcal{M}$ in such way that we assume all important dynamics happen inside this area. Note, however, that usually only those objects can be detected by investigations of symbolic images which are completed covered by M. Generally, there are no restrictions concerning the geometry of M and of the sets $M(i) \in C$, except that they have to be closed and compact. The investigated domain M could be any confined part of the phase space and has to be provided by the user. For the simplicity of the implementation we assume here that it is an d-dimensional rectangular region:

$$ M = \left[M_1^{\min}, M_1^{\max}\right] \times \cdots \times \left[M_n^{\min}, M_n^{\max}\right]. \qquad \text{(B.1)} $$

Then we can subdivide the area M into *uniform grid boxes*. In order to do this, the user has to define for each state space coordinate k $(k = 1, \ldots, d)$ the numbers $i_k^{\max}$ of subdivisions for the domain M. Then M can be subdivided into

$$m = \prod_{k=1}^{d} i_k^{\max} \tag{B.2}$$

d-dimensional rectangular boxes. The length of the edge k is given for each box by

$$d_k(M(i)) = \frac{1}{i_k^{\max}} \left(M_k^{\max} - M_k^{\min} \right). \tag{B.3}$$

Every box must hold the information about its position in the d-dimensional state space. In context of an efficient implementation, the positions of boxes are represented as d-dimensional *multi-indices* $i \in \mathbb{N}^d$,

$$i = (i_1, \ldots, i_d) \in \mathbb{N}^d \tag{B.4}$$
$$\text{with} \quad i_k \in [1, \ldots, i_k^{\max}], \quad \forall k = 1, \ldots, d,$$

so that for every box $M(i) \in C$ exists a unique multi-index I defining its position in M.

This representation was chosen because it allows a fast and easy mapping from $\mathbf{x} \in M(i)$ to i and vice versa. Furthermore, the mapping range within the domain of usual integer values is much larger if multi-indices are used instead of one-dimensional indices. If N_{int} is defined as the maximal integer value of a computer system and one would only use one-dimensional indices to identify the boxes $M(i)$ then each $i \in \{1, \ldots, N_{\text{int}}\}$. An area of investigation M could only be subdivided in $m = N_{\text{int}}$ boxes because no more values are available for the description of all possible positions. Note that there must be an index for every position i, no matter if $M(i)$ exists or not. This is a crucial restriction for higher-dimensional systems. If multi-indices are used instead, an d-dimensional domain M can be subdivided in $m = (N_{\text{int}})^d$ boxes which means we have an exponential growth for the number of possible box positions.

After the construction of the covering C is completed, an approximation of the symbolic image based on C can be constructed. It represents a graph G, whereby the vertices of G correspond to the boxes $M(i)$ of C. In the following, we denote the vertices of G as *cells* c_I. Each of these cells c_I is uniquely connected with a box $M(i)$. This correspondence represents the link between the domain M and the symbolic image G. The edges of the graph G are defined by *adjacency lists*. To each cell belongs such a list with its target cells. Note that we do not use an *adjacency matrix* for the representation of the edges. Reason for this is that the symbolic image graph is considered to be huge but sparse. Hence, an adjacency matrix would require by far more memory resources than the lists.

B.1.2 Construction of the Symbolic Image

The construction of a symbolic image based on numerical calculations is always only an estimation of the "real" symbolic image G. Besides the usual numerical errors which occur by the computation of a mapping $f(\mathbf{x})$ for $\mathbf{x} \in M$, another reason for this is the fact that the construction of the image

$$T(\mathrm{i}) = f(M(\mathrm{i})) = \{\mathbf{y} \mid \mathbf{y} = f(\mathbf{x}), \mathbf{x} \in M(\mathrm{i})\} \subset \mathbb{R}^d \qquad (\mathrm{B.5})$$

for a box $M(\mathrm{i})$ would involve the calculation of $f(\mathbf{x})$ for every $\mathbf{x} \in M(\mathrm{i})$. This is, of course, beyond the limits of every finite numerical computation.

In our implementation, the image $T(\mathrm{i})$ is approximated by a finite set of points. This technique was also used for the implementation of similar numerical methods [7, 12], and has proved to be an efficient and reliable approach in practice. From each box $M(\mathrm{i})$ a representative set of k points is selected,

$$S(\mathrm{i}) = \{\mathbf{x}_j \mid \mathbf{x}_j \in M(\mathrm{i}), \quad j = 1 \ldots k\} \qquad (\mathrm{B.6})$$

the so-called *scan points* of the box $M(\mathrm{i})$. Then the approximation $\tilde{T}(\mathrm{i})$ of the region $T(\mathrm{i})$ in the state space is calculated by

$$\tilde{T}(\mathrm{i}) = f(S(\mathrm{i})) = \{\mathbf{y}_j \mid \mathbf{y}_j = f(\mathbf{x}_j), \mathbf{x}_j \in S(\mathrm{i})\} \qquad (\mathrm{B.7})$$

As one can see, the continuous region $T(\mathrm{i})$ will be approximated by the discrete set $\tilde{T}(\mathrm{i}) \subset T(\mathrm{i})$ consisting of k points. The number k of scan points for the boxes as well as the positions of these points within the boxes are parameters of the described method, which must be set by the user. A general strategy how the scan points should be placed within the box $M(\mathrm{i})$ is that they are either uniformly distributed within $M(\mathrm{i})$ or that they are put into the neighborhood of the boundaries.

Besides the calculation of scan points, a mapping

$$p : M \mapsto \mathbb{N}^d, \quad \forall \mathbf{x} \in M(\mathrm{i}) \ \Rightarrow \ p(\mathbf{x}) = \mathrm{i} \qquad (\mathrm{B.8})$$

of a point $\mathbf{x} \in M(\mathrm{i})$ onto a box index i is required for further computations. Additionally, we need its inverse mapping

$$p^{-1} : \mathbb{N}^d \mapsto M, \quad \forall \mathrm{i} : \quad p^{-1}(\mathrm{i}) = \mathbf{x} \ \Rightarrow \ \mathbf{x} \in M(\mathrm{i}) \qquad (\mathrm{B.9})$$

which defines for every $M(\mathrm{i})$ the spatial coordinates of a point within this box. Note that $p^{-1}(\mathrm{i})$ is only defined if $M(\mathrm{i})$ exists for i.

Due to the fact that all uniform grid boxes have the same size, the mapping $p : M \mapsto \mathbb{N}^d$ can be simply defined as

$$p(\mathbf{x}) = \mathrm{i} = (i_1, \ldots, i_d) \qquad (\mathrm{B.10})$$

$$\text{with } i_k = \left\lfloor \frac{\mathbf{x}_k - M_k^{\min}}{i_k^{\max}} \right\rfloor + 1, \quad k = 1, \ldots, d$$

The inverse mapping can be described in a similar way. Note, however, that the inverse mapping is not unique and, in practice, requires the definition of an arbitrary reference point within the box $M(\text{i})$. If using, for instance, the minimal point of each box, one can get

$$p^{-1}(I) = \mathbf{x} = (x_1, \ldots, x_d)^T$$
$$\text{with } x_k = M_k^{\min} + (i_k - 1) \cdot d_k(M(\text{i})), \ k = 1, \ldots, d \qquad \text{(B.11)}$$

After the functions p and p^{-1} have been defined, the approximation $\tilde{C}(\text{i})$,

$$\tilde{C}(\text{i}) = \left\{ M(\text{i}') \mid M(\text{i}') \cap \tilde{T}(\text{i}) \neq \emptyset \right\}, \qquad \text{(B.12)}$$

of the covering $C(\text{i})$,

$$C(\text{i}) = \{ M(\text{i}') \mid M(\text{i}') \cap T(\text{i}) \neq \emptyset \}, \qquad \text{(B.13)}$$

can be computed. Obviously, we have the relation $\tilde{C}(\text{i}) \subseteq C(\text{i})$.

Proceeding this task, the following steps have to be performed for each box $M(\text{i})$ in the covering C :

1. The set of scan points $S(\text{i})$ is calculated using a set of k globally defined relative coordinates

$$S = \left\{ \xi_j \mid \xi_j = (\xi_{j,1}, \ldots, \xi_{j,d})^T, \xi_{j,i} \in [0,1], i = 1..d, j = 1..k \right\} \ \text{(B.14)}$$

with respect to the reference point $\mathbf{x}_0 = p^{-1}(\text{i})$:

$$S(\text{i}) = \{ \mathbf{x}_j \mid x_{j,i} = x_{0,i} + \xi_{j,i} \cdot d_i(M(\text{i})), i = 1..d, \ \xi_j \in S, j = 1..k \} \ \text{(B.15)}$$

This is necessary for scaling the relative coordinates ξ_j, which are defined within the hypercube $[0,1]^d$, onto the area $M(I)$.
2. For every point $\mathbf{x}_j \in S(\text{i})$ the target point $\mathbf{y}_j \in \tilde{T}(\text{i})$ will be calculated. Dealing with dynamical systems discrete in time, i.e. $\mathbf{x}_{k+1} = f(\mathbf{x}_k)$, the target point $\mathbf{y}_j$ can be found by a simple one step iteration of the point $\mathbf{x}_j$:

$$\mathbf{y}_j = f(\mathbf{x}_j) \qquad \text{(B.16)}$$

Note that also the n-th iterated function f^n can be used in Eq. B.16 instead of f. This is necessary if we want to apply symbolic images to dynamical systems continuous in time, see Sect. B.5.1, or if we want to work with higher iterated functions, see Sect. B.6 for a more detailed discussion.
3. For each target point $\mathbf{y}_j \in \tilde{T}(\text{i})$ the corresponding box object $M(\text{i}')$ with $\mathbf{y}_j \in M(\text{i}')$ must be found.
   The index i' of this box is given by

$$\text{i}' = p(\mathbf{y}_j) \qquad \text{(B.17)}$$

It is important to check, whether the conditions

$$1 \leq i'_k \leq i_k^{\max} \quad \forall k = 1, \ldots, d \tag{B.18}$$

hold for the components of the multi-index i'. If not, the index i' exceeds the dimension range and there is no box defined for this index.

4. Within the implementation context, a memory access function

$$\mathbf{g} : \mathbb{N}^d \mapsto M, \ \exists M(\mathbf{i}) \in C \ \Rightarrow \ \mathbf{g}(\mathbf{i}) = M(\mathbf{i}) \tag{B.19}$$

is required in order to access a box object $M(\mathbf{i})$ for an index i. In our approach, a hash map H is used which contains a key i iff there exists a corresponding $M(\mathbf{i})$. Hence, no memory space is wasted for indices without a corresponding box object. Note that a fast and proper implementation of H requires the definition of a *strict weak order* relation $\prec$ for the box indices i of a covering C. This can be achieved by applying an iterative comparison of the components i_k $(k = 1, \ldots, d)$ of an index I, starting with the largest "digit" i_d:

$$\mathbf{i} \prec \mathbf{i}' \quad \text{iff} \quad \exists k = 1, \ldots, d \tag{B.20}$$
$$\text{with} \quad i_k < i'_k \quad \text{and} \quad i_j = i'_j \quad \forall j > k.$$

5. If $M(\mathbf{i}')$ has been located, a reference to it will be added to the list of $\tilde{C}(\mathbf{i})$. When the location of the target boxes $M(\mathbf{i}')$ for all scan points $\mathbf{x}_j \in S(\mathbf{i})$ is completed, the list of $\tilde{C}(\mathbf{i})$ is an estimation of the covering $C(\mathbf{i})$ for the box object $M(\mathbf{i})$.

After the described steps were performed for all boxes of the state space discretization, an approximation G of the symbolic image has been constructed. The vertices of the graph are the cells c_I corresponding to the boxes $M(\mathbf{i})$. For each c_I the adjacency list of target cells is given by the cells corresponding to the boxes of the covering $\tilde{C}(\mathbf{i})$.

B.1.3 Subdivision Process

In the previous section, the construction of a symbolic image G was described. It was already mentioned that the precision of the state space discretization for such a symbolic image is increased by an iterative process. To describe this process, we introduce an index s which indicates the level of state space discretization, $s = 0, 1, \ldots$, or, in other words, the subdivision depth of a symbolic image. In the following, the notation for the symbolic image G is extended to G^s. The notation C^s, I^s and so on is introduced in the same way.

In order to get the new graph G^{s+1} for a G^s, a new covering C^{s+1} has to be calculated. This covering depends on a selection of cells $SV(G^s) \subseteq V(G^s)$ chosen by the application of an investigation method to the graph G^s. The covering C^{s+1} usually covers only parts of the area covered by C^s. Let $c_{\mathbf{i}^s}$ be the cell in G^s which matches to the box $M(\mathbf{i}^s)$. Then the new area of investigation is given by the joint of all boxes which belong to a selection of cells $SV(G^s)$:

$$M^{s+1} = \bigcup_{c_{i^s} \in SV(G^s)} M(i^s). \tag{B.21}$$

Each of these boxes will be divided into m sub-boxes $M(i^{s+1})$, see Eq. B.2, which build together the new covering C^{s+1}. For every $i^s = (i_1, \dots, i_d)$ the m new indices i^{s+1} are defined as follows:

$$i^{s+1} = \left(j_1 + (i_1 - 1) \cdot i_1^{\max}, \quad \dots \quad, \quad j_d + (i_d - 1) \cdot i_d^{\max} \right) \tag{B.22}$$

$$\text{with} \quad j_k = 1, \dots, i_k^{\max} \quad \forall k = 1, \dots, d$$

After the subdivision, the graph G^{s+1} is constructed for the covering C^{s+1} as described in Sect. B.1.2, and the whole calculation process is repeated.

It is a distinctive feature of the methods of symbolic analysis that only the covering of the current subdivision step is subject of investigation. Hence, after parts of a covering C^s are selected and subdivided into a more precise covering C^{s+1} then C^s can be deleted. This is essential for an efficient usage of the memory resources. It is furthermore assumed that each covering C^s, except the initial covering C^0, does only cover parts of the area of investigation M.

B.2 Basic Investigations on the Graph

In this section we discuss some basic investigation techniques that can be applied to the symbolic image graph. Note that we use variations of some standard graph algorithms. A discussion of the original algorithms is out of scope of this work, therefore we refer to [1, 4].

B.2.1 Localization of the Chain Recurrent Set

The most important kind of investigation technique on the graph is the localization of the recurrent cells. Applying this technique, we can determine a neighborhood of the chain recurrent set Q. Theorem 16 in Chapter 6 states that the chain recurrent set can be approximated as precisely as one likes by the methods of symbolic analysis. The symbolic image must be constructed and subdivided for several times. The cells which have to be selected for subdivision are those which are recurrent. Hence, we propose now an algorithm for the localization of the recurrent cells in a symbolic image. Note that almost all other investigation methods of symbolic analysis also require the detection of the recurrent cells. Therefore, this technique is considered as a general first computation step of all investigations on a symbolic image graph.

An efficient approach to detect the recurrent cells is the variation of Tarjan's algorithm for the calculation of strongly connected components in directed graphs [22]. This algorithm locates the strongly connected components of a directed graph G by a *depth-first search*. Two vertices a and b of G are said to be strongly connected ($a \sim b$) if there exists a path from a to b and from b to a. Furthermore, the relation $a \sim a$ (reflexivity) always holds by definition. It can easily be proved that $\sim$ is an equivalence relation and that therefore G will be partitioned by the relation $\sim$ into equivalence classes, the strongly connected components. Although recurrent cells of G and strongly connected vertices are not the same, they are closely related to each other. If

$$\gamma_a = \{b \mid a \sim b\} \tag{B.23}$$

is a strongly connected component for which there is a path from a to each b and vice versa with $a \neq b$ then, for a as well as b, exists a periodic path. It follows that, if $|\gamma_a| > 1$, then for all cells $c \in \gamma_a$ exists a periodic path and therefore all these cells are recurrent. The special case to look at is $|\gamma_a| = 1$. Due to reflexivity, if there is only one component in the set it could mean that this cell is either non-recurrent or, if there is an edge $a \to a$, its least period size is 1.

So, for the localization of recurrent cells, Tarjan's algorithm needs a minor extension. What has to be done in addition is to perform a test for each set γ_a if $|\gamma_a| = 1$ holds. In this case, it has to be checked for the single cell of this set whether it is one-periodic (or recurrent), which means one of its target cells is itself, or not. If the cell is not one-periodic, it is non-periodic (or non-recurrent). All cells belonging to a set γ_a with $|\gamma_a| > 1$ are periodic, i.e. recurrent. All recurrent cells that belong to the same set γ_a can be considered as a set of equivalent recurrent cells which we denote in the following by H_k. Furthermore, we define the union of these sets by $\zeta = \{H_k\}$.

B.2.2 Localization of Periodic Points

A related investigation is the localization of p-periodic points $P(p) = \{\mathbf{x} \mid f^p(\mathbf{x}) = \mathbf{x}, \mathbf{x} \in M\}$ for a given value p. Theorem 23 states that the set of p-periodic points can be approximated as precisely as one likes by the methods of symbolic analysis. The symbolic image must be constructed and subdivided for several times. The cells which have to be selected for subdivision are those which are p-periodic. Hence, we propose now an algorithm for the localization of the periodic cells in a symbolic image.

For a practical application of Theorem 23, we have to consider that there might be more than one admissible path to which a cell c_i belongs to, especially in case of a coarse phase space discretization. Indeed, the number of admissible paths to which a cell belongs can even be infinite. In that case, it is impossible to explicitly compute each periodic path to which the cell belongs. On the other hand, if each cell of a symbolic image represents exactly one periodic point in the phase space then this cell c_p belongs only to one p-periodic path $\{\ldots, c_{i_0}, \ldots\}$ with $c_{i_0} = c_p$. Although such a precise covering can not be achieved by numerical computation, a reasonably fine phase space discretization is usually sufficient to get a unique path for a cell which belongs to a covering of a periodic point. Considering these facts, the p-periodic points can be located by selecting all cells for subdivision which have a *shortest periodic path* for a period $p' \leq p$. Obviously, such a selection contains all cells which belong to a p-periodic path. After several subdivisions, we check that there exists a unique periodic path for each cell of the symbolic image. In case this can not be achieved, a higher precision of the symbolic image is required, i.e. more subdivisions must be applied.

Consequently, an algorithm is needed which is able to find the shortest periodic path $c_i \to \cdots \to c_i$ for every recurrent cell c_i on the symbolic image graph. Furthermore, the length of such a path must be detected. Note that Tarjan's algorithm is not capable to solve this task. Instead, we introduce a different algorithm which is based on the idea of Dijkstra's algorithm for calculation of shortest paths in directed graphs [8]. It belongs to the class of so-called *greedy algorithms* and performs a *breadth-first search*.

The Dijkstra algorithm does not only find the shortest paths from each cell c_i to all other cells of the graph, but also locates at the k-th step first the path $c_i \to \cdots \to c_u$ so that the following equation is fulfilled:

$$d(c_i, c_u) = \min\{d(c_i, c_v) \mid c_v \in V(G) \wedge (c_i \to \cdots \to c_v) \notin D_{k-1}\}, \quad \text{(B.24)}$$

where $d(c_i, c_u)$ is defined as the length n of the shortest path between c_i and c_u, and D_{k-1} is the set of all shortest paths which have already been detected in the previous steps. Then the shortest periodic path of a cell c_i can be found by checking for the first detected shortest path $c_i \to \cdots \to c_u$ whether the edge $c_u \to c_i$ exists. If so, the algorithm can be stopped because the path $c_i \to \cdots \to c_u \to c_i$ is the shortest periodic path for the cell c_i and the length of this path is the period of c_i. If not, then the next shortest path has to be detected and checked for the same condition until a periodic path has been found or until all shortest paths have been visited.

There are several improvements to speed up Dijkstra's algorithm within our context. First of all, the original Dijkstra algorithm is developed for weighted graphs while the edges of G are unweighted. This means that the edge weight $\gamma(c_i \to c_j)$ is 1 for all edges of G. Therefore, the outer edge of visited but not yet examined cells can be implemented as a queue. Every cell which is visited first time and becomes a part of the outer edge will be pushed into the queue, while the next cell which will be examined can be popped out of the queue. This works fine because our edges are unweighted and so the distance between c_i and the first element in the queue is always the minimum distance between c_i and every other element in the queue.

Next it should be considered that all periodic cells of G have to be inspected. So in worst case, the modified Dijkstra algorithm must be started once for each cell $c_i \in V(G)$. In order to spare out some of the cells, we can first run the Tarjan algorithm to detect the recurrent cells and the sets ζ of equivalent recurrent cells. The Dijsktra algorithm must then only be started for the recurrent cells $c_i \in RV(G)$. Furthermore, it is sufficient to check for each of these cells only the paths to equivalent recurrent cells, i.e. those cells which belong to the same set $H_k \in \zeta$. Cells which do not belong to the same set can not belong to the same shortest periodic path.

Despite all improvements, the modified Dijkstra's algorithm can not compete with the performance of the aforementioned Tarjan's algorithm. So it should only be chosen by the user if the additional information about the periodic paths and/or the least period sizes are really required for the calculation.

B.3 Performance Analysis

The performance of symbolic image construction as described above is analysed by studying the *worst-case scenario* for the time complexity. We use the standard *O-notation*. In the following, n_s denotes the number of cells in the symbolic image G^s.

Proposition 189. [2] *The construction of the symbolic image G^s with respect to the covering C^s is in $O(n_s \log(n_s))$.*

In order to construct G^s, we consider a function $getBoxMapping(M(i))$ which is called for every box $M(i) \in C^s$. This function calculates an estimation $\tilde{C}(i)$ for $C(i)$ by first locating all indices i' with $M(i') \in \tilde{C}(i)$ and afterwards accessing these boxes by calling the function $g(i')$, see Eq. B.19. Each call of $getBoxMapping(M(i))$ is in $O(\log(n_s))$. Hereby, the computation time depends on a constant c which is governed by the number of scan points, $|S|$, and a number m_f so that $c = m_f |S|$. Note that m_f describes the number of function iterates, see Sects. B.5.1 and B.6. The size of the constant c is significant for the practical computation.

These results are closely related to the proposed method for the approximation of the covering $C(I)$. The number of edges for a cell is limited by the scan points per box, and in our case this is a constant. If another approach is chosen for the approximation of the covering like, e.g. interval arithmetic [11] or the calculation of the Lipschitz constant [15], the number of edges per cell is not longer bounded by a constant. For the performance analysis this means that the time complexity must possibly be multiplied by n_s.

Proposition 190. [2] *The construction process and investigation of a symbolic image for a subdivision phase s can be done in time $O(n_s \cdot \log(n_s))$ if only recurrents cells are located using Tarjan's algorithm, and in time $O(n_s^2)$ if the least period size and the shortest periodic path for each cell should also be calculated.*

The time required for the computation of recurrent cells is within $O(n_s \cdot \log(n_s))$, and therefore almost ideal from the theoretical point of view, especially for large n_s. Hereby, the construction of the symbolic image requires the main computational effort. The effort for the subdivision of cells as well as the localization of recurrent cells can be neglected. The time complexity changes in case the least period sizes and shortest periodic paths are computed by application of the algorithm proposed in Sect. B.2.2. Then the investigation of the graph can require by far more computational effort than the graph construction.

Performance analysis shows that the computation time of the algorithms is no major obstacle for the construction of the symbolic image. Instead, the crucial factor is the size of the input value n_s, i.e. the memory resources required for a computation. Note that n_s could grow almost exponential during the subdivision process. The size and growth rate of n_s depend hereby not only on the investigation task, i.e. the dimension of those objects which are the subjects of investigation, but also on the specific properties of the focused dynamical system. Practical application has shown that a high growth rate of n_s is a limitation for many computations. Hence, it should be the main concern to keep the number of cells n_s low and avoid an high growth rate during subdivision. Often this can be achieved by appropriate parameter settings or tunings of the method, see Section B.6.

B.4 Accuracy of the Computations

Let us next consider the accuracy of the numerical calculation. One should recall that we do not calculate specific points in the domain space but boxes $M(i)$ with some extent. These boxes are always only an outer covering of a solution. In our implementation, a box $M(i)$ is defined as a uniform grid box. The size of such a grid box defines the accuracy ϵ of the calculation. Let us denote by d_k the edge length of a grid box $M(i)$ on the dimension axis k, and by L^s the union of boxes which correspond to the selected cells $SV(G^s)$ in the symbolic image G^s constructed after the s-th subdivision. Then these boxes in L^s are neighborhoods (or an outer covering) of a solution S. The basic principle of symbolic analysis is that the sequence of embedded neighborhoods $L^0 \supset L^1 \supset \cdots \supset L^m$ gets for every subdivision step $s = 0, \ldots, m$ closer to S in the way that, if the largest edge length tends to zero as s becomes infinite, then,

$$\lim_{s \to \infty} L^s = \bigcap_s L^s = S. \tag{B.25}$$

See also Theorem 4 in Chapter 3 and Theorem 16 in Chapter 6 as examples of this principle. Unfortunately, for practical numerical calculations there is a minimal edge length which limits the accuracy. Reason for this is that the n-dimensional phase space covered by M, see Eq. B.1, gets divided into regions which are identified by multi-indices $i \in N^n$, Eq. B.4. As mentioned above, a value i_k of the k-th component of the multi-index i is represented by an integer value, and for every computation machine there exists a constant N_{int} giving the largest number which can be represented as an integer value. Consequently, we have the limitation $i_k^{max} \leq N_{int}$. Therefore, every edge length d_k is limited to

$$d_k^{min} = \frac{M_k^{max} - M_k^{min}}{N_{int}} \tag{B.26}$$

which means that the minimal error ϵ_k can only shrink down to $\epsilon_k \geq d_k^{min}$.

Note that this limit is not specific for the presented method but only for the implementation presented in this work. Furthermore, it is possible to extend the limit to any size by taking a different representation of a number i_k, though this implies a higher memory consumption.

B.5 Extensions for the Graph Construction

We consider two important extensions of the proposed implementation. One is the integration of dynamical systems continuous in time, the other a technique for the better approximation of the image of a box. Both of these extensions aim to improve the construction of the symbolic image graph and extend the possible field of application for the methods of symbolic analysis.

B.5.1 Dynamical Systems Continuous in Time

Only dynamical systems discrete in time have been discussed so far. The symbolic image for such a system

$$\mathbf{x}_{k+1} = f(\mathbf{x}_k), \ x_k \in M$$

can be constructed by performing one iteration which means simply applying the system function $f(\mathbf{x})$ on the points $\mathbf{x} \in M(i)$ lying in a box $M(i)$ of a certain covering, see Eq. B.16. If we are dealing now with systems continuous in time given by an ODE, i.e. $\dot{\mathbf{x}} = \mathbf{F}(t, \mathbf{x})$, $t \in \mathbb{R}$, some kind of mapping is required which transforms an orbit continuous in time into one discrete in time. As already mentioned in Chapter 1, a shift operator along trajectories is needed. Such a shift operator $\phi(t, t_0, \mathbf{x}_0)$ is considered to be the solution of the vector field $\mathbf{F}$ with an initial condition $\phi(t_0, t_0, \mathbf{x}_0) = \mathbf{x}_0$.

In our implementation, we use a stroboscopic mapping with a fixed discretization time t. This approach is applicable if the underlying differential equation is autonomous. Such a shift operator has the form $f(\mathbf{x}) = \phi(t, \mathbf{x})$ with $\phi(0, \mathbf{x}) = \mathbf{x}$. It can be calculated by solving the equation

$$\dot{\mathbf{x}}(t) = \mathbf{F}(\mathbf{x}(t)) \tag{B.27}$$

for the time t and initial conditions $\mathbf{x}(0) = \mathbf{x}$. We assume a fixed $t > 0$. In that case, $\phi(t, \mathbf{x})$ is also called a *time-t map*. Such a time-t map is a restriction of ϕ to $M \times tZ$ and, hence, a discretization of the dynamical system continuous in time. Note that similar approaches for a discretization were also proposed in [14, 20].

Consider now that in the context of a computer implementation it is suitable to use a small integration step size Δt for the applied integration method in order to minimize numerical errors. Hence, we do not calculate $\phi(t, \mathbf{x})$ explicitly. Instead, we use an integration step size $\Delta t = t/n$ with $n \in N$ and iterate ϕ for n times so that

$$\phi(t, \mathbf{x}) = \phi(\Delta t \cdot n, \mathbf{x}) = \phi^n(\Delta t, \mathbf{x}).$$

This approach allows us the numerical computation of time-t maps for any precision, independently of the chosen discretization time t. In the following, we use the notation $f(\mathbf{x}) = \phi(\Delta t, \mathbf{x})$ so that the time-t map for a $t = \Delta t \cdot n$ is given by $f^n(\mathbf{x})$. Hence, the symbolic image is constructed assuming $f^n(\mathbf{x})$ as the system function instead of $f(\mathbf{x})$, see also the comments to Eq. B.16.

B.5.2 Error Tolerance for Box Images

As already stated, the construction of symbolic images requires the approximation $\tilde{C}(\mathrm{i})$ of the covering $C(\mathrm{i})$, see Eqs. B.12 and B.13. Note that for our proposed implementation there is $\tilde{C}(\mathrm{i}) \subseteq C(\mathrm{i})$, and so some boxes may get lost. This behavior can be reduced by usage of a large number of scan points for each box. A disadvantage of this approach is that the computation time may become inappropriately large. As another solution of the problem one can extend the covering $\tilde{C}(\mathrm{i})$ by boxes which correspond to boxes in its neighborhood. We define a small constant ϵ and introduce the extended covering

$$C^{\mathrm{ext}}(\mathrm{i}) = \big\{ M(\mathrm{i}') \mid \exists M(J) \in \tilde{C}(\mathrm{i}), \exists \mathbf{x} \in M(J), \exists \mathbf{y} \in M(\mathrm{i}'), \\ \rho(\mathbf{x}, \mathbf{y}) \leq \epsilon \big\}. \tag{B.28}$$

Note that a suitable setting of the constant ϵ depends on the edge length $d_k(M(\mathrm{i}))$ of the boxes, Eq. B.3, and hence on the subdivision level. In our implementation, the user can define a parameter e, which is in the following denoted as the error tolerance, so that

$$\epsilon_k = e \cdot d_k \tag{B.29}$$

for the k-th phase space component. The condition $|x_k - y_k| \leq \epsilon_k$ is used instead of $\rho(\mathbf{x}, \mathbf{y}) \leq \epsilon$ in Eq. B.28. Note that d_k is the generalized edge length of the boxes in a covering C.

In practice, this parameter was used to detect p-periodic trajectories if they can not be found otherwise, whereby a setting of $e = 0.1$ proved to be sufficient in our simulations. One should keep in mind that the use of this parameter usually increases the size of the symbolic image and, therefore, it should only be applied if necessary. Furthermore, it is often a good alternative to increase the number of scan points $S(i)$ for a more precise calculation instead of applying error tolerance. Some of the scan points should then be placed close to the corners of the boxes.

Note that this approach was motivated by the technique described in Junge [15]. Actually, if ϵ is calculated by Lipschitz constants as proposed by Junge, it could be guaranteed that $C(i) \subseteq C^{\text{ext}}(i)$.

B.6 Tunings for the Graph Investigation

Performance analysis has shown that the computation time of the algorithms is no major obstacle for the construction of the symbolic image. Assuming n_s is the number of cells belonging to a symbolic image G^s, i.e. $n_s = |V(G^s)|$, all computations can be performed in $O(n_s \cdot \log(n_s))$, see Prop. 190. Instead, the crucial factor is the size of the input value n_s or, in other words, the memory resources required for a computation. Note that n_s could grow almost exponential for $s \to \infty$, i.e. during the subdivision process. Ideally, this growth rate should only depend on the investigation task or, more precisely, the dimension of those objects which are the subjects of investigation. However, due to the complexity of the underlying dynamics, this is often not the case. Instead, we observed that in many computations much more cells are selected for subdivision than necessary. The major problem we come across is that not only those cells are selected which correspond to boxes containing parts of the solution, but also several more cells which correspond to boxes in the neighborhood of the solution. Reason for this is a too coarse discretization of the phase space. In the following, we will refer to this phenomenon as *clustering*. Due to clustering, the growth rate of the number of cells in a symbolic image increases during the subdivision process, and the accuracy of the computation shrinks. Moreover, the analysis of computed data is more difficult.

Taking the theoretical point of view, the selection of too many cells for subdivision does not matter. By successive application of the subdivision process the discretization of the phase space gets finer. Eventually, those boxes which do not contain a solution will be deleted and the solution is detected as precisely as one likes. However, taking the practical point of view, one has to deal with limited resources. That means that the number of applicable subdivisions is limited by the memory space of the computation machine which only allows the storage of a symbolic image graph of a limited size. Therefore, it is our strong concern to avoid clustering, i.e. the selection of cells for subdivision which do not contain a solution. This aim can only be achieved by a change of paradigm. The target is not anymore the rigorous construction of the symbolic image graph for a phase space discretization. We are not interested in providing all existing edges between the cells as requested in the theoretical approach, but rather more only those edges which are necessary for the detection of the solution. By doing so, we are also aware of the fact that some important information might get lost. However, empirical studies have shown that numerical investigations are mostly limited by performance resources instead of an insufficient approximation of the symbolic image. A significant reason for this is that the method is typically quite robust.

B.6.1 Use of Higher Iterated Functions

When dealing with dynamical systems discrete in time $x_{n+1} = f(x_n)$, the points $y \in \tilde{T}(i)$, which represent the images of x, are calculated as direct successors of the scan points: $y = f(x)$. However, in some cases it is more suitable to use an iterated function of f and calculate the image points by

$$y = f^n(x), \quad n > 1 \tag{B.30}$$

In other words, the symbolic image is not constructed for the function f but for the n-th iterated function f^n. In the following, we will denote a symbolic image constructed for f by G_f and for f^n by $G_{f[n]}$.

Obviously, the symbolic image graph $G_{f[n]}$ with $n > 1$ differs from G_f. More precisely, $G_{f[n]}$ might have less edges than G_f. However, $G_{f[n]}$ is still useful for investigations. In order to clarify this, we introduce some theorems about the relations of f^n and f with regard to invariant sets.

Proposition 191. [9] *If $Q \subset M$ is an invariant set for f, then also for any f^n, $n \in N$, i. e.*

$$f(Q) = Q \Rightarrow f^n(Q) = Q.$$

Considering this result, we can conclude that all invariant sets of a dynamical system generated by f can also be found in a dynamical system generated by f^n.

Proposition 192. [9] *If $Q' \subset M$ is an invariant set for f^n, $n \in N$ then $Q = \bigcup_{0 \leq k < n} f^k(Q')$ is an invariant set for f, i.e.*

$$f^n(Q') = Q' \Rightarrow \bigcup_{0 \leq k < n} f^k(Q') = Q = f(Q).$$

Proposition 193. [9] *If $Q' \subset M$ is an invariant set for f^n, $n \in N$ then there is an invariant set Q for f with $Q' \subseteq Q$, i.e.*

$$f^n(Q') = Q' \Rightarrow \exists Q \supseteq Q' : f(Q) = Q.$$

An important conclusion of Proposition 193 is that a dynamical system generated by f^n still consists of the same invariant sets than the one generated by f. Each invariant set Q' found for f^n belongs to an invariant set Q of f. Furthermore, according to Proposition 191, all sets Q of f can be found in the dynamical system of f^n.

Recall now that most of our investigations aim to detect specific types of invariant sets. If all invariant sets of f are preserved in the dynamical system of f^n then, obviously, they can also be detected in $G_{f[n]}$. However, note that the characteristics of the sets might change. Let us look, for instance, on the invariant sets of periodic points $P(p)$. The invariant set $P(6)$ of f is then equivalent to the invariant set $P(2)$ of f^3 but the points belonging to these sets have a different periodicity with respect to f and f^3. Hence, one has to be careful when analyzing the results of $G_{f[n]}$. However, although the periodicity might change, every periodic point of f is also periodic for f^n, and no other periodic points than for f are found for f^n. The same is true for points belonging to quasiperiodic trajectories (without proof).

Each edge in the graph $G_{f[n]}$ represents a longer part of a trajectory than in G_f. In terms of tuning this is of interest because transient dynamics can then be better distinguished from asymptotic ones. Less cells which do not contain a solution are selected for subdivision, and the growth rate of cells during the subdivision process is lower. However, the tuning has also some drawbacks. First of all, the computation time for the construction of $G_{f[n]}$ increases by factor n in comparison to G_f. Furthermore, it is more likely that unstable parts of the solution, e.g. unstable periodic or quasiperiodic points, might not be detected because the forward iterates $\mathbf{y} = f^n(\mathbf{x})$ diverges stronger from these objects than $\mathbf{y} = f(\mathbf{x})$, see also the discussion in Sect. B.6.2. Last but not least, taking the analytical point of view, one must be aware about the change of characteristics regarding the invariant sets of $G_{f[n]}$ in G_f.

B.6.2 Reconstruction of Fragmented Solutions

In the former sections we have discussed the usage of higher iterated functions. In many calculations, this option turned out to be an adequate technique to tune investigation methods. However, it was also mentioned that unstable parts of a solution might not be detected if the number of iterations n or the time t is chosen too large. In practice, we observed, that for crucial settings of the parameters, some unstable invariant sets do not completely disappear at once, but rather more fall apart. Some parts of them are still recognized while others vanish, as it can be seen, for instance, in Fig. B.6(a) in case of a computation of the chain recurrent set for the Lorenz system.

Such a phenomenon is a result of taking only a limited number of scan points per box in combination with following a relatively long run of trajectories in order to construct the edges of the symbolic image graph. This leads to a loss of information about the structure of unstable invariant sets. It is not our intention to give here a detailed analysis of this problem, but rather more a solution for the reconstruction of such unstable objects. Nevertheless, one should keep in mind that not every structure that looks like a disappearing unstable invariant set is necessarily a fragment of the solution. In some cases, it turned out that objects which seemed to be parts of unstable limit cycles belonged to non-cyclic orbits. So, after the application of the method of reconstruction, further tests have to be applied to approve the correctness of obtained results.

The method, as introduced here, aims only on the reconstruction of the chain recurrent set. For other investigations, slight changes might be necessary. The reconstruction can be done by application of an extension to the symbolic image construction algorithm. The basic idea here is to add and/or select all cells belonging to boxes $M(i)$ of the symbolic image $G_{f[n]}$ which will be passed by the forward iterates $f^1(\mathbf{x}), \ldots, f^{n-1}(\mathbf{x})$ on its way from $\mathbf{x}$ to its image $\mathbf{y} = f^n$. Therefore, first the symbolic image $G_{f[n]}$ will be constructed according to the standard approach. Then the investigation method is applied in order to get the set of recurrent cells $RV(G_{f[n]})$. Afterwards, the following extension must be applied before the next subdivision. For every recurrent cell $c_I \in RV(G_{f[n]})$ its corresponding box $M(i)$ is detected. Then, for every scan point $\mathbf{x} \in S(i)$ it has to be checked, whether its target point $f^n(\mathbf{x}) = \mathbf{y} \in \tilde{T}(i)$ lies in a selected cell which is equivalent, i.e. belongs to the same set of strongly connected components. If so, we locate for each value $f^k(\mathbf{x})$, $k = 1, \ldots, (n-1)$, the box $f^k \in M(i')$. If the box $M(i')$ and its corresponding cell c_i' do not exist for a visited area, they will be added to the symbolic image. Furthermore, the cell c_I' will be marked as recurrent, no matter if it already existed or was just added.

If this extension is applied, the course of a trajectory, which connects recurrent cells, will be reconstructed. Note that the symbolic image can only become more precise by this extension. If a source cell c_i and its target cell c_i' are recurrent and equivalent, then, consequently, all the cells corresponding to boxes which are passed by the connecting trajectory are also recurrent. In fact, if all numerically computed symbolic images of an investigation would have been exact and no approximation, reconstruction would not change them. As already mentioned, this operation might add new cells to the symbolic image. So it can still be applied in a stage of subdivision when the fragmented invariant set has already fallen apart to a large extent. This can be seen in Fig. B.6(b), where unstable limit cycles of the Lorenz system are reconstructed.

B.7 Numerical Case Studies

In order to demonstrate the capabilities of symbolic analysis we present some typical examples of global analysis. The main aim hereby is to demonstrate what kind of results can be obtained with the basic investigation techniques presented in Sect. B.2, and how the parameters of the method have to be adjusted for specific investigation tasks. Also, the application of the extensions and tunings is shown. As examples we chose systems discrete as well as continuous in time. The reference machine for all these calculations was an

Asus L3000D laptop with an AMD Athlon XP-M 1400+ processor and 512MB SDRAM.

B.7.1 Ikeda Map

We start with a 2-dimensional map, namely the Ikeda map [13]. The system is defined as

$$\mathbf{x}(n+1) = f_I(\mathbf{x}(n)), \tag{B.31}$$

$$f_I(\mathbf{x}) = (r + a\, (x\, \cos g(x,y) - y\, \sin g(x,y)),\, b\, (x\, \sin g(x,y) + y\, \cos g(x,y)))$$
with $g(x,y) = c_1 - \frac{c_2}{1+x^2+y^2}$.

The numerical simulations have been carried out for the parameter values $a = b = 0.9$, $c_1 = 0.4$, $c_2 = 6.0$ and $r = 0.9$.

We start the global analysis of the system by localization of the chain recurrent set. As already mentioned, the chain recurrent set contains all kind of return trajectories and, hence, should give an overview about the areas of interest for further investigation. We set the area of investigation in the domain space to $M = [-5.0; 5.0] \times [-5.0; 5.0]$. This area M is initially divided into a covering C^0 of 20×20 boxes. Then the symbolic image G^0 is constructed for C^0. We apply the Tarjan algorithm, see Sect. B.2.1, to detect the recurrent cells. The construction and cell detection process was repeated for 4 subdivisions. In each step the boxes are subdivided into 4×4 new smaller boxes. After four subdivision steps, the distinct features of the Ikeda mapping can be found in the different recurrent sets of the symbolic image. Three areas in the state space are detected. One of them represents the stable point, the other one the unstable saddle point, and the last one contains all cells representing the chaotic attractor. It is worth mentioning that chaotic attractors can be found by computation of the chain recurrent set because their skeleton is typically build up from unstable periodic cycles [5, 6] and, therefore, recurrent points. The calculation of the symbolic image takes less than 2 minutes. In Table B.1 the number of cells in the symbolic image and the number of located recurrent cells for every subdivision level are shown. We observe that the number

Table B.1. Ikeda system: Computation of the recurrent cells. The number s marks the level of subdivision. The phase space discretization of the area $M = [-5.0; 5.0] \times [-5.0; 5.0]$ is shown. Furthermore, the number of cells belonging to a symbolic image, $|V(G^s)|$, and the number of localized recurrent cells $|RV(G^s)|$

Subdivision level s	Phase space discretization	$|V(G^s)|$	$|RV(G^s)|$
0	20×20	400	79
1	80×80	1 264	415
2	320×320	6 640	3 018
3	$1 280 \times 1 280$	48 288	27 878
4	$5 120 \times 5 120$	446 048	284 727

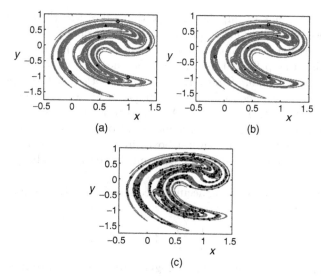

Fig. B.1. Ikeda system: (a) A fixed point (square), 2-periodic (triangles) and the two period-3 orbits (empty and filled circles). (b) Some detected unstable limit cycles with periods 5 (empty circles) and 6 (points). (c) All detected unstable 6-periodic (empty circles) and 13-periodic (points) points. Note that the chaotic attractor $\mathcal{A}$ in the background is visible for better orientation but was not calculated by the same computation

of recurrent cells grows during the subdivision process by factor ≈ 10. This high growth rate is due to the fact that it depends on the dimension of those objects which are the subject of localization, see also [14,19]. In this case, one of the subjects of localization is the chaotic attractor which has a dimension close to 2.

Our next target is the localization of periodic points. This requires, of course, the usage of the time consuming variant of the cell location algorithm based on the Dijkstra algorithm, see Sect. B.2.2. We took the same area of investigation, $M = [-5.0; 5.0] \times [-5.0; 5.0]$, than for the last computation but select only the cells with period size $p' < p = 3$ for further subdivisions. After an initial subdivision of M into 20×20 boxes, the boxes get subdivided into 8×8 new smaller boxes in every subdivision. The construction and cell detection process was repeated for 9 subdivisions. The results of the calculation in the area $[-1.5; 2.5] \times [-1.5; 2.0]$ can be seen in Fig. B.1(a), namely three fixed points, a period-2 and two period-3 orbits. It turns out that the usage of a sufficient number of scan points is important. In this example, every box contained 8×8 scan points scattered over the box, four more points close to the box corners and another one in the center. Such a high number is needed to acquire all the periodic orbits. If less scan points are chosen, another parameter must be set for error tolerance, see Sect. B.5.2, or some of the periodic cycles will not be detected. Although this calculation uses the more time consuming period detection algorithm, the computation takes less than 30 seconds on the reference machine. This is due to the fact that only very few cells have a period size smaller or equal than 3. In our calculations, not more than 97 cells per subdivision step fit to this criterion. So the size of the symbolic images can be kept very small. However, one should notice that the performance time can increase exponentially if the parameter p is set to a higher value and more such cycles with $p' \leq p$ exist. A serious problem we come across in this computation is clustering. For most points not only one box corresponding to a periodic cell is found, but several boxes in the neighborhood. In this case we get up to 5 boxes as an outer covering for each periodic point instead of one box.

Theoretically, the following accuracy, compare Sect. B.4, could be achieved for the calculated points:

$$\epsilon \leq \frac{1}{20} \cdot \frac{1}{8^9} \cdot 4 \approx 2 \cdot 10^{-9}. \tag{B.32}$$

However, in practice, the error is higher because of clustering. If taking this into account and analyzing the computed results, the error increases to $\epsilon \leq 1 \cdot 10^{-7}$. One can expect to find in the vicinity of the chaotic attractor some unstable limit cycles with periods higher than 3. So first we increased p to 6 and then to 14. Some results of these calculations are presented in Fig. B.1(b), which shows two of the detected unstable 5- and 6-periodic orbits, and Fig. B.1(c), an overview of all detected 6- and 13-periodic points. Remarkably, the symbolic images for $p = 6$ contained not more than 325 cells, for which the corresponding boxes got subdivided and thus the calculation did not take much more computation time than in the first case (≈ 30 seconds). But the location of cells with a period size ≤ 14 consisted of up to 27 000 selected cells. Boxes corresponding to each of them get subdivided into 8×8 new smaller boxes, so that the symbolic images had up to 1 700 000 cells. Therefore, the calculation took around eight hours in this case.

Until now we investigated the Ikeda system for fixed parameter values, as described above. Using the methods of symbolic analysis under variation of some parameters, interesting results can be obtained as well. For instance, one can observe the bifurcations which causes the emergence of unstable periodic orbits. These periodic orbits determine the structure of the chaotic attractor discussed above. Performing this task, we consider the area $M = [-0.4; 1.5] \times [-1.7; 1]$ in the state space and calculate the periodic orbits up to period six. Using an initial subdivision into 20×20 boxes and performing 4 subdivision steps, whereby each box is divided into $2 \times 8 \times 8$ smaller boxes, we obtain the results shown in Fig. B.2. The parameter a is varied in the interval $[0; 0.9]$. The other parameters are kept fixed to the same values as above. In this experiment we observe a period doubling bifurcation scenario and a large number of saddle-node bifurcations.

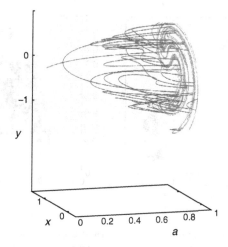

Fig. B.2. Ikeda system: Periodical points up to period 6 under variation of parameter a

B.7.2 Coupled Logistic Map

We take a look at another 2-dimensional map, the coupled logistic map defined by:

$$\mathbf{x}(n+1) = f_C(\mathbf{x}(n)), \tag{B.33}$$
$$f_C(\mathbf{x}) = ((1-r)\, g(x,a) + r\, g(y,b), r\, g(x,a) + (1-r)\, g(y,b) \tag{B.34}$$

with $g(x,m) = m\, x\, (1-x).$

The system, as presented here, can be considered as a 2-dimensional case study of coupled map lattices [16] for the logistic map [10]. For all our investigations, we fixed the parameter settings to $a = b = 3.8$ and $r = 0.07$. Analytically, it is easy to show that, due to $a = b$, we have symmetric behavior with respect to the diagonal $y = x$. This means that orbits become symmetric if one interchanges the x- and y-coordinates, and that all points on the diagonal at $y = x$ form an invariant set $\mathcal{D}$. By numerical analysis based on forward iterations and calculation of Lyapunov exponents, one can find out, that the system is governed by a single attractor $\mathcal{A}$ which consists of two symmetric parts in the phase space, see Fig. B.3(a).

Our first investigation of the system by symbolic image analysis was the computation of the chain recurrent set. We initially divided the area $M = [0.0; 1.0] \times [0.0; 1.0]$ into 5×5 boxes. In each subdivision step, a box gets divided into 3×3 new ones. After 5 subdivisions the outer covering of the chain recurrent set consists of $430\,000$ boxes with a side length $\approx 1 \cdot 10^{-3}$. It is important to mention that a high number of scan points is required. If taken too little, large parts of the chain recurrent set get lost during the first subdivisions. Hence, for our investigation we covered each box with a regular

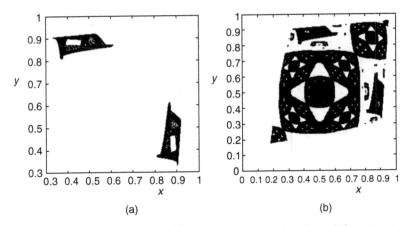

Fig. B.3. Coupled logistic map: (a) Numerical approximation of the attractor $\mathcal{A}$. (b) Numerical approximation of the chain recurrent set. One of the components is an approximation of the attractor $\mathcal{A}$

grid of 100 scan points. Applying these settings, the computation takes around 8 minutes, and its results can be seen on Fig. B.3(b). The chain recurrent set does not only consist of the chaotic attractor but also of fractal structures which are symmetric with respect to the diagonal. Note that these fractal structures are unstable entities. Orbits started in a neighborhood of the chain recurrent set are attracted by the attractor $\mathcal{A}$. We observed that even orbits started in the area covered by the computed fractal structures are attracted by $\mathcal{A}$. However, this can be explained by the fact that our numerical computation produced an outer covering of the real chain recurrent set and, hence, covers also the chain recurrent set's neighborhood. Note that the chain recurrent set consists of 4 distinct components of equivalent recurrent cell sets, one of them represents $\mathcal{A}$, and another one a 2-periodic unstable orbit in the holes of $\mathcal{A}$.

In order to verify our results, we also computed periodic orbits. We used the cell location algorithm based on the Dijkstra algorithm, see Sect. B.2.2, and computed all periodic points with a periodicity ≤ 8. We applied 17 subdivisions so that the error $\epsilon \leq 1 \cdot 10^{-8}$. The computation took around 25 minutes, and we got 614 periodic points. These points belong to periodic orbits which are scattered over the whole area designated by the approximation of the chain recurrent set. Furthermore, we checked that each of these periodic orbits is unstable.

Combining the results of our numerical computations so far, we find strong evidence for the hypothesis that the computed fractal structure of the chain recurrent set is an outer covering of a set of unstable periodic orbits of any size. This reminds us of the hypothesis of Cvitanović [5] regarding periodic orbits as the skeleton of chaotic attractors. However, the fractal structure we observe here is not an attractor.

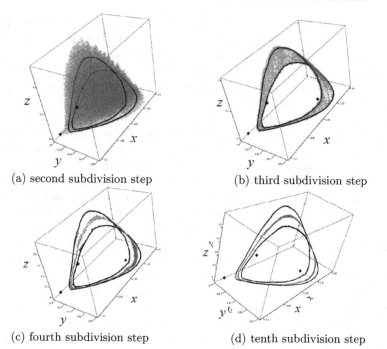

(a) second subdivision step   (b) third subdivision step

(c) fourth subdivision step   (d) tenth subdivision step

Fig. B.4. Discrete food chain model: Numerically calculated fixed points and four subdivision steps of the symbolic image construction. The outer covering of the chain recurrent set (gray), the attractor (black) and the unstable fixed points (points) are shown. The fourth fixed point at $(0,0)$ can not be seen. Note that the attractor was separately computed by forward iterates but is also covered by the approximation of the chain recurrent set

B.7.3 Discrete Food Chain Model

Next we analyzed a discrete system of mathematical biology. The 3-dimensional dynamical model describes a discrete food chain model, studied by Lindström in [17]. The system is defined by

$$\mathbf{x}(n+1) = f_{fc}(\mathbf{x}(n)), \qquad (B.35)$$

$$f_{fc}(\mathbf{x}) = \left(\frac{\mu_0\, x\, e^{-y}}{1 + x\, \max(e^{-y}, g(z)g(y))},\ \mu_1\, x\, y\, e^{-z} g(y) g(\mu_2\, y\, z),\ \mu_2\, y\, z \right)$$

$$\text{with}\quad g(s) = \begin{cases} \frac{1 - e^{-s}}{s}, & \text{if } s \neq 0, \\ 1, & \text{if } s = 0. \end{cases}$$

We only focus on the following parameter setting: $\mu_0 = 3.4001$, $\mu_1 = 1$ and $\mu_2 = 4$. The analytic results of Lindström showed, that the system possesses at most four fixed points.

Our target is the computation of the chain recurrent set. We chose the area $M = [-1.0; 4.0] \times [-1.0; 4.0] \times [0.0; 1.6]$ for investigation. It turned out that it is not possible to get an appropriate approximation for the chain recurrent set by means of usual symbolic image construction. The tuning techniques must be applied to get satisfiable results. By doing so, the equilibrium points, and maybe some other information, get lost in the symbolic image after several subdivision steps. On the other hand, two invariant manifolds can be detected which belong to different components of the chain recurrent set, see Fig. B.4(d). By application of forward iteration, it can be verified that both of them consist of quasiperiodic trajectories, and that one is a stable invariant set, namely an attractor (black), while the other is an unstable invariant set (gray). Hereby, the unstable entity is not a repeller but of saddle type. For this reason, it could not be approximated by backward iterates. Such a calculation takes around one hour and the symbolic image grows up to $\approx 1\,100\,000$ cells. The long calculation time is mainly caused by the application of the tuning-techniques. Note that the localization of the unstable quasiperiodic manifold is, from the computational point of view, a nontrivial task.

In order to get a better impression how the construction process works, Fig. B.4 shows the results of several subdivision steps. Hereby, 17 scan points per box are taken. The rough position of the attractor can be located after the second subdivision of the domain space M into $200 \times 200 \times 32$ regions, see Fig. B.4(a), then, in the third subdivision, see Fig. B.4(b), the principal shape of the attractor becomes visible. But only after the fourth subdivision into $1\,200 \times 1\,200 \times 192$ regions, see Fig. B.4(c), the symbolic image splits into two different sets of equivalent recurrent cells, which correspond to the stable and unstable invariant manifolds. In order to achieve these results, it is necessary to compute the symbolic image graph for the iterated function f^{40} in the third subdivision and for f^{80} in the fourth subdivision step. Otherwise, the principal shape of the cone, see Fig. B.4(a), would persist during further subdivisions. Additionally, reconstruction of the fragmented parts must be applied in order to avoid that the cycles vanish. The final result, see Fig. B.4(d), is computed after the sixth subdivision. Note that in the subdivisions 5 and 6, also the function f^{40} is used and reconstruction of the cycles applied.

B.7.4 Lorenz System

As an example for a dynamical system continuous in time, we consider the well-known system introduced by Lorenz in [18] which is defined by

$$\dot{\mathbf{x}}(t) = f(\mathbf{x}(t))$$
$$f(\mathbf{x}) = (\sigma(y - x),\ x(r - z) - y,\ xy - bz). \tag{B.36}$$

We use the standard values of the parameters $\sigma = 10$, $b = 8/3$ and investigate the Lorenz system at two values of the parameter r, namely $r_1 \approx 14.6$ and $r_2 \approx 20$. As shown in [21], for these settings exist an unstable fixed point

$P = (0,0,0)^T$ and two stable ones C_1 and C_2, each of them accompanied by an unstable limit cycle. The value r_1 is chosen close to the so-called homoclinic explosion which occurs at $r \approx 13.926$, where the unstable manifolds of P return to the origin. Furthermore, at parameter value r_2, the both unstable limit cycles around C_1 and C_2 are situated close to each other and to C_1 and C_2.

In order to reproduce these results with methods of symbolic analysis, we compute the chain recurrent set. We define for r_1 and r_2 the domain spaces $M_1 = [-35.0; 35.0] \times [-35.0; 35.0] \times [0.0; 30.0]$ and $M_2 = [-20.0; 20.0] \times [-20.0; 20.0] \times [0.0; 30.0]$ as the area of investigation. The division of these spaces is initially set to $4 \times 4 \times 2$ and $2 \times 2 \times 2$ boxes. In the following subdivision stages each box is divided into $2 \times 2 \times 2$ smaller boxes. The integration step Δt is set to 0.001, and the number of iteration steps to $n_1 = 100$, $n_2 = 200$. In order to compute the integration step $\phi(\Delta t, \mathbf{x})$, the Runge-Kutta method was applied.

Figs. B.5(a) and B.5(b) show the results of the calculations for the parameters r_1 and r_2. Remarkably, one can see that the limit cycles for r_1 still touch each other, which is due to some numerical inaccuracy, while for r_2 the cycles

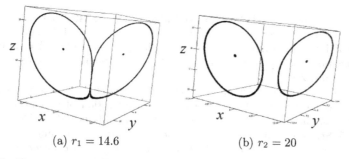

(a) $r_1 = 14.6$ (b) $r_2 = 20$

Fig. B.5. Lorenz system: Computation of an outer covering of the chain recurrent set at positions $r_1 = 14.6$ and $r_2 = 20$

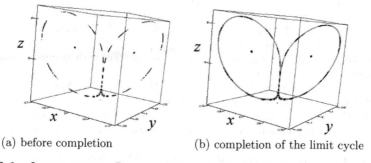

(a) before completion   (b) completion of the limit cycle

Fig. B.6. Lorenz system: Reconstruction of unstable limit cycles at parameter $r_1 = 14.6$ with a large discretization time. The limit cycles fall apart and vanish by time (black), but will be completed (gray)

shrinked closer around C_1 and C_2. The computations took 30 minutes for r_1 and 2 hours for r_2. Ten subdivision steps were computed, and the symbolic images contained up to $1\,400\,000$ cells. Hereby, the high computation time is mainly due to the relative high setting of the iteration time t. Furthermore, the unstable fixed point P can not be computed by this setting. However, if t would be set to a lower value, the limit cycles could not be detected at all because too many cells would be selected for subdivision and the memory resources would be exceeded after a few subdivisions.

The reconstruction of fragmented solutions is illustrated for the parameter setting $r_1 = 14.6$. In Fig. B.6(a) the computed approximation of the chain recurrent set after 10 subdivisions is shown. As can be seen, parts of the unstable limit cycles got lost. For this reason, the method for reconstruction of the fragmented solutions must be applied. The results are shown in Fig. B.6(b). We see that the final computation produces a precise outer covering of the unstable limit cycles.

References

1. Home page of the AnT 4.669 project, http://www.AnT4669.de, 2005
2. V. Avrutin, D. Fundinger, P. Levi, G.S. Osipenko, and M. Schanz, Investigation of dynamical systems using symbolic images: Efficient implementation and applications, accepted for publication by International Journal of Bifurcation and Chaos, 2005
3. V. Avrutin, R. Lammert, M. Schanz, G. Wackenhut, and G.S. Osipenko, On the software package AnT 4.669 for the investigation of dynamical systems, in *Fourth International Conference on Tools for Mathematical Modelling*, v.9, 24–35, St. Petersburg State Polytechnic University, Russia, June 2003
4. T. Cormen, C. Leiserson, and R. Rivest, *Introduction to algorithms*, The MIT electrical engineering and computer science series, MIT Press, 2000
5. P. Cvitanović, Periodic orbits as the skeleton of classical and quantum chaos, *Physica D*, 51, 1991
6. P. Cvitanović, Focus issue on periodic orbit theory, *Chaos*, 2, 1992
7. M. Dellnitz and A. Hohmann, A subdivision algorithm for the computation of unstable manifolds and global attractors, *Numerische Mathematik*, 75, 293–317, 1997
8. W. Dijkstra, A note on two problems in connection with graphs, *Numerische Math.*, 1:269–271, 1959
9. D. Fundinger, *Investigating Dynamics by Multilevel Phase Space Discretization*, PhD thesis, University of Stuttgart, 2006
10. Mat Gyllenberg, Gunnar Söderbacka, and Stefan Ericsson, Does migration stabilize local population? Analysis of a discrete metapopulation model, *Math. Biosciences*, 118, 25–49, 1993
11. S.L. Hruska, *On the numerical construction of hyperbolic structures for complex dynamical systems*, PhD thesis, Cornell University, 2002
12. C.S. Hsu, *Cell-to-Cell Mappings*. Springer, N.Y., 1987
13. K. Ikeda, Multiple-valued stationary state and its instability of the transmitted light by a ring cavity system, *Opt. Commun.*, 30, 57–261, 1979

14. O. Junge, *Mengenorientierte Methoden zur numerischen Analyse dynamischer Systeme*, PhD thesis, University of Paderbonn, 1999
15. O. Junge, Rigorous discretization of subdivision techniques, in *Proceedings of Equadiff '99, Berlin*, 2000
16. K. Kaneko, editor, *Theory and Applications of Coupled Map Lattices*, Wiley, New York, 1993
17. T. Lindström, On the dynamics of discrete food chains: Low- and high-frequency behavior and optimality of chaos, *Journal of Mathematical Biology*, 45, 396–418, 2002
18. E.N. Lorenz, Deterministic nonperiodic flow, *J. Atmos. Sci.*, 20, 130–141, 1963
19. K. Mischaikow, Topological techniques for efficient rigorous computations in dynamics, *Acta Numerica*, 2002
20. P. Pilarczyk, Computer assisted method for proving existence of periodic orbits, *TMNA*, 13(2), 365–377, 1999
21. C. Sparrow, *The Lorenz Equations: Bifurcations, Chaos, and Strange Attractors*, Springer, N.Y., 1973
22. R. Tarjan, Depth-first search and linear graph algorithms, *SIAM J. Comput*, 1, 146–160, 1972

Index

Lecture Notes in Mathematics

For information about earlier volumes
please contact your bookseller or Springer
LNM Online archive: springerlink.com

Recent Reprints and New Editions